PROGRESS IN FOOD SCIENCE AND TECHNOLOGY

PROGRESS IN FOOD SCIENCE AND TECHNOLOGY

VOLUME 1

PROGRESS IN FOOD SCIENCE AND TECHNOLOGY

Additional books in this series can be found on Nova's website under the Series tab.

Additional E-books in this series can be found on Nova's website under the E-books tab.

Progress in Food Science and Technology

Progress in Food Science and Technology

Volume 1

Anthony J. Greco
Editor

Nova Science Publishers, Inc.
New York

Copyright © 2011 by Nova Science Publishers, Inc.

All rights reserved. No part of this book may be reproduced, stored in a retrieval system or transmitted in any form or by any means: electronic, electrostatic, magnetic, tape, mechanical photocopying, recording or otherwise without the written permission of the Publisher.

For permission to use material from this book please contact us:
Telephone 631-231-7269; Fax 631-231-8175
Web Site: http://www.novapublishers.com

NOTICE TO THE READER

The Publisher has taken reasonable care in the preparation of this book, but makes no expressed or implied warranty of any kind and assumes no responsibility for any errors or omissions. No liability is assumed for incidental or consequential damages in connection with or arising out of information contained in this book. The Publisher shall not be liable for any special, consequential, or exemplary damages resulting, in whole or in part, from the readers' use of, or reliance upon, this material. Any parts of this book based on government reports are so indicated and copyright is claimed for those parts to the extent applicable to compilations of such works.

Independent verification should be sought for any data, advice or recommendations contained in this book. In addition, no responsibility is assumed by the publisher for any injury and/or damage to persons or property arising from any methods, products, instructions, ideas or otherwise contained in this publication.

This publication is designed to provide accurate and authoritative information with regard to the subject matter covered herein. It is sold with the clear understanding that the Publisher is not engaged in rendering legal or any other professional services. If legal or any other expert assistance is required, the services of a competent person should be sought. FROM A DECLARATION OF PARTICIPANTS JOINTLY ADOPTED BY A COMMITTEE OF THE AMERICAN BAR ASSOCIATION AND A COMMITTEE OF PUBLISHERS.

Additional color graphics may be available in the e-book version of this book.

LIBRARY OF CONGRESS CATALOGING-IN-PUBLICATION DATA

ISSN: 2157-863X

ISBN: 978-1-61122-314-9

Published by Nova Science Publishers, Inc. † New York

CONTENTS

Preface		**vii**
Chapter 1	Flavour Compounds in Legumes: Chemical and Sensory Aspects *Sorayya Azarnia and Joyce I. Boye*	**1**
Chapter 2	Micronutrient Disorders and Various Mitigation Options in Indian Agriculture *M. P. S. Khurana, K. P. Patel, V. P. Ramani G. Mahajan and Kuldip-Singh*	**35**
Chapter 3	Legumes: Properties, Nutrition, Consumption and Health *Tzi Bun Ng and Randy Chi Fai Cheung*	**65**
Chapter 4	Role of the Processing Industry for Food Security in India *S. Mangaraj, M. K.Tiripathi and S. D.Kulkarni*	**89**
Chapter 5	Characteristics and Utilization of Black Rice *Bipin Vaidya and Jong-Bang Eun*	**113**
Chapter 6	In-Season Wheat Yield Prediction in the Semiarid Pampa of Argentina using Artificial Neural Networks *A. Bono, J. De Paepe and R. Álvarez*	**133**
Chapter 7	Legumes as Cover Crops or Components of Intercropping Systems and Their Effects on Weed Populations and Crop Productivity *I. S. Travlos*	**151**
Chapter 8	Optimizing Barley Grain use by Dairy Cows: A Betterment of Current Perceptions *A. Nikkhah*	**165**
Chapter 9	Bioactivity of Oleanane Triterpenoids from Soy (Glycine Max Merr) Depends on the Chemical Structure *Wei Zhang and David G. Popovich*	**179**

Chapter 10	Enhanced Production of β – Carotene and L – Asparaginase from a Tribal Food Alga, *Vaucheria Uncinata* *Usha Pandey and J. Pandey*	**195**
Chapter 11	Wheat Grain a Unique Prepartal Choice for Transition Holstein Heifers *A. Nikkhah and F. Ehsanbakhsh*	**209**
Chapter 12	Germinated Legumes: A Boon to Human Nutrition *Geetanjali Kaushik*	**221**
Chapter 13	Sucrose Metabolism and Grain Development of Contrasting Spikelets Located at Different Positions of the Rice Panicle *Pravat Kumar Mohapatra*	**231**
Chapter 14	Wheat Grain in Dairy Rations: A Major Feasible Commodity Overlooked *A. Nikkhah, F. Amiri, H. Khanaki and H. Amanlou*	**243**
Chapter 15	World Food Demand: Coping with its Growth *Anil Hira, Hayley Jones and James Morfopoulos*	**253**
Index		**271**

PREFACE

This book reviews research on new developments in all areas of food science and technology. Topics discussed herein include the role of the processing industry for food security in India; the characteristics and utilization of black rice; the formation of flavours and off-flavours in legumes; legumes as cover crops and their effects on weed populations and crop productivity; optimizing barley grain use by dairy cows; and enhanced production of B-carotene and L-asparginase from a tribal food alga.

Chapter 1 - Flavour is an important criterion for food selection as it influences taste and can be perceived even before the food is ingested. As different flavour compounds have different characteristics, changes in their concentrations could affect the taste and flavour of foods. Legumes such as soybeans, beans, peas and lentils are good sources of energy, protein and fibre and are considered to be healthy legumes for human consumption. They are utilized in the whole seed form, as split seeds and as flour, for soup mixes, purees and processed products.

One of the major challenges facing the legume industry is the maintenance of a level of production which is adequate to meet markets need. Thus, the majority of the published literature has focused on improving yields and quality traits such as seed size and colour without paying much attention to processing traits such as flavour and organoleptic quality and the impact of seed composition on these traits. Knowledge of the flavour profile of legumes and the impact of variety, processing conditions, ingredients and storage are important in selecting and marketing the right varieties for targeted food applications.

This chapter addresses the major factors involved in the formation of flavours and off-flavours in legumes. It also covers different flavour compounds identified in different legumes along with their characteristics. Additionally, recent developments in the sensory and instrumental analyses of flavour compounds in legumes are detailed.

Chapter 2 - The deficiency of micronutrient in Indian soils is attributed to several factors which include: adoption of high-yielding crop cultivars during last two decades that removed large quantities of micronutrients in the harvested crop resulting in mining of nutrients from the soil, particularly those not applied in required quantities through fertilizers and manures, increased cropping intensity, increased irrigation facility, increased use of high analysis fertilizers, reduced recycling of micronutrient contained in crop residues and organic manures. Micronutrient deficiencies in India are widely reported across the different soil types and crop species. As much as 48, 12, 5, 4 and 33 per cent soils in India are afflicted with deficiency of zinc (Zn), iron (Fe), manganese (Mn), copper (Cu) and boron (B) respectively (Singh, 1999). This causes failure of normal crop growth thereby resulting in low crop yields. Ample evidences has become available on hand from hundreds of experiments that micronutrient fertilization of crops is essentially required to ensure balanced nutrition for producing optimum yield, enhancing fertilizer use efficiency of other nutrients besides improving quality of crops. It is further noticed that crop response were not related to total content but the size of available pool of particular micronutrient.

Chapter 3 - A legume is a plant or a fruit belonging to family Fabaceae (or Leguminosae). A legume fruit or pod is a simple dry dehiscent fruit. Alfalfa, beans, carob, clover, peas, lentils, lupins, mesquite, peanuts, locust trees (Gleditsia or Robinia), wisteria, and the Kentucky coffeetree (*Gymnocladus dioicus*) are examples of legumes. Legumes made their appearance early in America, Europe and Asia around 6,000 BC. They became a staple food, serving as in important protein supplement when there was an inadequate supply of meat.

Chapter 4 - Agro-processing development could make a significant contribution to the vocation of agriculture and hence, in national and international development. Agro-processing at village/rural level can expand the local markets for primary agricultural products, add value by vertically integrating primary production and food processing systems and minimize post harvest losses. Being mindful of the pitfalls and obstacles to agro-processing related development, it may be instructive to understand the concept of establishing agro-processing center and its economic perspective associated to increase income generation and sustainable rural development. The country has achieved breakthrough in agricultural production. However, level of productivity is low compared to many neighboring countries, let alone developed world. As a result margin of profit to the farmers in most of the agricultural commodities is very low. Post harvest loss prevention, value addition, and entrepreneurship development are important for higher income and rural employment generation (Shukla, 1993). Agriculture contributes 25% to annual gross domestic products and provides livelihood to more than 76% of the people. The agricultural processing sector has immense potential for value addition and employment generation. Sun drying, winnowing, paddy hulling, pulse milling, oil expelling, wheat milling, pickle making, gur and khandsari, ghee & khoa etc. are the major processing activities undertaken by the farmers. The traditional processing equipment used by the farmers include supa, chalni, chakiya, janta, silbatta, Okhli, mathani, puffing pan, mini oil ghanis/kolhus, rice hullers and flour chakkis etc. The quality of the products made by the traditional methods has been accepted and popular in the rural markets but have little scope for marketing in urban areas due varied consumers preferences. High capacity modern machines introduced in urban and sub-urban areas for processing of agricultural produce have helped in increasing the income of the processors. But, many of the agro-processing activities earlier being undertaken in the rural areas are now gradually being shifted to urban areas as a result rural workers are deprived of their due share of modernized agriculture. Farmers and industries mutually depend on each other for inputs and raw materials for processing. A balance has to be maintained in the agricultural development so that the farmers equally share the fruits of the higher production by involving them in primary processing and value addition. Also agro industries lead to the creation of forward and back ward linkages on large scale by maximizing complementarities of agriculture and industries (Desai, 1986).

Chapter 5 - Rice cultivation feeds more than half of the global population, as it is found on all continents except Antarctica. Among the 24 recognized species of rice, *Oryza sativa* and *Oryza glaberrima* are the most cultivated. Asian cultivated rice, *Oryza sativa*, has migrated across the globe with the human population, whereas *O. glaberrima* is cultivated only in certain areas of western Africa (1). A few varieties of *Oryza sativa* have unique characteristics in terms of their aroma, kernel color, and chemical composition (2). Among them, pigmented rice is a unique variety and has been consumed for a long time in Asia, especially in China, Japan, Korea Laos, Thailand, and Burma. Pigmented rice with black and red colored pericarp is also planted in places such as the United States, Italy, Greece, and Africa. Among these varieties, the most famous is black rice, which is generally used as an ingredient in different snacks and desserts. Rice with colored pericarp is considered to be a health food. For example, whole grain pigmented rice is categorized as a potent functional food since it contains high amounts of phenolic compounds, especially anthocyanins in pericarp (*3*) (*4*).

Chapter 6 - More than 40 % of wheat is produced in the Argentinean Pampa under semiarid conditions, in the so called Semiarid Pampean Region. In this region, seeded area rounded 2.2 Mha with and average crop yield of 2000 kg ha^{-1} during the last years. Yield varied depending on soil fertility, texture, water content, rainfall and management. The objectives of this study were 1) to generate an explicative wheat yield model that accounted for the effect of environmental and cultural variables affecting yield in the Semiarid Pampa and 2) to generate a predictive model for in-season crop yield estimation. An analysis was performed from results produced in 85 field experiments from 1996 to 2004, in which the combination of soils x climate x management conditions resulted in 912 yield data. Surface regression and artificial neural networks (ANN) methodologies were tested and compared for data analysis. A surface regression response model was developed that included soil moisture, organic nitrogen, nitrate nitrogen and fertilizer nitrogen, soil phosphorus, soil depth, tillage system, previous crop and rainfall as independent variables that accounted for 78 % of yield variance. An ANN was also fitted to the data accounting for 90 % of the yield variance which used as inputs the same variables included in the regression model. Comparing predicted vs. observed yields resulted in a lower (P = 0.05) RMSE using the ANN than using the regression method (381 vs. 552 kg ha^{-1}). A predictive ANN model that accounted for 88 % of the yield variability could also be developed excluding rainfall as input. In-season prediction of wheat yield can be performed with this later ANN model by farmers and decision-makers.

Chapter 7 - Legumes are the second family after the Poaceae in importance to humans. The more than 18,000 species of legumes include important species. In recent years, there has been remarkably increased interest in alternative agricultural production systems in order to achieve high productivity and promote sustainability over time. Among them, intercropping and cover crop mulches seem quite promising since they affect several agronomic traits. Cover crops have long been used to reduce soil erosion and water runoff and improve several soil parameters, while in many cases are also used to manage weeds in several crops. Several annual legume species have already been studied for their weed control benefits, while in the same time they biologically fix nitrogen that subsequently becomes available during residue decomposition. This further improves crop yield, making the use of cover crops economically profitable by reduced inputs and increased yield. Additionally, intercropping legumes for forage or grain production with cereals or other crops is widely used in many parts of the world. The presence of legumes as a major component of these systems contributes into soil conservation, efficient resource utilization, higher forage protein level and better weed management. It is also noticeable that some legumes, such as sweetclover, suppress weeds both during growth and after death. In addition to competition-based or physical weed suppression; certain legumes used as cover crops or in intercropping (hairy vetch, red clover etc) are also known to suppress weeds through their allelopathic activity. Future research should focus on legumes which can suppress weeds without reducing main or next crop yield, in order to maintain and enhance the profitability of these sustainable agriculture systems.

Chapter 8 - Dairy cows are irreplaceable major suppliers of amino acids, unique unsaturated fatty acids, energy, and calcium to humans. Optimizing ruminal and small intestinal starch and protein assimilation has been a challenge in modern dairy farms. Dietary barley inclusion rate and processing type manipulation leads our efforts in improving starch utilization. Grinding has traditionally been considered a risk to diet palatability and healthy rumen conditions whilst steam-processing has been rationalized to improve diet palatability and reduce barley fermentation rate, aiming to reduce the risk of subacute rumen acidosis. However, no conclusive research has thoroughly compared processing techniques at different dietary inclusion rates. The main objective of the first experiment was to determine effects of feeding either (1) finely ground, (2) steam-rolled, (3) finely dry-rolled, or (4) coarsely dry-rolled barley grain on rumen fermentation, digestibility and milk production. Eight multiparous midlactation Holstein cows were used in a replicated 4×4 Latin square design study with four periods of 21 d. Diets contained 256 g barley grain/kg DM. Processing type did not affect milk yield and composition, DM intake, rumen pH and volatile fatty acids (VFA), fecal and urine pH, and apparent total tract nutrient digestibility. Results suggested that finely ground barley grain is no different than dry-rolled and steam-rolled barley grains in stimulating feed intake and productivity of midlactation cows, with 256 g/kg barley grain. The objective of the second experiment was to compare effects of grinding versus steam-rolling of barley grain at 30% or 35% of diet dry matter on feed intake, chewing behavior, rumen fermentation, and milk production. Eight multiparous Holstein cows (85 ± 9 d in milk) were used in a replicated 4 × 4 Latin square design experiment with four 21-d periods. Treatments included grinding (GB) or steam-rolling (SB) of barley grains at either 35% or 30% of dietary DM. Diets were prepared as a mixed ration and delivered twice daily at 0730 and 1600 h. Neither processing method nor dietary barley grain use affected dry matter intake, daily eating, ruminating and chewing times, rumen pH and major VFA molar percentages, or milk percentages and yields of fat and protein. Energy-corrected milk yield increased for SB compared with GB at 35% but not at 30% barley grain. Feed efficiency was increased by SB, but was unaffected by dietary barley grain level. Results suggest that at 30% dietary barley grain, GB resulted in similar lactation performance as SB and that SB did not affect productivity when dietary barley grain increased from 30 to 35%. Regardless of barley grain level, grinding effectively maintained feed intake and rumen pH at 4 h post-feeding, whereas steam-rolling increased feed efficiency. Increasing barley grain from 30% to 35% of diet dry matter did not improve feed intake and milk production.

Chapter 9 - Soyasaponins are oleanane triterpenoids molecules that are found in Soy (*Glycine Max* Merr) and other legumes such as green peas (*Pisum sativum* L) and lentils (*Lens culinaris*). Soyasaponins are diverse, structurally complex and possess amphiphilic properties, with polar water soluble sugar moieties attached to a non polar, water insoluble pentacyclic ring structure. In legumes, soyasaponins are present primarily as glycosides which are storage forms of saponins and are secondary plant metabolites. These molecules serve as defensive molecules for the plant to defend against biological predators such as fungi and herbivores. A resurgence of interest into secondary plant metabolites has emerged as many of these compounds possess diverse biological activity which includes the potential to act as chemo-preventive agents. Soyasaponins are classified into five groups based on the chemical structure of their respective aglycones but the two main groups are A and B. Group B soyasaponins are the most abundant group of saponins found in soy and are thought to contribute to bulk of soy's chemo-preventative properties attributed to soyasaponins.

Chapter 10 - The growing concern towards the exponential increase in human population, environmental stresses and associated health risks during recent years has enhanced the demand for potential biodiversity resources with high nutritional and therapeutic values. In India, rural and tribal people use a wide variety of wild plants including algae for maintaining vigor and treatment of diseases.

We investigated enhanced production of β – carotene and L – asparaginase in a yellow green alga, *Vaucheria uncinata* in a full factorial design using 1mM each of nitrate (KNO_3^-), ammonium {$(NH_4)_2SO_4$}, glutamine and NaCl in free and immobilized cell culture condition. Glutamine was found to be the most suitable N – source for the production of β – carotene and L – asparaginase. Maximum production of β – carotene (7.02 $\mu g\ ml^{-1}$) and about five fold increase in L – asparaginase (5.89 IU mg^{-1} protein) was recorded when glutamine grown cells were exposed to 1mM NaCl under immobilized cell and free cell culture respectively. Exposure to NaCl also leads to a substantial increase in proline and glutathione contents. Our study indicates *V. uncinata* to be a promising bioresource for production of β – carotene and L – asparaginase. Use of immobilized and free cell cultures supplemented with glutamine and NaCl could be an effective approach for scaling – up production of β – carotene and L – asparaginase, respectively for commercial application.

Chapter 11 - Wheat grain is unique in possessing extensive starch and nitrogen fermentation, low cation-anion difference and reasonable palatability, making it a potentially useful prepartal dietary choice. The objective was to determine effects of WG provision to prepartum first-calf heifers on metabolic, health, and productive criteria during the transition period. Wheat grain replaced barley grain as commonly used. Fifteen Holstein heifers at 31 ± 6 days prepartum were blocked based on expected calving date and assigned to three treatments, or feeding totally mixed rations containing either 1) barley grain (13.8%) and wheat bran (6.1%) (BGW), 2) 10% wheat grain (WG10), or 3) 18% WG (WG18) (DM basis) from 31 days prepartum until calving. Prepartal diets contained no supplemental anionic salts. Cows were monitored until 21-day postpartum and received the same early lactation diet. The prepartal provision of WG reduced urine pH at 7-day prepartum, and elevated blood calcium and glucose at 7-day prepartum and at 3-day postpartum. Milk fat and protein yields increased during 21-day postpartum by prepartal WG provision. Blood albumin, globulins, total proteins, and urea concentrations were similar among treatments. Feeding WG did not affect body condition score, calving difficulty, calf weight and health, placenta weight, and the time interval from calving to placenta expulsion. It is suggested that prepartal provision of WG led to simultaneous improvements in energy and calcium status of the heifers experiencing their first periparturient phase without compromising parturition and health.

Chapter 12 - Food legumes or pulses are high in protein, carbohydrates, and dietary fibre and are a rich source of other nutritional components viz B group vitamins and minerals and their consumption and production extends world-wide. Pulses used for human consumption include peas, beans, lentils, chickpeas, and faba beans. On account of their high nutritive value they are the main sources for human and animal nutrition especially in the developing countries. While in the developed countries, legumes are increasingly being used in dietetic formulations in the treatment and prevention of diabetes, cardiovascular diseases and cancer of colon. Frequent legume consumption (four or more times compared with less than once a week) has been associated with 22% and 11% lower risk of coronary heart disease (CHD) and cardiovascular disease (CVD), respectively.

Chapter 13 - Individual grain weight varies in rice panicle, because some of the caryopses located mostly on the basal part of the panicle fill poorly or partially. Physiological mechanism responsible for premature cessation of grain growth in an inferior caryopsis, thereby resulting in a heterogeneous architecture of panicle morphology, is elusive. During grain filling sucrose is the major phloem solute, which moves symplastically in the sieve tube from the source to the pericarp of grain caryopsis by an osmotically driven pressure gradient. Pericarp comprising cells of maternal tissue is however, not connected symplastically to the filial endosperm cells. Carrier mediated efflux of sucrose from the inner most nucellar tissue of pericarp into an apoplasm and its subsequent influx there from into the outermost aleurone tissue of the endosperm are energy dependent. Entry of sucrose into endospermal cells primes the starch biosynthesis pathway for grain filling; the activity is controlled by a number of enzymes. Empirical evidences ruled out deficiency of sucrose as a possible cause of decreased growth of the inferior caryopsis and instead identified sink limitation responsible for the down regulation of starch filling capacity of grains. The review examines the functions of the pericarp, apoplasmic solutes and enzymes of the starch biosynthesis pathway in sink control of grain filling in rice caryopsis for the purpose of improving yield.

Chapter 14 - Wheat grain (WG) is in many regions less common than corn and barley grains as dietary starch sources for dairy cows, due to climatic, nutritional and economical constraints. Wheat starch and proteins usually possess high rumen degradation rates that, if used at high levels, could predispose cows to subacute rumen acidosis, laminitis and milk fat depression. Beliefs in such risks as well as competition for human uses of WG have restricted major WG use in dairy diets. Nevertheless, when and where is WG widely accessible, optimizing its dietary use and processing becomes economically and nutritionally critical. A paucity of in vivo data exists on feeding lactating cows different levels of differently processed WG. The objective was to determine ground WG (GW) level and particle size (PS) effects on blood metabolites, nutrient digestibility, and production. Eight midlactation cows (176 ± 8 days in milk; 554 ± 13 kg body weight (BW), 3.12 ± 0.14 body condition score (BCS); mean $\pm$ SE) were used in a 4×4 replicated Latin square design study with 21-d periods. Treatments were feeding either 20% or 10% WG, ground either finely or coarsely. Alfalfa hay-based total mixed rations with forage to concentrate ratio of 47.5:52.5 were offered individually at 0900, 1600 and 2300 h. Tail veins blood was sampled at 0 and 2 h relative to feeding. WG at 10% vs. 20% of diet DM tended to decrease blood BHBA (0.64 vs. 0.54 mmol/L), increased blood total proteins (8.28 vs. 8.46), and tended to increase blood albumin (3.72 vs. 3.83) levels. Dry matter intake increased (19.9 vs. 19.4 kg/d) when WG replaced half of dietary barley grain (i.e., 10% WG). Milk NE_L yield and the ratio of milk NE_L to NE_L intake, milk solids content and yield as well as urine pH were unaffected. Fecal pH tended to increase (6.9 vs. 6.7, P=0.10) by increasing WG from 10% to 20%. Total tract apparent DM digestibility was greater for coarse than for fine WG (70 vs. 65, P<0.01). Findings suggest feasible major uses of ground wheat grain in midlactation mixed rations.

Chapter 15 - This chapter will examine the growing concern about possible future crises in global food production, from sources such as population growth, climate change, and biofuels production. Some estimates indicate that global farm output might need to double to meet projected demand to avoid price increases. The authors demonstrate that there is a large degree of latent productivity increase possible, well beyond what is needed to alleviate the aforementioned threats, if agricultural productivity in the developing world is improved. They then examine the various impediments to achieving such targets, both on an international and domestic level. The authors close with some lessons about the role of government institutions in both Brazil and Argentina in terms of major productivity improvements in agriculture over the past two decades.

In: Progress in Food Science and Technology, Volume 1 ISBN: 978-1-61122-314-9
Editor: Anthony J. Greco © 2012 Nova Science Publishers, Inc.

Chapter 1

FLAVOUR COMPOUNDS IN LEGUMES: CHEMICAL AND SENSORY ASPECTS

Sorayya Azarnia and Joyce I. Boye

Agriculture and Agri-Food Canada, Food Research and Development Centre,
St-Hyacinthe, QC, Canada

ABSTRACT

Flavour is an important criterion for food selection as it influences taste and can be perceived even before the food is ingested. As different flavour compounds have different characteristics, changes in their concentrations could affect the taste and flavour of foods. Legumes such as soybeans, beans, peas and lentils are good sources of energy, protein and fibre and are considered to be healthy legumes for human consumption. They are utilized in the whole seed form, as split seeds and as flour, for soup mixes, purees and processed products.

One of the major challenges facing the legume industry is the maintenance of a level of production which is adequate to meet markets need. Thus, the majority of the published literature has focused on improving yields and quality traits such as seed size and colour without paying much attention to processing traits such as flavour and organoleptic quality and the impact of seed composition on these traits. Knowledge of the flavour profile of legumes and the impact of variety, processing conditions, ingredients and storage are important in selecting and marketing the right varieties for targeted food applications.

This chapter addresses the major factors involved in the formation of flavours and off-flavours in legumes. It also covers different flavour compounds identified in different legumes along with their characteristics. Additionally, recent developments in the sensory and instrumental analyses of flavour compounds in legumes are detailed.

1. INTRODUCTION

Flavour is one of the most important criteria in the quality and acceptability of foods (Pattee et al., 1982). Among other things, flavour can be affected by food ingredients, processing and storage conditions. Food components (e.g., proteins, carbohydrates) can interact with flavour compounds and influence flavour perception by altering the chemistry of flavour compounds in food or by increasing or decreasing their intensities (Heng et al., 2004). Various chemical families with different chemical composition and physical properties contribute to food flavours. Although carbohydrates, lipids and proteins are important in the overall flavour of foods, alcohols, aldehydes, ketones and different heterocyclic compounds play a major role in the flavour of different foods (Sessa et al., 1977; O'Connor & O'Brien, 1991; Liu, 1997; Heng et al., 2004).

Legumes such as peas, beans, lentils, soybeans have an important role in human nutrition (Sathe, 2002; Boye et al., 2010). They are sources of carbohydrates, protein, dietary fibre, vitamins, minerals, and calories (Sathe, 2002; Tharanathan & Mahadevamma, 2003; de Almeida Costa et al., 2006). Moreover, inclusion of legumes in the daily diet may have an effect on controlling and preventing diseases such as diabetes, heart disease and colon cancer (Wang et al., 1998; Sathe, 2002; Boye et al., 2010).

Legumes are used as either whole seeds, split seeds, or as flour for soup mixes, purees, and processed products. They are also used as protein sources to formulate novel protein foods which could be used as substitutes for meat or meat products. The taste and flavour of these new food protein sources is therefore an important factor to ensure consumer acceptance (Heng et al., 2004).

One of the major challenges facing the legume industry is the maintenance of a level of production to meet markets need. Seed composition and quality affect price and seed selection. The majority of studies conducted on seed quality have focused mainly on seed size and color without paying much attention to flavour and organoleptic quality. Data on the flavour properties of legumes is therefore unfortunately lacking in the literature. Detailed understanding of the flavour profile of legumes is important to ensure that the right cultivars are selected and marketed for different food applications.

Flavour analysis of foods can be studied either with sensory panels or with laboratory equipments such as gas-chromatography. This chapter will provide a review of the major factors involved in the formation of flavours and off-flavours in legumes, the different flavour compounds identified in legumes along with their characteristics, and finally synthesize some of the recent developments in the sensory and instrumental analyses of flavour compounds in legumes.

2. OVERVIEW OF FLAVOUR COMPOUNDS FORMATION IN LEGUMES

Flavour has been a restrictive factor for the utilization of legume products in foods destined for human consumption (Tsukamoto et al., 1995; Torres-Perranada & Reitmeier, 2001). A variety of factors contribute to the formation of flavour compounds found in legumes and legume products. Fresh-legume flavours, for example, are associated with the normal metabolism of the plant. On the other hand, flavour associated with cooked legumes occurs due to enzymatic, non-enzymatic and chemical reactions induced during different thermal processes (Sessa, 1979).

Legumes contain unsaturated fatty acids such as linoleic, linolenic, and arachidonic acids which are degraded by either enzymatic or non-enzymatic reactions during harvesting, storage and processing (Rackis et al., 1979). The by-products of these reactions contribute both desirable and undesirable flavours to foods. Additionally, bitterness and astringency, which occur due to the presence of saponins, phenolic acids, oxidized phospholipids, oxidized fatty acids, and isoflavones, may also impart undesirable characteristics to legumes (Arai et al., 1966; Sessa et al., 1976; Okubo et al., 1992).

Table 1 provides a list of some of the different flavour compounds identified in legumes. Some of these flavour compounds have a pleasant aroma. In legumes, however, the flavours that are of most concern are the ones contributing to the off-flavours often described as beany, grassy and/or rancid (Sessa & Rackis, 1977; Rackis et al., 1979).

2.1. Role of Enzymes in the Formation of Off-Flavour Compounds in Legumes

Legumes are susceptible to oxidative degradation by enzymatic reactions (Eskin et al., 1977) resulting in the generation of a variety of off-flavours. Enzymatic degradation of fatty acids in legumes is facilitated as a result of cell wall breakdown as well as thermal processing. The substrate of the enzyme reaction is the fatty acids present in the legume seeds. Thus, in general the higher the fat content of the legume the greater the risk of off-flavour production, especially when high amounts of unsaturated fatty acids are present. The role of lipoxygenases and lipases in generating undesirable flavours in legumes is detailed below.

2.1.1. Lipoxygenases

Lipoxygenase (LOX), an iron-containing dioxygenase is an important factor in the generation of off-flavour compounds in legumes (Wolf, 1975). This enzyme catalyzes the oxidation of polyunsaturated fatty acids containing *cis,cis*-1,4-pentadiene units to the conjugated *cis,trans*-dienoic monohydroperoxides (Rackis et al., 1979). Hydroperoxides are the initial products of LOX activity, and they are further degraded into a variety of products such as aldehydes, ketones and alcohols (Gardner, 1975; Itoh & Vick, 1999; Pérez et al., 1999). Some of the end-product flavour compounds formed have low flavour threshold values, and they play a significant role in the formation of off-flavours in many fruits and vegetables (Eriksson, 1967; Eriksson & Svensson, 1970; St-Angelo & Ory, 1975; Eskin et al., 1977; O'Connor & O'Brien, 1991; Liu, 1997).

In general, compounds produced by LOX could cause either desirable or undesirable flavours in legumes depending on their chemical structure (O'Connor & O'Brien, 1991). Kalbrener et al. (1974) described the flavour associated with linoleic acid hydroperoxide in water as grassy-beany, musty-stale and bitter, whereas that of linolenic acid hydroperoxide was described as grassy-beany, bitter and astringent.

Table 1. Some flavour compounds identified in legumes

Flavour compounds Bean[1]	Bean[1]	Pea[2]	Soybean[3]	Lentil[4]
Aldehydes	Ketones	Aldehydes	Aldehydes	Aldehydes
Acetaldehyde	Acetone	Acetaldehyde	Ethanal	Acetaldehyde
Benzaldehyde	2-Butanone	Ethanal	2-Methylpropanal	Methylpropanal
Decanal	6-Methyl-5-hepten-	Methylpropanal	Pentanal	Methylbutanal
2-Methylpropanal	2-one	Methylbutanal	Hexanal	Hexanal
Pentanal	3,5-Octadien-2-one	Hexanal	Heptana	Benzaldehyde
Hexanal	1-(2-furanyl)-1-	Benzaldehyde	(E)-2-Hexenal	Phenylacetaldehyde
2-Hexenal	propanone	Phenylacetaldehyde	Octanal	Hydrocarbons
(E)-2-hexanal	Acetophenone	2-Methylbutanal	Benzaldehyde	Toluene
Heptanal	3-Methyl-2-	Propanal	Phenylacetaldehyde	Benzene
(E)-2-Heptenal (E, E)-butanone		Pent-2-enal	Deca-2,4-dienal	Dichlorobenzene
2,4-Heptadienal	2-Pentanone	Heptanal	Nonanal	Trichlorobenzene
Octanal	3-Penten-2-one	(E)-2-Hexenal	Decanal	Ethylbenzene
Nonanal	2-Hexanone	(E)-2-Heptenal	Hydrocarbons	Xylene
(E)-2-Nonenal	2-Heptanone	(E)-2-Octenal	Trichloromethane	Styrene
Cuminyl aldehyde	2-Octanone	(E,E)-2,4-Heptadienal	(Chloroform)	Naphtalene
Hydrocarbons	Pentane-2,3-dione	Octanal	Pentane	Methyl naphtalene
Benzene	2,3-Butanedione	Nonanal	Methylcyclopentane	Dodecane
Ethyl benzene	3-Hydroxybutan-2-	Non-2-enal	Hexane	Trichloromethane
Propylbenzene	one	Hept-2,4-dienal	Benzene	(Chloroform)
1,3,5-	Pent-3-en-2-one	Non-2,4-dienal	Dichlorobenzene	Alcohols
trimethylbenzene	Octan-3-one	Dec-2,4-dienal	Trichlorobenzene	Methanol
1,2,3-	β-Ionone	Hydrocarbons	Toluene	Ethanol
trimethylbenzene	Lactones	Toluene	Octane	2-Propanol
Toluene	γ-Butyrolactone	Benzene	Ethylbenzene	1-Propanol
Xylene	γ-Hexalactone	Dichlorobenzene	Xylene	2-Butanol
Styrene	Terpenoids	Trichlorobenzene	Naphtalene	2-Methylpropanol
α-Metylstyrene	Pinene	Ethylbenzene	Heptadecane	1-Butanol
Decane	3-Carene	Xylene	Nonadecane	1-Pentanol
4-Methyldecane	Limonene	Styrene	Methyl naphtalene	1-Hexanol
Dodecane	Camphene	Naphtalene	Ketones	Ketones
Tridecane	Myrcene	Methyl naphtalene	Acetone	Acetone
Tetradecane	Terpinene	Dodecane	Methyl ketone	2-Butanone
Pentadecane	p-Cymene	Trichloromethane	2-Heptanone	2-Pentanone
Hexadecane	β-Phellandrene	(Chloroform)	3-Octanone	Acetophenone
Heptadecane	Terpinolene	Alcohols	2-Octanone	Sulfur compounds
Octadecane	α-Pinene oxide	3-Pentanol	Lactones	Dimethyl sulfide
Nonadecane	4-Terpineol	1-Penten-3-ol	γ-Butyrolactone	Carbon disulfide
Hexane	Furans	2-Methylbutanol	ß-Methyl-	γ-
Heptane	2-Ethylfuran	3-Methylbutanol	butyrolactone	
Octane	2-propyl furan	(Z)-3-Hexen-ol	γ-Valerolactone	
2, 6-Dimethyloctane	2-Pentylfuran	(E)-2-Hexanol	Δ- Valerolactone	
Nonane	3-(4-Methyl-3-	1-Octen-3-ol	γ-Hexalactone	
4-Methylnonane	pentenyl)-furan	Heptanol	γ-Nonalactone	
3-Methylnonane	Esters	Octanol	Terpenoids	
Butylcyclohexane	Ethyl acetate	Methanol	Pinene	
Undecane	2-Propyl acetate	Ethanol	Terpinene	
Octylcyclopropane	2-Butyl acetate	2-Propanol	p-Cymene	
2,6,10-Trimethyl	3-Methyl-1-butyl	1-Propanol	ß-Phellandrene	
Alcohols	acetate	2-Butanol	Limonene	
1-Penten-3-ol	2-Heptyl acetate	2-Methylpropanol	Terpinolene	
Ethanol	Bornyl acetate	1-Butanol	α-Pinene oxide	
2-Propanol	1-Hexyl acetate	1-Pentanol	4-Terpineol	
1-Propanol	Isoamylacetate	(Z)-2-Pentenol	Alcohols	

Flavour Compounds in Legumes: Chemical and Sensory Aspects

Flavour compounds Bean[1]	Bean[1]	Pea[2]	Soybean[3]
2-Butanol	(Z)-Hex-3-enylacetate	(E)- 2-Pentenol	Ethanol
2-Methyl-1-propanol	Hexyl acetate	(Z)-3-Hexenol	Methanol
1-Butanol	Methyl-2-ethyl-	(E)-3-Hexenol	2-Propanol
3-Methyl-2-butanol	hexanoate	(E)-2-Octenol	1-Propanol
3-Pentanol	(Z)-Hex-3-enyl	1-Hexanol	2-Butanol
2-Pentanol	propionate	Ketones	2-Methyl-1-propanol
3-Methyl-1-butanol	Hexyl propanoate	Acetone	1-Butanol
2-Methyl-1-butanol	Z)-Hex-3-enyl	2-Butanone	iso-Amylalcohol
3-Hexanol	butyrate	2-Pentanone	2-Methyl-1-Butanol
2-Hexanol	Sulfur compounds	2-Propanone	3-Methyl-1-Butanol
(Z)-3-hexen-1-ol	Dimethyl disulfide	Acetophenone	1-Penten-3-ol
2-Heptanol	Carbon disulfide	3-Octanone	1-Pentanol
1-Pentanol	Thiophene	Terpenoids	2-Pentanol
3-Methyl-1-butanol	Acids	Pinene	3-Pentanol
1-Hexanol	Acetic acid	Sabinene	2-Hexanol
1-Octen-3-ol	Octadienoic acid	3-Carene	1-Hexanol
2-Ethyl-1-hexanol	Nonanoic acid	Limonene	2-Heptanol
1-Octanol	Dodecanoic acid	Esters	1-Octen-3-ol
1-Nonanol	Tetradecanoic acid	Ethyl acetate	1-Octanol
Benzyl alcohol	Palmitoleic acid	Hexyl acetate	3-Octanol
Phenol	Pyrazines, pyridines,	(Z)-3-Hexenyl acetate	2-Octanol
Methanol	thiazoles	Methyl-3-	1-Nonanol
Linalool	2,6-	methylbutanoate	Benzylalcohol
Geraniol	Dimethylpyrazine	Ethyl-3-	Furans
Nerolidol	2,5-	methylbutanoate	2-Pentylfuran
α-Bisabolo	Dimethylpyrazine	Sulfur compounds	3-(4-Methyl-3-
Fernesol	2,3-	Dimethyl sulfide	pentenyl-furan)
Phytol	Dimethylpyrazine	Carbon disulfide	Esters
Furfurol	Pyridine	Methyl propyl	Methyl acetate
3-Penten-1-ol	2-Isobutyl-3-	disulfide	Ethyl acetate
(Z)-3-Hexenol	methoxypyrazine	Dipropyl disulfide	2-Propyl acetate
(E)-2- Hexenol	2-Methylpyridine	Pyrazines	1-Butyl acetate
	2-Methyl-5-	3-Isopropyl-2-	1-Pentyl acetate
	ethylpyridine	methoxypyrazine	1-Hexyl acetate
	2-Acethylpyridine	5-or6-Methyl-3-	2-Heptyl acetate
	2,4,5-	isopropyl-2-	Sulfur compounds
	Trimethylthiazole	methoxypyrazine	Carbon dioxide
	2,5-Dimethyl-4-	3-sec-Butyl-2-	Acids
	ethylthiazole	methoxypyrazine	Acetic acid
	2,4-Dimethyl-5-	3-Isobutyl-2-	Pyrazines
	ethylthiazole	methoxypyrazine	Methylpyrazine
	2-Isopropyl-4,5-		
	dimethylthiazole		
	2-Isobutyll-4,5-		
	dimethylthiazole		
	2-Acethylthiazole		
	2,4,5-Trimethyl-2-		
	thiazole		

[1](Buttery et al., 1975; de Lumen et al., 1978; del Rosario, et al., 1984; Mtebe & Gordon, 1987; Oomah & Liang, 2007; Barra et al., 2007). [2](Whitfield & Shipton, 1966; Murray et al., 1968; Lovegren et al., 1979). [3](Lovegren et al., 1979; Honig, et al., 1979; Maga et a., 1973; Kato et al., 1981; del Rosario, et al., 1984). [4](Lovegren et al., 1979).

Among plants, the highest activity of LOXs is observed in legumes (St-Angelo & Ory, 1975), and among them, soybean seeds are the richest source of these enzymes. Three to four different types of lipoxygenase isozymes have been identified in soybean (LOX 1, LOX 2, LOX 3) and additional isozymes have been identified during seed germination (Aspelund & Wilson, 1983; Matoba et al., 1985; Davies et al., 1987; Hildebrand et al., 1990; Kato et al., 1992).

As a result, the most studied legume in terms of off-favour production is soybean. Soybean contains a considerable amount of lipid (17-23%) as well as protein (35-45%) and is considered to be highly nutritious with great functionality. Although widely accepted today, soybean utilisation in foods in Western Societies met with great resistance initially due to the beany, bitter taste which remained even after processing of whole soybeans into other derived products (Rackis et al., 1979; Walker & Kochhar, 1982; O'Connor & O'Brien, 1991; Saxby, 1993; Torres-Perranada & Reitmeier, 2001; Achouri et al., 2006).

Aliphatic aldehydes and alcohols such as propanal, hexanal, 3-*cis*-hexenal, 2-*trans,6-cis*-nonadienal, pentanol, hexanol, and heptanol are some of the major compounds identified which have been linked to quality loss and off-flavours in soybean products (Morita & Fujimaki, 1973; Kato et al., 1981; Matoba et al., 1985; O'Connor & O'Brien, 1991; Wang et al., 1998). Among these compounds, hexanal, which can be formed in disrupted soybean seeds during processing is the main compound responsible for the formation of the greeny flavour in soybean products due to its low flavour threshold value (Wilkens & Lin, 1970; Hildebrand et al., 1990; Liu, 1997). Hexanol, hexenal, ethyl vinyl ketone, and 2-pentylfuran have been isolated from soymilk and described as green, grassy and/or beany (Wilkens & Lin, 1970).

Bitterness is another limiting factor in the consumption of soybeans and derived products. Oxidized soybean phosphatidylcholine has been shown to be a major source of bitterness in soybeans (Sessa et al., 1976; Baur et al., 1977; Sessa et al., 1977). Bitter phosphatidylcholines containing oxygenated fatty acids were isolated from defatted soybean flakes. These oxygenated fatty acids derive from polyunsaturated fatty acid hydroperoxides generated by the action of LOX. 5-Pentenyl-2-furaldehyde was isolated from a bitter-tasting soybean phospholipid fraction which was formed from 9-hydroperoxide of linolenic acid (Sessa, 1979). Non-volatile compounds such as 9,12,13-trihydroxy-octadec-10-enoic acid and 9,10,13- trihydroxy-octadec-11-enoic acid, resulting from the action of LOX on linoleic acid also contribute to the formation of bitterness in soybeans (Baur et al., 1977; O'Connor & O'Brien, 1991).

LOX in untreated peas can cause undesirable flavours after harvesting (Bengtsson & Bosund, 1964; Eriksson, 1967). Volatile content of peas increases rapidly during the interval between harvesting and enzyme inactivation by blanching in boiling water (Murray et al., 1968; Shipton & Last, 1968). As with soybeans, degradation of fatty acids to hydroperoxides by LOX resulted in the formation of aldehydes responsible for the off-flavour in intact peas (Whitfield & Shipton, 1966). Hexanal with a green-grassy character and 3-*cis*-hexenal with a green-beany character have been identified in raw peas (Keppler, 1977).

Degradation products of fatty acids are also responsible for the hay-like off-odour in frozen green pea (Bengtsson & Bosund, 1964; Whitfield & Shipton, 1966; Murray et al., 1968; Murray et al., 1976; Jakobsen et al., 1998). The flavour of pea flour and protein concentrate were altered from a fresh pea flavour to a musty, fishy flavour after storage at high temperature and humidity. The development of these off-flavours was a result of the degradation of pea lipids by LOX (Pattee et al., 1982). Off-flavours in stored unblanched frozen peas have also been attributed to the activity of LOX (Whitfield & Shipton, 1966).

Additionally, hexanal and 2-pentylfuran generated by the action of LOX have been reported to be responsible for beany flavours in winged beans (Saxby, 1993).

2.1.2. Lipases

Lipolytic enzymes contribute to the formation of undesirable flavours in legumes (Sessa et al., 1976; Sessa & Rackis, 1977; Pattee et al., 1982). These enzymes play a critical role in the initial hydrolysis of lipids to the corresponding fatty acids rendering them more susceptible to oxidation (Desnuell & Savary, 1963; Pattee et al., 1982). Methyl, ethyl, *p*-nitrophenyl and naphthyl esters of fatty acids, water-soluble triglyceride, short- or long-chain triglycerides and natural oils are substrates for plant lipases (Desnuell & Savary, 1963). Lipases attack the fatty acid ester bonds of triglycerides and phospholipids leading to the formation of free fatty acids. The type of flavours produced depends on the fat composition and the fatty acids liberated. Thus, formation of some types of short chain fatty acids causes undesirable flavours, whereas the long chain fatty acids (C12 and above) produce candle-like or soapy flavours (Forss, 1969; Sessa & Rackis, 1977). Furthermore, fatty acids produced by the lipases can be precursors of bitter substances (Sessa, 1985).

2.2. Non-Enzymatic Oxidation

Lipid oxidation can be caused by non-enzymatic reactions (Pattee et al., 1982; Dörnenburg & Davies, 1999). Non-enzymatic lipid oxidation is a free radical chain process which occurs in the presence of molecular oxygen (Pattee et al., 1982; Brannan et al., 2001). Hydroperoxides are the major products of oxidative reactions which in turn cause degradation of other components in the foods resulting in the formation of low molecular weight volatile compounds, particularly carbonyl compounds. These hydroperoxide degradation products contribute to the generation and development of rancid off-flavours in peas, lentils, beans, and soybeans during processing and storage (Pattee et al., 1982; Dörnenburg & Davies, 1999; Brannan et al., 2001). Although some legumes are low in fat, legumes are in general particularly prone to non-enzymatic oxidation due to their high content of unsaturated fatty acids (Choudhury & Rahman, 1973, Sessa et al., 1976; Sessa & Rackis, 1977; Pattee et al., 1982).

Aldehydes such as *cis*-pentenal, hexanal, *cis*-hexenal, heptenal, 2,4-heptadienal, 3-nonenal, 2,4-nonadienal, 2,4-decadienal are examples of some flavour compounds generated by the autoxidative degradation of fatty acids in peas (Whitfield & Shipton, 1966). Additionally, 2-pentylfuran and *cis*-3-hexenyl acetate derived from lipid autoxidation were isolated from peas (Jakobsen et al., 1998). Wilkens & Lin (1970) reported the formation of 2,4-decadienal from autoxidation of soybean oil linoleic acid. The formation of bitter flavours in soybeans has also been attributed to autoxidative degradation of fatty acids attached to phosphatidylcholine (Sessa et al., 1976; Sessa et al., 1977).

2.3. Other Compounds Influencing the Flavour of Legumes

2.3.1. Protein-Saponin Interaction

Proteins can influence the taste perception of food by binding to flavour compounds, and controlling their release during mastication. Interaction between proteins and flavour compounds can affect either the overall flavour intensity or it can result in the generation of new desirable or non-desirable flavour compounds in foods including legumes (Sessa et al., 1977; Curl et al., 1985). Soybean proteins, for example, have a stronger binding affinity to aldehydes compared to ketones. They, however, do not have binding affinity to alcohols containing six carbons or more (Franzen & Kinsella, 1974). Acid-sensitive fractions of soybean protein have binding affinity to grassy, beany, bitter, astringent flavour compounds in soybeans (Anderson & Warner, 1976).

Binding sites on proteins can be occupied by saponins (Curl et al., 1985; Potter et al., 1993) which are sterol or triterpenoid glycosides with non-volatile, amphiphilic, and surface-active characteristics (Fenwick & Oakenfull, 1983; Hostettmann & Marston, 1995; Messina, 1999; Heng et al., 2004). The name "saponins" is derived from the ability of these compounds to form stable, soap-like foams in aqueous solutions (Hostettmann & Marston, 1995; Shi et al., 2004). Saponins can therefore interfere with the interaction of flavour compounds with proteins (Curl et al., 1985; Potter et al., 1993) resulting in the formation of bitterness and astringency in legumes and their derived products (Curl et al., 1985; Okubo et al., 1992; Hostettmann & Marston, 1995; Heng et al., 2006). In milled whole pea seeds, astringent, bitter, and metallic off-flavours have been attributed to saponins (Price et al., 1985).

2.3.2. Phenolic Compounds

Phenolic compounds can contribute to bitterness and astringency in legumes (Arai et al., 1966; Sessa et al., 1976; Curl et al., 1985; Okubo et al., 1992; Hostettmann & Marston, 1995; Messina, 1999). The extent to which these compounds affect the flavour of legumes will depend on their concentration and changes to their chemical structure induced by processing (Curl et al., 1985; Price et al., 1985).

High concentrations of phenolic compounds have been reported in legumes with a dark seed coat such as soybeans, beans, peas and lentils (Dueñas et al., 2004). Bitter and astringent flavours have also been associated with soybean products (Curl et al., 1985; Tsukamoto et al., 1995; Aldin et al., 2006), soybean protein concentrates and isolates as well as traditional soybean foods (Robinson et al., 2004) which have been linked to bitter and astringent phenolic acids in soybeans (How & Morr, 1982). Furthermore, an unpleasant cooked odour in soybean products was related to the thermal decarboxylation of phenolic (p-coumaric and ferulic) acids (Pattee et al., 1982). Non-specific binding of phenolic compounds to proteins during processing of beans also resulted in the formation of a bitter taste in beans and bean products (Deshpande et al., 1984).

2.3.3. Carbohydrates

Carbohydrates in legumes occur as natural constituents. Besides their nutritional and technological functions, they have functions such as flavour carrier, and flavour and colour precursor through the Maillard reaction (Voragen, 1998) which often occurs during food processing and storage (Whitfield, 1992).

The Maillard reaction, which is a non-enzymatic reaction between the primary amino group of an amino acid and the carbonyl group of a reducing sugar, leads to the formation of heterocyclic compounds with low threshold values which can contribute to different flavour compounds in foods (Vernin & Parkanyi, 1982).

Examples of Maillard reaction products include 2-oxopropanal, 2,3-butanedione, 1,2-ethanedial, 1-hydroxy-2-propanone, 2-hydroxyethanal, 2,3-dihydroxypropanal, and 1,3-dihydroxy-2-propanone (van den Ouweland et al., 1978). Furthermore, interactions between the Maillard reaction products and other components can lead to the formation of compounds such as pyrazines, pyridines, oxazoles, thiophenes, thiazoles (Murray et al., 1970; Murray & Whitfield, 1975; van den Ouweland et al., 1978; Zhang & Ho, 1989) which are important in the flavour of different legumes as described in sections 3.4, 3.7.

3. CHEMICAL CHARACTERISTICS OF FLAVOUR COMPOUNDS IDENTIFIED IN LEGUMES

Volatile compounds such as low-molecular weight acids, alcohols, aldehydes, amines, esters, ketones, pyrazines, sulphur compounds, and terpenes contribute to the flavour of legumes. These compounds produce either the characteristic flavour associated with normal metabolism of a plant or the flavour derived during harvesting, processing and storage (Sessa et al., 1977).

Volatile compounds associated with the sensation of flavour can be formed from lipids, proteins, and carbohydrates. The different flavour compounds formed can generally be classified into different chemical families with each compound having different characteristics (Table 2) which can impart peculiar taste and flavour to legumes.

3.1. Alcohols

Oxidation of lipids by LOX leads to the formation of aliphatic and unsaturated alcohols in legumes (Oomah & Liang, 2007). Absence or very small amounts of free fatty acids has been reported in fresh legumes; however, during storage or freezing, their concentration increases due to enzymatic action (O'Connor & O'Brien, 1991; Liu, 1997) which increases the likelihood of oxidation. Alcohols can also be produced due to physical damage or during processing of plant materials (de Lumen et al., 1978).

Each alcohol has distinct characteristic which could affect the taste or flavour of legumes (Table 2). 1-Octanol, as an example, has a fresh, orange-rose odour and an oily, sweet taste. 1-Propanol, on the other hand, has an alcoholic odour, ripe and fruity flavour. 1-Heptanol has an aromatic and fatty odour and a pungent spicy taste, and 3-octanol has a fruity odour (Burdock, 2002).

Trans-2-hexenol also occurs in plant tissue, and it is the most abundant alcohol in green peas. 3-Hexenol has a strong green leafy odour. Ethanol and methanol contribute to the flavour of peas as well. The higher molecular weight saturated alcohols such as hexanol have also been shown to influence the flavour of green peas (Murray et al., 1976).

In dry beans, 1-penten-3-ol originating from linoleate oxidation has been shown to have a penetrating grassy etheral odour, and 3-methyl-1-butanol reportedly has a penetrating green aroma (Oomah & Liang, 2007).

Table 2. Description of some flavour compounds identified in legumes

Compounds	Description	Natural occurrence[1]
Aldehydes		
Benzaldehyde	Aromatic taste similar to bitter almond	Bean (Oomah & Liang, 2007; del Rosario et al., 1984; Barra et al., 2007) Pea (Lovegren et al., 1979)
Hexanal	Fatty, green, grassy, fruity odour and taste	Soybean
2-Hexenal (E, 2-hexenal)	Sweet, fragrant, almond, fruity green, leafy, apple, plum, vegetable odour	Soybean, pea
2-Heptenal (E, 2-heptenal)	A pungent green fatty odour	Soybean, bean, pea
2-Octenal (Trance-2-octenal)	Green-leafy odour, orange, honey-like, cognac-like aroma	Soybean, pea
2, 4-Nonadienal	Fatty, floral odour	Soybean, bean, frozen pea
n-Valeraldehyde (n-Pentanal)	Powerful, acrid, pungent odour, warm, slightly fruity and nut-like taste	Soybean, pea
n-Octanal (Octyl aldehyde)	Fatty, citrus, honey odour	Soybean, bean, pea
Trans-2-cis-6-Nonadienal	Reminiscent of green cucumber odour	Pea
2-Nonenal (Trans-2-nonenal)	Powerful, penetrating fatty odour in concentrated form, orris-like, waxy and pleasant odour on dilution, reminiscent of dried orange peels odour	Soybean, bean, pea
3-Methyl-2-butenal	Almond odour	Bean (Barra et al., 2007) Pea (Lovegren et al., 1979; Murray et al., 1968)
3-Methylbutyraldehyde (3-methylbutanal, Isovaleraldehyde)	Choking, powerful, acrid, pungent, apple-like odour, fruity, fatty, animal, almond odour	Soybean, bean, pea
Hydrocarbons		
Styrene	Sweet, balsamic, floral odour	Soybean, bean, pea
Alcohols		
2-Pentanol	Mild green, fusel oil, winey, ethereal odour	Soybean, bean
2-Heptanol	Brassy, herbaceous odour reminiscent of lemon, fruity, green, bitter taste	Soybean, bean
1-Octen-3-ol	Powerful, sweet, earthy odour, herbaceous note reminiscent of lavender-lavandin, rose and hay, sweet, herbaceous taste	Soybean, bean
1-Octanol	Fresh, orange-rose odour, sweet, oily, herbaceous taste	Soybean, bean, pea
Isopropyl alcohol (2-propanol)	Alcoholic, unpleasant odour, burning taste	Soybean, bean, pea
1-Propanol (Propyl alcohol)	Alcoholic odour, ripe, fruity flavour	Soybean, bean, pea
Isobutyl alcohol (2-Methyl propanol)	Wine-like disagreeable odour	Soybean, pea
1-Butanol (Butyl alcohol)	Fusel-like sweet, pleasant odour, dry, burning taste	Bean (del Rosario et al., 1984; Lovegren et al., 1979) Pea (Murray et al., 1968)
1-Penten-3-ol	Bitter, mild green odour	Soybean, bean, pea
Heptyl alcohol (1-heptanol)	Fragrant, woody, heavy, oily, faint, aromatic, fatty odour, pungent spicy taste	Soybean, pea
Linalool	Floral odour free from camphoraceous and terpenic notes	Pea

Compounds	Description	Natural Occurence[1]
3-Hexanol	Alcoholic, ethereal, medicinal odour	Soybean, bean
Guaiacol	Sweet odour, slightly phenolic	Soybean, bean
Ketones		
2-Pentanone	Wine, acetone-like odour, penetrating, buttery taste	Soybean, bean, pea
2, 3-Pentanedione	Sweet odour similar to quinine	Soybean, bean, pea
3-Octanone	Strong, penetrating, fruity odour reminiscent of lavender	Soybean, bean, pea
2- Octanone	Floral and bitter, green, fruity odour, bitter, camphoraceous taste	Soybean, bean, pea
2-Nonanone	Rue odour, rose and tea-like flavour	Soybean
2-Butanone	Sweet apricot-like odour	Natural products Bean (Oomah & Liang, 2007; del Rosario et al., 1984; Lovegren et al., 1979)
1-Octen-3-one	Mushroom odour	Soybean, pea
Terpenoids		
3-Carene	Sweet, pungent turpentine-like taste	Natural products Bean (Oomah & Liang, 2007)
α -Pinene	Pine-like, turpentine-like odour, resin-like odour in oxidized material	Soybean
Limonene	Lemon-like odour free from camphoraceous and turpentine-like notes	Bean (Oomah & Liang, 2007; Buttery et al., 1975; del Rosario et al., 1984; Barra et al., 2007)
Esters		
Ethyl acetate	Ethereal-fruity, brandy-like odour, reminiscent of pineapple, fruity sweet taste	Soybean
Methyl-2-methylbutyrate	Sweet, fruity, apple-like odour, apple-like taste	Pea
Sulfur compounds		
Methyl disulfide (Carbon disulfide)	Diffuse, intense onion odour	Soybean, bean, pea
Acids		
Hexanoic acid (Caproic acid)	Sickening, sweaty, rancid, sour, sharp, pungent, cheesy, fatty, unpleasant odour reminiscent of copra oil, acid taste	Soybean, bean
Pyrazines		
2-Ethyl pyrazine (Ethyl pyrazine)	Peanut butter, musty, nutty, woody, buttery odour	Soybean
2-Methoxy-3-isopropylpyrazine (2-Isopropyl-3-methoxyparazine)	Earthy, bell pepper, raw potato, galbanum aroma	Bean, pea
2-Ethyl-3,5(6)-dimethylpyrazine	Rroasted coca odour	Soybean products
2-Ethyl-5-methyl pyrazine	Odour threshold of 100 ppb in water	Soybean products
2-Ethyl-3-methylpyrazine	Strong, raw potato, roasted, earthy odour	Soybean products

[1]All description of compounds and their natural occurrence are cited from G. A. Burdock (2002), except those for which other references are provided.

3.2. Aldehydes

Aldehydes are derived from either enzymatic or autoxidative decomposition of fatty acids, mainly linoleic and linolenic acids (Gaddis et al., 1961; Buttery et al., 1975; Barra et al., 2007).

Physical damage, processing and storage can lead to the formation of aldehydes in legumes. Blending and frost damage could cause enzymatic oxidation of fatty acids present in legumes leading to the formation of aldehydes such as, hexanal, *trans*-2-hexenal, heptanal, *trans*-2-heptenal, 2,4-heptadienal, *trans*-2-octenal, and 2-nonenal (Buttery et al., 1975; Oomah &Liang, 2007). Furthermore, 2-methylpropanal and 2,3-methylbutanal are typical aldehydes in dried vegetables (Whitfield, 1992) and have been isolated from French beans and dried beans (Lovegren et al., 1979). Ethanal, propanal, hexanal, 2-pentenal, 2-hexenal, 2-heptenal, 2-octenal, 2-nonenal, 2,4-heptadienal, and 2,4-nondienal have been identified in stored unblanched frozen peas (Whitfield & Shipton, 1966) and are responsible for off-flavour formation in frozen peas (Bengtsson & Bosund, 1964; Whitfield & Shipton, 1966; Barra et al., 2007). Formation of these aldehydes occurs as a result of the decomposition of hydroperoxides derived from enzymatic oxidation of unsaturated fatty acids.

The presence of aldehydes can affect the flavour of legumes and derived products as they have different characteristics. For example, hexanal has a fatty, green, grassy, fruity odour and taste whereas pentanal has an acrid, pungent odour and a warm, slightly fruity and nut-like taste. *Trans*-2-octenal has a green-leafy odour, and *trans*-2-heptenal has a pungent green fatty odour (Burdock, 2002). *cis*-4-Heptanal and 2,4-pentadienal, isolated from soybean oil have, respectively, fish- or potato-like flavours (Seals & Hammond, 1970). Alka-2,4-dienals, generated by enzymatic oxidation in legumes, have been described as oxidized, cardboard-like, oily, and painty (Sessa et al., 1977; Rackis et al., 1979; Sessa, 1979).

3.3. Hydrocarbons

Alkanes derive from oxidative decomposition of lipids in foods, and they contribute to desirable flavour characteristics of green beans and peas (Perkins, 1989). Aromatic hydrocarbons such as toluene, ethylbenzene, 1-methylethylbenzene, xylene, styrene, furans and 1,3,5-trimethylbenzene are also derived from oxidation of unsaturated fatty acids (Märk et al., 2006). They affect the characteristic aroma and flavour of legumes due to their distinct characteristics (Table 2), (e.g., styrene with a sweet and floral odour) (Burdock, 2002). Trichloromethane (chloroform) a chlorinated hydrocarbon is considered to be a natural compound in plants and it is produced in plants even with minimal exposure to chlorinated organic compounds (Buttery et al., 1975; Lovegren et al., 1979). Some aliphatic and aromatic hydrocarbons, however, cause off-flavours in legumes (Murray et al., 1976).

3.4. Sulphur Compounds

Volatile sulphur compounds occur either naturally in foods or are formed during processing and storage. Sulphur compounds contribute to the overall flavour and sensory properties of foods (Shankaranarayana et al., 1974; Maga & Katz., 1975). Most sulphur compounds have very low odour threshold values and are important in the aroma perception of foods. Heterocyclic sulphur compounds such as thiazoles, thiophenes, trithiolanes, trithianes are important flavour compounds in cooked foods (Mottram et al., 1996). Low boiling sulphur compounds such as hydrogen sulphide, methanethiol, ethanethiol, propanethiol and dimethyl sulphide have been reported in cooked peas and beans (Self et al., 1963). Dimethyl sulfoxide has also been identified in peas (Shankaranarayana et al., 1974).

Thiophenes are a group of sulphur compounds which contribute to the undesirable sensory properties of cooked foods. Thiophene has been reported in heated soybean protein isolate (Maga & Katz., 1975). Formation of sulphur compounds such as dipropyl disulfide and methyl propyl disulfide have been reported in blanched green peas (Buttery, 1981; Jakobsen et al., 1998).

3.5. Ketones

Ketones are derived from lipid oxidation. Seals & Hammond (1970) associated 2,3-pentanedione to the buttery flavour found in oxidized soybean oil. Ketones such as acetone, 6-methyl-5-hepten-2-one and 3,5-octadien-2-one are abundantly found in dry beans (Oomah & Liang, 2007). Ketones have distinct sensory characteristics. For example, cyclohexanone, 2-pentanone, and 2-butanone have, respectively, a peppermint or acetone-like odour, wine or acetone-like odour, and sweet apricot-like odour (Burdock, 2002).

3.6. Terpenes

Terpenes such as 3-carene, α- and β-pinene and limonene are found in essential oils and headspace of flowers and leaves (Jakobsen et al., 1998; de Schutter et al., 2008). The presence of these compounds in legumes may result from the degradation of carotenes by either legume LOXs or hydroperoxides generated by these enzymes (Arens et al., 1973; Weber & Grosch, 1976; Jakobsen et al., 1998).

Murray et al. (1976) reported the presence of cineole in green peas and suggested that this terpenoid was adsorbed from the soil.

3.7. Pyrazines

Pyrazines are heterocyclic nitrogen containing compounds (Maga et al., 1973; Müller & Pappert, 2010). Natural pyrazines have been isolated from non-heat treated foods. Pyrazines are associated with desirable food flavours and contribute to the characteristic aroma of vegetables (Maga et al., 1973; Alberts et al., 2009; Müller & Pappert, 2010). They are readily evaporated due to their low vapour pressure and have intense smells and contribute to the aroma and flavour of many foods (Maga et al., 1973; Alberts et al., 2009; Müller & Pappert, 2010). As an example, pyrazines contribute directly to flavour of fresh foods such as peas, tomatoes, green bell papers as well as to the flavour of roasted or cooked foods. The formation of pyrazines in heat-treated foods is due to the Maillard reaction (Koehler et al., 1969; Murray et al., 1970; Murray & Whitfield, 1975).

3-Alkoxy-2-, 3-isopropyl-, 3-*sec*-butyl- and 3-isobuty-2-methoxypyrazines are found in many vegetables and natural products (Murray & Whitfield, 1975). 3-Alkyl-2-methoxypyrazines has been suggested to contribute significantly to the characteristic green pea aroma (Murray et al., 1970; Murray & Whitfield, 1975; Murray et al., 1976). 2-Isobutyl-3-, 2-butyl-3- and 2-isopropyl-3-methoxypyrazines were isolated in green peas (Murray et al., 1970; Murray & Whitfield, 1975), and 3-isopropyl-2-methoxypyrazine has been reported in raw and cooked French beans (Hinterholzer et al., 1998). Murray & Whitfield (1975) also isolated methoxypyrazines in the juice of unblanched peas.

4. FLAVOUR IMPROVEMENT IN LEGUMES

Undesirable flavours such as beany, grassy, bitter, and astringent associated with legumes and legume products are the most important barrier in the utilization of these resources for human consumption. Removal or reduction of these undesirable flavours has therefore been the subject of many studies. Inhibition and/or inactivation of LOX, removal of components responsible for the undesirable flavour and flavour masking are the three techniques which have been extensively explored for improving the flavour of legumes and derived products (Macleod et al., 1988).

4.1. Inhibition and/or Inactivation of Lipoxygenase

Most methods developed commercially for controlling LOX activity are based on heat treatments such as dry or steam heating (Wolf, 1975; Sessa, 1979; Gardner, 1980; Jakobsen et al., 1998). For example, soymilk having a good flavour has been produced by grinding soybeans in hot water (Wilkens et al., 1967), and off-flavour formation in peas has been prevented by blanching to inactivate LOX (Jakobsen et al., 1998).

To provide mild heating condition as well as to minimize insolubilization of the legume proteins, a combination of heating with the use of aqueous alcohol has been explored (Borhan & Snyder, 1979; Brown et al., 1982; Ediriweera et al, 1987). Soaking of soybeans, full-fat soybean flour, defatted flour, concentrate, or isolate in 50% ethanol for 24 h at 25°C decreased soybean LOX activity as well as reduced intensities of grassy, beany, bitter and astringent flavours (Eldridge et al., 1977).

Furthermore, the activity of LOX can be inhibited by natural and synthetic antioxidants and inhibitors. Application of α-tocopherol, propyl gallate, and nordihydroguaiaretic acid inhibit soybean LOX activity (Gardner, 1980). Cysteine treatment reversibly inhibits LOX activity as well (Gardner et al., 1977).

Genetic removal of LOX has also been attempted to improve legume flavours (Kobayashi et al., 1995; King et al., 1998). Production of soymilk with LOX-null soybean mutants resulted in the reduction of beany, rancid, and oily flavours in the product (Matoba et al., 1985; Davies et al., 1987).

Processing of legumes in the absence of oxygen has also been proposed as a technique to prevent off-flavour development (Wolf, 1975; Sessa et al., 1976; Rackis et al., 1979; Pattee et al., 1982).

4.2. Removal of Undesirable Flavour Compounds from Legumes

A variety of approaches using heat, solvents, pH adjustment, enzymes or physical methods have been applied to remove undesirable flavour compounds from legumes and legume products.

Flavour attributes of legumes can be affected by different heat treatment techniques such as toasting and hot-water blanching (Iwuoha & Umunnakwe, 1997). Toasting, for instance has been used by the soybean industry to improve soybean and soybean product flavours. Using this technique resulted in replacing the beany flavour of soybean with sweet, nutty, and toasted flavours in defatted soybean meal (Wolf, 1975; Kato et al., 1981). Strong residual beany flavours in soymilk prepared with unblanched soybeans was not observed in soymilk prepared from either hot-water blanched or toasted beans (Iwuoha & Umunnakwe, 1997).

Vacuum distillation has also been used by many workers to eliminate unwanted compounds responsible for beany, greeny flavours from legumes (Arai et al., 1970; Sugawara, et al., 1985).

Binding of other components in legumes to flavour compounds may cause undesirable flavours, thus, using enzymatic treatments to break down these bonds can be useful for improving flavour quality (Abdo & King, 1967; Fujimaki, et al., 1968; Dijkstra et al., 2003). Abdo & King (1967) extracted defatted soybean flakes with water containing carbohydrases, proteases, lipases, oxidases and pectinases which improved the flavour of the product (Abdo & King, 1967). However, using proteolytic enzymes to break down the bonds between soybean proteins and flavour components may also cause bitterness mainly due to the formation of bitter peptides (Fujimaki, et al., 1968; Cho et al., 2004).

Solvent extraction and pH adjustment techniques are also used to reduce objectionable flavours from the legumes. The former method involves in the fractionation of legume proteins based on their isoelectric point values (Che Man et al., 1989; Dijkstra et al., 2003; Cordero-de-los-Santos et al., 2005). However, acid degradation may sometimes cause bitterness in the product (Maga, 1973).

In comparison with other methods used to improve the legume flavour, solvent extraction has been widely used due to its better efficiency (Rackis et al., 1975; Honig et al., 1976). Although using organic solvents such as pentane and hexane are useful from the economic viewpoint as well as their efficiency in extracting oil from legumes, they are not able to remove undesirable flavours from soybeans and derived products (Rackis et al., 1979). Among the different solvents suggested as substitution for pentane and hexane, alcohols have caused better flavour outcome in soybean products (Honig et al., 1976).

Solvent extraction has also been applied for extracting and identifying bitter compounds like saponins and isoflavones from legumes (Achouri et al., 2005; Aldin et al., 2006; Xu & Chang, 2007). Various isoflavones and saponins have been identified in defatted soybean flakes, soybean protein isolate and soybean germ extracted and fractionated with ethanol (Aldin et al., 2006). The result of fractionation of the different soybean extracts showed that malonyl-ß-glucoside isoflavone and 2,3-dihydro-2,5-dihydroxy-6-methyl-4H-pyran-4-one-saponin contributed to bitter and astringent flavours more than other isoflavones and saponins found in the different soybean extracts.

Combination of solvent extraction with other treatments may cause better flavour in products compared to using only the solvent. In this regard, the use of steam in conjunction with alcohol extraction resulted in the removal of greeny, beany and bitter tastes from defatted soybean flakes (Honig et al., 1976). Honig and coworkers (1976) reported that alcohol extraction of toasted defatted flakes improved flavours of concentrates and isolates prepared from them. Extraction of black bean, lentil, black soybean, and red kidney bean with acidic 70% acetone yielded the highest total phenolic content and total flavonoids content compared to either 50% or 80% acetone alone (Xu & Chang, 2007). The impact of this latter technique on flavour and taste was, however, not reported.

4.3. Masking of Undesirable Flavour Compounds

There are different approaches to mask undesirable flavours in legume products. One of these approaches, for example, is the addition of meat flavourings to soybean proteins to give a desirable meaty flavour to the product (MacLeod & Seyyedain-Ardebili, 1981). Another approach attempted is to add sugar and amino acids to soybean proteins to impart desirable flavour on heat processing due to the Maillard reaction (Moll et al., 1979). Furthermore, addition of this masking agent causes reduction in bitter and astringent flavours in soybean products (Calvino et al., 1990; Aldin et al., 2006).

Additionally, off-flavour in soymilk has been masked with either adding different flavourings (e.g., strawberry, chocolate, vanilla) (Wang et al., 2001) or with lactic acid fermentation (Pinthong et al., 1980; Buono et al., 1990). Addition of chocolate and almond flavourings to soymilk improved sensory properties and partially masked the beany flavour in the product (Wang et al., 2001). Preparation of soymilk with chocolate reduced the beany flavour of soymilk as well (Gatade et al., 2009). Also, Pinthong et al. (1980) reported that production of soymilk fermented by adding commercial yoghurt culture containing lactic acid bacteria (i.e., mixture of *Lactobacillus bulgaricus* and *Streptococcus thermophilus*) resulted in partially masking the beany flavour in the product.

4.4. Other Methods

Centrifugation, filtration (Ohta et al., 1980; Ohren, 1981; Dijkstra et al., 2003), foam formation (Baur & Grosch, 1977; Sathe & Salunke, 1984), soaking (Sathe & Salunke, 1984; Uzogara & Ofuya, 2007), and the use of activated charcoal or ion-exchange resins (Arai et al., 1966; Wolf & Cowan, 1971; Dijkstra et al., 2003) have been investigated as physical methods to improve legume flavours. The use of activated charcoal and ion-exchange resins led to the removal of phenolic compounds, but bitter and astringent flavours remained in the soybean isolate (How & Morr, 1982). Soybean isolate treated with activated carbon had better flavour compared to the sample treated with ion-exchange resin (How & Morr, 1982).

5. INSTRUMENTAL ANALYSIS OF FLAVOURS IN LEGUMES

Flavour analysis of foods can be studied with either sensory panels or laboratory equipments and could be either qualitative or quantitative. Some of the most recent analytical techniques used to analyse volatile compounds in legumes are reported below.

5.1. Electronic Nose

Flavour of foods can be evaluated by the perception of volatile compounds either with the human nose or with instruments based on the same principle as the human nose (Schaller et al., 1998; Stephan et al., 2000). Electronic nose is a rapid and simple technique for detecting volatile compounds (Craven et al., 1996; Dewettinck et al., 2001). The principle of this technique is to mimic the human olfaction system (Craven et al., 1996).

An odour sensor array, a data pre-processor, and a pattern recognition engine are the major parts of an electronic nose (Craven et al., 1996). Metal oxide semiconductors, metal oxide semiconductor field effect transistors, conducting organic polymers, and piezoelectric crystals (bulk acoustic wave) are the most common sensors used in the electronic nose (Schaller et al., 1998).

In electronic nose, volatilization of compounds is based on the static headspace technique. Volatile compounds are concentrated by heating in the vapour phase which is on the liquid or solid sample, and then they are introduced into the detection system. The detection system comprises of a sensor that measures the different physical and chemical properties of aroma compounds and then converts them into a measurable signal which is processed by a computer using chemometric techniques. The results are plotted on a chart that represents the fingerprint of this odour. Most electronic noses use sensor arrays that react with volatile compounds. The adsorption of volatile compounds on the sensor surface causes a physical change of the sensor. A specific response is recorded by the electronic interface and the signal is tranferred into a digital value (Schaller et al., 1998; Stephan et al., 2000).

Electronic nose is used in the food and beverage industry (Craven et al., 1996) by research and development laboratories and quality control laboratories for various purposes (Schaller et al., 1998; Stephan et al., 2000; Dewettinck et al., 2001).

An electronic nose is a multi-sensor array analytical technique not a real nose (Schaller et al., 1998) so human sensory analysis of foods is still evaluated by the human nose (Stephan et al., 2000).

5.2. Solid Phase Microextraction Gas Chromatography

Gas chromatography (GC) is used in analytic chemistry for separating and analyzing compounds of a mixture that can be vaporized without decomposition. In GC, the mobile phase which is the carrier phase is an inert gas, usually helium and the stationary phase is a very thin layer of liquid or polymer on an inert solid support inside a column. The volatile compounds being analyzed interact with the walls of the column, which is coated with different stationary phases. Each compound elutes from the column at a different time, known as the retention time of the compound (Willet, 1987; Grob & Barry, 2004). The eluted compounds are then identified using an appropriate detector. Flame ionization detector and mass spectrometry are the most commonly used detectors for flavour analysis (Vas & Vékey, 2004).

Most flavour compounds in foods are volatile. This characteristic is used for their separation from food matrices. As flavour compounds usually occur at concentrations too low to be detected by GC, headspace extraction or some form of concentration of volatiles is required prior to GC (Werkhoff et al., 1998; Deibler et al., 1999; Prosen & Zupančič-Kralj, 1999; Zambonin, 2003).

Different methods such as purge and trap, static headspace, liquid-liquid, solid phase extraction, and solid phase microextraction are used for extraction and concentration of the headspace. The composition of a flavour extract will therefore depend on the isolation method.

Among different isolation methods, solid phase microextraction (SPME) has gained increased attention as a new technique for sample preparation (Arthur & Pawliszyn, 1990; Pawliszyn, 1995; Deibler et al., 1999; Kataoka et al., 2000; Ouyang & Pawliszyn, 2008). In comparison with conventional extraction techniques, SPME is a solvent-free, less expensive, fast, simple and easily automated method with different analytical techniques for the analysis of volatile, semi-volatile, polar and non-polar compounds in different food systems.

The principle of SPME is the adsorption of volatile compounds onto an adsorbent. Adsorption is based on the equilibrium partitioning of the analytes between the adsorbent and the sample matrix. Adsorbed analytes are desorbed into an analytical instrument such as a gas chromatograph (GC), high performance liquid chromatograph (HPLC), liquid chromatograph or capillary electrophoresis (CE) equipment (Pawliszyn, 1995; Eisert & Pawliszyn, 1997; Li & Weber, 1997; Pawliszyn, 1997; Prosen & Zupančič-Kralj, 1999; Zambonin, 2003).

Most SPME methods are coupled with GC due to the ease of automation (Penũalver et al., 1999; King et al., 2003). In SPME-GC, the injection port of the GC is used for thermal desorption of analytes from a fibre. When the temperature increases, the affinity of analytes towards the fibre is reduced and they are liberated into the GC column (Pawliszyn, 1995; Penũalver et al., 1999; King et al., 2003; Vas & Vékey, 2004; Anli et al., 2007). Different SPME techniques are available as follows:

5.2.1. Fibre Solid Phase Microextraction

In fibre solid phase microextraction (Fibre-SPME), a fused-silica fibre coated with an appropriate stationary phase is used. Analytes are extracted from the sample onto a polymeric stationary phase coated onto the fibre. The solid phase allows the concentration of the organic analytes in the sample matrix. Fibre-SPME can be used in either headspace mode or direct extraction mode to extract the analytes from food matrices (Pawliszyn, 1997; Prosen & Zupančič-Kralj, 1999).

In the headspace mode, sample is placed into a vial, which is sealed with a septum-type cap. The fibre is exposed in the vapour phase above a gaseous, liquid or solid sample. Volatiles adsorb onto the fibre. After extraction, the SPME fibre is placed into the injection port of a GC where the analytes are thermally desorbed (Pawliszyn, 1995; Prosen & Zupančič-Kralj, 1999; Vas &Vékey, 2004).

In the direct mode, analytes in aqueous samples are extracted by immersion of the fibre in the liquid sample. In direct extraction, adsorption is based on the equilibrium partitioning of the analytes between the aqueous matrix and the fibre coating. The analyte is desorbed from the fibre by heating the fibre into the injection port of the SPME-GC, or by loading solvent into the desorption chamber of the SPME-HPLC (Pawliszyn, 1995; King et al., 2003; Vas & Vékey, 2004).

In headspace extraction, equilibrium is obtained more rapidly than in direct extraction. Analytes with a high vapour pressure are extracted by immersion or headspace whereas analytes with a low vapour pressure are extracted only by immersion (Wan et al., 1994; King et al., 2003).

Several different fibres are used in SPME. Polydimethylsiloxane (PDMS) is apolar and has a high affinity for non-polar compounds such as benzene, toluene, ethylbenzene and xylene. Polyacrylate (PA) has a more polar coating material and extracts polar compounds such as phenols and their derivatives. Bipolar fibres also exist such as PDMS-divinylbenzene (DVB), PDMS-carboxen and carbowax-DVB which have greater holding capacity than monopolar fibres (Chen & Pawliszyn, 1995; Penüalver et al., 1999; Vas & Vékey, 2004; Januszkiewicz et al., 2008). PDMS-DVB (60 µm) and CW-TPR (carbowax-templated resin) are used for SPME coupled with HPLC (Penüalver et al., 1999).

In fibre-SPME the extraction of analytes by the fibre can be affected by the characteristic of the coating materials, as well as the extraction temperature and time, presence of salt or an organic solvent, pH, agitation, sample volume, matrix effects and derivatization (King et al., 2003).

5.2.2. In-Tube Solid Phase Microextraction

In-tube solid phase microextraction (In-tube SPME) is used as an SPME device coupled with HPLC or LC (Chen & Pawliszyn, 1995; Vas & Vékey, 2004). In this sample preparation technique, organic compounds in aqueous samples are directly extracted from the sample into a capillary column. These compounds are then desorbed by introducing a stream of mobile phase, or by using a static desorption solvent. The desorbed compounds are subsequently injected into an analytical instrument for analysis. In-tube SPME technique is used for extraction of polar and non-polar compounds in liquid samples (Chen & Pawliszyn, 1995).

5.2.3. Solid Phase Dynamic Extraction

In solid phase dynamic extraction (SPDE), stainless steel needles coated with polydimethylsiloxane and 10% activated carbon are used. Analytes in SPDE are concentrated onto the polydimethylsiloxane and activated carbon coated onto the inside wall of the stainless steel needle of a gas tight syringe. The trapped analytes are then thermally desorbed into a GC injection port (Kataoka et al., 2000).

5.2.4. Stir Bar Sorptive Extraction

In stir bar sorptive extraction (SBSE), a magnetic stir bar coated with polydimethylsiloxane is used, similar to fibre-SPME but in a thicker layer. In this technique, the sample is put into a 20-mL headspace vial, which is stirred with a stir bar coated with the polydimethylsiloxane. After sampling, the stir bar is removed with tweezers. After removing the residual water, it is placed in an empty glass tube for thermal desorption. Using SBSE results in higher recovery of the sample compared to Fibre-SPME based on PDMS (Kataoka et al., 2000).

In summary, SPME is widely used as an extraction and concentration step prior to analysis by analytical methods. This technique has advantages in sample preparation, increasing reliability, selectivity, sensitivity and reducing the cost and time of analysis (Vas & Vékey, 2004). SPME has been applied in the analyses of different flavour compounds in foods such as legumes and other vegetables, fruits, beverages and dairy products (Prosen & Zupančič-Kralj, 1999; Kataoka et al., 2000).

5.3. Gas Chromatography-Olfactometry

In gas chromatography-olfactometry (GCO), the human nose is used as a sensitive and selective biosensor to detect volatiles eluting from the GC column. Volatile compounds in food have different aroma threshold values. Volatile compounds with high concentration may have little or no aroma activity, whereas volatile compounds with low concentrations may produce intense aroma activity. In general, GCO allows the selection of odour-active compounds from non-odourous volatile compounds (Deibler et al., 1999; Stephan et al., 2000; Dewettinck et al., 2001; Frank et al., 2001; van Ruth & O'Connor, 2001) based on their odour threshold values (Gocmen et al., 2004). This technique can also be used to estimate volatile flavour intensity. The description of compounds (using a lexicon) as well as estimation of the detection time is carried out by a trained human assessor (Gocmen et al., 2004).

6. SENSORY EVALUATION METHODS

In addition to instrumental techniques, sensory evaluation using human panels can provide valuable information to determine consumer acceptance (Piggott, 1984). In general, sensory evaluation methods are divided into affective and analytical sensory tests. The affective method evaluates consumer preference and/or opinion about a product. This test is based on employing a large number of untrained consumers, whereas the analytical method is based on employing trained panelists. The most important analytical tests are discriminating tests (difference and threshold), as well as descriptive sensory test. Analytical methods are suitable for both identifying flavour compounds in a product and discriminating sensory properties between different products. Popular difference tests are triangle tests and duo-trio tests. Descriptive sensory tests discriminate between different products based on their sensory characteristics and provide a quantitative description of the sensory differences (Piggott, 1984; Marsili, 2007). The most important sensory evaluation tests are described below:

6.1. Analytical Sensory Tests

6.1.1. Discriminating Tests
The objective of a discriminating test is to determine differences between 2 or more products (Piggott, 1984; Drake, 2007; Marsili, 2007). Popular difference tests are paired-comparison, duo-trio, and triangle tests. The paired comparison test is a two-sample test to

determine whether the products are the same or different. In the duo-trio test, the subject is presented a reference sample as well as two coded samples. One of the coded samples is the same as the reference sample. The subject is asked to indicate which sample is the same as the reference or different from it. In the triangle test, all three products are coded and the subject is asked to determine which two samples are similar or which one is different from the others (Piggott, 1984; Stone & Sidel, 1993; Drake, 2007; Marsili, 2007).

6.1.2. Threshold Tests

A threshold is the lowest concentration at which a sensory response is detectable. Threshold tests are applied to both undesirable and desirable compounds in foods (Drake, 2007). Taste dilution analysis (TDA) was developed for screening of taste-active compounds. This technique is based on the determination of the relative taste threshold of compounds in a serial dilution of a sample (Frank et al., 2001; Seo et al., 2008).

TDA has been used to evaluate the taste threshold of hydrolysed soybean protein isolate. Soybean protein isolates are applied as functional and nutritional ingredients in many foods. Partially hydrolyzed soybean protein isolate, however, has a bitter taste (Matoba & Hata, 1972). Seo et al. (2008) evaluated bitterness of enzyme-hydrolyzed soybean protein isolate by TDA. In their study, soybean protein isolate was hydrolyzed with different commercial proteases, i.e., flavourzyme, alcalase, neutrase, protamex, papain, and bromelain at the same condition. Taste dilution (TD) factor was defined as the dilution at which a taste difference between the diluted sample and 2 blanks could be detectable. As enzymatic hydrolysis increased, the bitterness increased as well. Alcalase showed the highest TD factor, whereas flavourzyme showed the lowest TD factor. The authors indicated that TDA could be applied as an alternative test to the hedonic scale in evaluation of the bitterness in soybean isolates (Seo et al., 2008).

6.1.3. Descriptive Tests

Descriptive tests provide a qualitative and/or quantitative description of products based on the perception of a group of qualified panels. The panel members are highly trained, and they are also capable of using a standardized terminology (a lexicon) to describe their perception (Piggott, 1984; Price et al., 1985; Stone & Sidel, 1993; Drake, 2007). Evaluation during training and on actual product is carried out individually and usually in an isolated sensory booth (Drake, 2007; Marsili, 2007).

The most frequently described flavours used for description of legumes include beany, bitter, and nutty flavours. Other flavours attributed to legume products are green, grassy, sweet, toasted, musty, stale, spoiled, raw beany, cardboard, and chalky (Cowan et al., 1973; Rackis et al., 1979). Aldehydes formed from oxidized unsaturated fatty acids are responsible for oxidized flavours described as painty, oily, tallowy, or cardboardy. Ketones such as 1-octen-3-one and 1-penten-3-one are, respectively, responsible for metallic and oily flavours in legumes (Pattee et al., 1982).

In dry beans, the following aldehydes have been used for aroma descriptors: hexanal (green-grassy), octanal (green-fresh), nonanal (green), benzaldehyde (bitter almond-like, fragrant, aromatic), heptenal (green, fatty, pungent), decanal (musty, fruity), 2-nonenal (stale, musty, buttery oily), 2-hexenal (green, bitter almond-like) (Oomah & Liang, 2007). Furthermore, flavours associated with soybean products with undesirable taste characteristics have been described with terms such as beany, green, nutty, astringent and bitter flavours (Huang et al., 1982).

Quantitative descriptive analysis was used to evaluate flavour of rehydrated diced French beans using a panel of 21 assessors. They scored for taste attributes such as cooked vegetables, grassy, sweet, spicy, sour, sharp and bitter. French beans were scored high in cooked vegetables, grassy, sweet and spicy attributes (van Ruth et al., 1995).

Price et al. (1985) evaluated undesirable sensory properties of dried pea using two groups of assessors. In preliminary studies, a hedonic rating was obtained on the pea sample by sixty assessors. Assessors rated the samples on a 9-point scale from "like extremely" through "neither like nor dislike" to "dislike extremely". They also described the sample, using their own words. Subsequently, nine of these assessors were trained to familiarise them with words such as bitter, astringent and metallic. Price et al. (1985) concluded that a significant part of the overall taste defects of the pea flours was due to the presence of bitter, astringent and metallic-tasting compounds. The role of saponins in the undesirable sensory properties of the dried pea was reported.

6.2. Affective (Consumer) Tests

The objective of affective tests is to evaluate preference and/or acceptance and/or opinion of a product with a large number of consumers (Piggott, 1984). In these sensory tests, individual consumers are used to identify and quantify the attribute of products. However, this group of tests is expensive, diverse, and complex. Consumers are variable due to differences in age and the effect of advertising. Large companies have sensory or market research departments that conduct such tests regularly with a large number of representative consumers (Piggott, 1984; Frank et al., 2001; Drake, 2007; Marsili, 2007).

The paired-comparison test and the nine-point hedonic scale are the most frequently used methods to evaluate, respectively, preference and acceptance of consumers. In the paired-comparison test the subject is asked to indicate which one of the two coded products is preferred. The option of "no preference" or "dislike both equally" can be included in this test. The nine-point hedonic scale is the most useful method for evaluating and quantifying an attribute. This scale is bipolar and the anchors are dislike and like. The middle choice is neutral "neither like nor dislike". Choices above the neutral are positive, and choices below the neutral are negative. This scale is easily understandable by consumers of all ages. For sensory evaluation, the result obtained using this method is informative, and the data can statistically be analyzed (Piggott, 1984; Stone & Sidel, 1993; Drake, 2007; Marsili, 2007).

In preference test, consumers are presented with 2 or more samples, and they are asked to indicate which sample they prefer or they are asked to express the degree of preference on a scale. The most commonly used scale is the 9-point hedonic scale (Drake, 2007).

As an example, the ranking procedure for individual quality, colour, texture, flavour and overall acceptability was used to evaluate the best cooking method for Syrian lentils using 18 panelists. The panelists preferred the taste of the lentils of Syrian type cooked in both water and in the presence of $NaHCO_3$ to that obtained by the traditional method (Moharram et al., 1986).

Selection of an appropriate sensory analysis test, testing conditions, and data analysis is helpful in the development of a new product and for quality control and marketing purposes.

CONCLUSION

Overall, literature on the sensory and flavour properties of legumes is sparse. Legumes are recognized around the world as a healthy food providing good supply of energy, protein and fibre (Sathe, 2002; Tharanathan & Mahadevamma, 2003; de Almeida Costa et al., 2006). In spite of their known health and nutritional benefits, legume consumption in some populations (especially in Western Societies) remains low. Flavour is an important criterion for food selection, and undesirable flavours in legumes could be one of the barriers to consumer acceptance. Thus, knowledge of the factors influencing the formation of desirable and/or undesirable flavours in legumes will be helpful in identifying ideal production and processing practices to obtain acceptable legume products for today's discriminating and health conscious consumer. In general, many of the volatile compounds resulting from fatty acid decomposition or other degradative processes which occur after harvesting or during storage and processing, could contribute to either desirable or undesirable flavours of plants (Rackis et al., 1979; Yousif et al., 2007). As these flavour compounds have different characteristics, their presence at different concentrations could affect the taste and flavour of legumes. Additionally, interactions of flavour compounds with food components can affect flavour characteristics, and it can also increase the difficulty of their removal from processed legume products (Sessa et al., 1977). Beany and bitter flavours of raw, full-fat, and defatted soybean flours, as an example, remain detectable in soybean flours, concentrates, and isolates (Sessa et al., 1976).

A number of workers have evaluated techniques to reduce or remove undesirable flavours from legumes. Rapid harvesting of legumes has been suggested as a way to prevent the formation of undesirable flavours caused by LOX (Hornostaj & Robinson, 2000). LOX-null legume varieties have also been developed in an attempt to reduce the generation of beany flavours during processing (Kobayashi et al., 1995; King et al., 1998). Attempts to control off-flavour generation by inactivation of LOX with heat, acid, alcohol, or antioxidants have been explored (Sessa et al., 1976; Rackis et al., 1979; Iwuoha & Umunnakwe, 1997; Dijkstra et al., 2003). These processes have been successful to different degrees. Thus, further research to identify ways to decrease off-flavour generation while enhancing the desirable flavours of legumes will be very useful.

REFERENCES

Abdo, K. M., & King, K. W. (1967). Enzymic modification of the extractability of protein from soybeans, *Glycine max*. *Journal of Agricultural and Food Chemistry*, 15, 83-87.

Achouri, A., Boye, J. I., & Belanger, D. (2005). Soybean isoflavones: Efficacy of extraction conditions and effect of food type on extractability. *Food Research International*. 38, 1199-1204.

Achouri, A., Boye, J. I., & Zamani, Y. (2006). Identification of volatile compounds in soymilk using solid-phase microextraction-gas chromatography. *Food Chemistry*, 99, 759-766.

Alberts, P., Stander, M. A., Paul, S. O., & de Villiers, A. (2009). Survey of 3-Alkyl-2-methoxypyrazine content of South African Sauvignon Blanc wines using a novel LC−APCI-MS/MS method. *Journal of Agricultural and Food Chemistry*, 57, 9347–9355.

Aldin, E., Reitmeier, H. A., & Murphy, P. (2006). Bitterness of soy extracts containing isoflavones and saponins. *Journal of Food Science*, 71, S211-S215.

Anderson, R.L., & Warner, K. (1976). Acid-sensitive soy proteins affect flavor. *Journal of Food Science*, 41, 293-296.

Anli, E., Vural, N., Vural, H., & Gucer, Y. (2007). Application of solid-phase micro-extraction (SPME) for determining residues of chlorpyrifos and chlorpyrifos-methyl in wine with gas chromatography (GC*). Journal of the Institute of Brewing*, 113, 213-218.

Arai, S., Suzuki, H., & Fujimaki, M. (1966). Studies on flavor components of soybean: Part II. Phenolic acids in defatted soybean flour. *Agricultural and Biological Chemistry*, 30, 364-369.

Arai, S., Kaji, M., & Fujimaki, M. (1970). n-Hexanal and some volatile alcohols: Their distribution in raw soybean tissues and formation in crude soy protein concentrate by lipoxygenase. *Agricultural and Biological Chemistry*, 34, 1420-1423.

Arens, D., Seilmeier, W., Weber, F., Kloos, G., & Grosch, W. (1973). Purification and properties of a carotene co-oxidizing lipoxygenase from peas. *Biochimica et Biophysica Acta*, 327, 295-305.

Arthur, C. L., & Pawliszyn, J. (1990). Solid phase microextraction with thermal desorption using fused silica optical fibers. *Analytical Chemistry*, 62, 2145-2148.

Aspelund, T. G., & Wilson, L. A. (1983). Adsorption of off-flavour compounds onto soy protein: A thermodynamic study. *Journal of Agricultural and Food Chemistry*, 31, 539–545.

Barra, A., Baldovini, N., Loiseau, A. M., Albino, L., Lesecq, C., & Lizzani Cuvelier, L. (2007). Chemical analysis of French beans (*Phaseolus vulgaris* L.) by headspace solid phase microextraction (HS-SPME) and simultaneous distillation/extraction (SDE). *Food Chemistry*, 101, 1279-1284.

Baur, C., & Grosch, W. (1977). Investigation about the taste of di, tri- and tetrahydroxy fatty acid. *Zeitschrift für Lebensmittel-Untersuchung und -Forschung*, 165, 82-84.

Baur, C., Grosch, W., Wieser, H., & Jugel, H. (1977). Enzymatic oxydation of linoleic acid: Formation of bitter tasting fatty acids. *Zeitschrift für Lebensmittel-Untersuchung und -Forschung,* 164, 171-176.

Bengtsson, B., & Bosund, I. (1964). Gas chromatographic evaluation of the formation of volatile substances in stored peas. *Food Technology*, 18, 179-182.

Borhan, M., & Snyder, H. E. (1979). Lipoxygenase destruction in whole beans by combinations of heating and soaking in ethanol. *Journal of Food Science,* 44, 586-590.

Boye, J, Zare, F., & Pletch, A. (2010). Pulse proteins: Processing, characterization, functional properties and applications in food and feed. Food Research International, 43, 414-431.

Brannan, R. G., Connolly, B. J., & Decker, E. A. (2001). Peroxynitrite: A potential initiator of lipid oxidation in food. *Trends in Food Science and Technology*, 12, 164-173.

Brown, B. D., Wei, L. S., Steinberg, M. P., & Villota, R. (1982). Minimizing protein insolubilization during thermal inactivation of lipoxygenase in soybean cotyledons. *Journal of the American Oil Chemists' Society*, 59, 88-92.

Buono, M. A., Setser, C., Erickson, L. E., & Fung, D. Y. C. (1990). Soymilk yoghurt: Sensory evaluation and chemical measurement. *Journal of Food Science*, 55, 528-531.

Burdock, G. A. (2002). *Handbook of flavour ingredients*. Boca Raton: CRC PRESS.

Buttery, R. G., Seifert, R. M., & Ling, L. C. (1975). Characterization of some volatile constituents of dry red beans. *Journal of Agricultural and Food Chemistry*, 23, 516–519.

Buttery, R. G. (1981). Vegetable and fruit flavors. In R. Teranishi, R. A. Flath, & H. Sugisawa (Eds.*), Flavor research: Recent advances* (pp. 175-216). New York: Marcel Dekker.

Calvino, A.M., García-Medina, M. R., & Cometto-Muniz, J. E. (1990). Interactions in caffeine- sucrose and coffee-sucrose mixtures: Evidence of taste and flavor suppression. *Chemical Senses*, 15, 505-519.

Chen, J., & Pawliszyn, J. B. (1995). Solid phase microextraction coupled to high-performance liquid chromatography. *Analytical Chemistry*, 67, 2530–2533.

Che Man, Y. B., Wei, L. S., & Nelson, A. I. (1989). Acid inactivation of soybean lipoxygenase with retention of protein solubility. *Journal of Food Science*, 54, 963-967.

Cho, M. J., Unklesbay, N., Hsieh, F., & Clarke, A. D. (2004). Hydrophobicity of bitter peptides from soy protein hydrolysates. *Journal of Agricultural and Food Chemistry*, 52, 5895-5901.

Choudhury, K., & Rahman, M. M. (1973). Fatty acids in different pulses produced and consumed in Bangladesh. *Journal of the Science of Food and Agriculture*, 24, 471-473.

Cordero-de-los-Santos, M. Y., Osuna-Castro, J. A., Borodanenko, A., & Paredes-López, O. (2005). Physicochemical and functional characterisation of amaranth (*Amaranthus hypochondriacus*) protein isolates obtained by isoelectric precipitation and micellisation. *Food Science and Technology International*, 11, 269-280.

Cowan, J. C., Rackis, J. J., & Wolf, W. J. (1973). Soybean protein flavor components. *Journal of the American Oil Chemists' Society*, 50, 426A-435A & 444A.

Craven, M. A., Gardner, J. W., & Bartlett, P. N. (1996). Electronic noses: Development and future prospects. *Trends in Analytical Chemistry*, 15, 486-493.

Curl, C. L., Price, K. R., & Fenwick, G. R. (1985). The quantitative estimation of saponin in pea (*Pisum sativum* L.) and soya (*Glycine max*). *Food Chemistry,* 18, 241-250.

Davies, C. S., Nielsen, S. S., & Nielsen, N. C. (1987). Flavor improvement of soybean preparations by genetic removal of lipoxygenase-2. *Journal of the American Oil Chemists' Society*, 64, 1428-1433.

de Almeida Costa, G. E., da Silva Queiroz-Monici, K., Pissini Machado Reis, S. M., & de Oliveira, A. C. (2006). Chemical composition, dietary fibre and resistant starch contents of raw and cooked pea, common bean, chickpea and lentil legumes. *Food Chemistry*, 94, 327–330.

Deibler, K. D., Acree, T. E., & Lavin, E. H. (1999). Solid phase microextraction application in gas chromatography/olfactometry dilution analysis. *Journal of Agricultural and Food Chemistry,* 47, 1616–1618.

Deshpande, S. S., Sathe, S. K., & Salunkhe, D. K. (1984). Chemistry and safety of plant polyphenols. In M. Friedman (Ed.), *Nutritional and toxicological aspects of food safety* (pp. 457-495). New York: Plenum Press.

de Lumen, B. O., Stone, E. J., Kazeniac, S. J., & Forsythe, R. H. (1978). Formation of volatile flavor compounds in green beans from linoleic and linolenic acids. *Journal of Food Science,* 43, 698-702.

del Rosario, R., de Lumen, B. O., Habu, T., Flath, R. A., Mon, T. R., & Teranishi, R. (1984). Comparison of headspace of volatiles from winged beans and soybeans. *Journal of Agricultural and Food Chemistry, 32*, 1011-1015.

de Schutter, D. P., Saison, D., Delvaux, F., Derdelinckx, G., Rock, J. M., Neven, H., & Delvaux, F. R. (2008). Release and evaporation of volatiles during boiling of unhopped wort. *Journal of Agricultural and Food Chemistry, 56*, 246-254.

Desnuell , P., & Savary, P. (1963). Specificities of lipases. *Journal of Lipid Research, 4*, 369-384.

Dewettinck, T., van Hege, K., & Verstraete, W. (2001). The electronic nose as a rapid sensor for volatile compounds in treated domestic wastewater. *Water Research, 35*, 2475–2483.

Dijkstra, D. S., Linnemann, A. R., & van Boekel, T. A. J. S. (2003). Towards sustainable production of protein-rich foods: Appraisal of eight crops for Western Europe. PART II: Analysis of the technological aspects of the production chain. *Critical Reviews in Food Science and Nutrition, 43*, 481-506.

Dörnenburg, H., & Davies, C. (1999). The relationship between lipid oxidation and antioxidant content in postharvest vegetables. *Food Reviews International, 15*, 435-453.

Drake, M. A. (2007). Invited Review: Sensory analysis of dairy foods. Journal of Dairy Science, 90, 4925–4937.

Dueñas, M., Estrella, I., & Hernández, T. (2004). Occurrence of phenolic compounds in the seed coat and the cotyledon of peas (*Pisum sativum* L.). *European Food Research and Technology, 219*, 116-123.

Ediriweera, N., Akiyama, Y., & Saio, K. (1987). Inactivation of lipoxygenase in soybeans with retention of protein solubility. *Journal of Food Science, 52*, 685-690.

Eisert, R., & Pawliszyn, J. (1997). Automated in-tube solid phase microextraction coupled to high performance liquid chromatography. *Analytical Chemistry, 69*, 3140-3147.

Eldridge, A. C., Warner, K., & Wolf, W. J. (1977). Alcohol treatment of soybeans and soybean protein products. *Cereal Chemistry, 54*, 1229-1237.

Eriksson, C. E. (1967). Pea lipoxidase, distribution of enzyme and substrate in green peas. *Journal of Food Science, 32*, 438-441.

Eriksson, C. E., & Svensson, S. G. (1970). Lipoxygenase from peas: Purification and properties of the enzyme. *Biochimica et Biophysica Acta, 198*, 449-459.

Eskin, N. A. M., Grossman, S., & Pinsky, A. (1977). Biochemistry of lipoxygenase in relation to food quality. *Critical Reviews in Food Science and Nutrition, 9*, 1-40.

Fenwick, D. E., & Oakenfull, D. (1983). Saponin content of food plants and some prepared foods. *Journal of the Science of Food and Agriculture, 34*, 186-191.

Forss, D.A. (1969). Role of lipids in flavors. *Journal of Agricultural and Food Chemistry, 17*, 681-685.

Frank, O., Ottinger, H., & Hofmann, T. (2001). Characterization of an intense bitter-tasting 1H,4H-quinolizinium-7-olate by application of the taste dilution analysis, a novel bioassay for the screening and identification of taste-active compounds in foods. *Journal of Agricultural and Food Chemistry, 49*, 231-238.

Franzen, K.L., & Kinsella, J. E. (1974). Parameters affecting the binding of volatile flavor compounds in model food systems: I. Proteins. *Journal of Agricultural and Food Chemistry, 22*, 675-678.

Fujimaki, M., Kato, H., Arai, S., & Tamaki, E. (1968). Applying proteolytic enzymes on soybean. I. Proteolytic enzyme treatment of soybean protein and its effect on flavour. *Food Technology*, 22, 889-893.

Gaddis, A. M., Ellis, R., & Currie, G. T. (1961). Carbonyls in oxidizing fat: V. The composition of neutral volatile monocarbonyl compounds from autoxidized oleate, linoleate, linoolenate esters and fats. *Journal of the American Oil Chemists' Society*, 38, 371-375.

Gardner, H. W. (1975). Decomposition of linoleic acid hydroperoxides: Enzymic reactions compared with non-enzymic. *Journal of Agricultural and Food Chemistry*, 23, 129-136.

Gardner, H. W., Kleunan, R., Weisleder, D., & Inglett, G. E. (1977). Cysteine adds to lipid hydroperoxide. *Lipids*, 12, 655-660.

Gardner, H. W. (1980). Lipid enzymes: Lipases, lipoxygenases, and hydroperoxidases. In M. G. Simic, & M. Karel (Eds.*), Autoxidation in food and biological systems* (pp. 447-504). New York: Plenum Press.

Gatade, A. A., Ranveer, R. C., & Sahoo, A. K. (2009). Physico-chemical and sensorial characteristics of chocolate prepared from soymilk. *Advance Journal of Food Science and Technology*. 1, 1-5.

Gocmen, D., Gurbuz, O., Rouseff, R. L., Smoot, J. M., & Dagdelen, A. F. (2004). Gas chromatographic-olfactometric characterization of aroma active compounds in sun-dried and vacuum-dried tarhana. *European Food Research and Technology*, 218, 573–578.

Grob, R. L., & Barry, E. F. (2004). *Modern practice of gas chromatography*. Hoboken: John Wiley & Sons.

Heng, L., van Koningsveld, G. A., Gruppen, H., van Boekel, M. A. J. S., Vincken, J. P., Roozen, J. P., & Voragen, A. G. J. (2004). Protein-flavour interactions in relation to development of novel protein foods. *Trends in Food Science and Technology*, 15, 217-224.

Heng , L., Vincken, J. P., van Koningsveld, G., Legger, A., Gruppen, H., van Boekel, T., Roozen, J., & Voragen, F. (2006). Bitterness of saponins and their content in dry peas. *Journal of the Science of Food and Agriculture*, 86, 1225 – 1231.

Hildebrand, D. F., Hamilton-Kemp, T. R., Loughrin, J. H., Kadum, A., & Anderson, R. A. (1990). Lipoxygenase 3 reduces hexanal production from soybean seed homogenates. *Journal of Agricultural and Food Chemistry*, 38, 1934-1936.

Hinterholzer, A., Lemos, T., & Schieberle, P. (1998). Identification of the key odorants in raw French beans and changes during cooking. *Zeitschrift für Lebensmittel-Untersuchung und –Forschung,* 207, 219–222.

Honig, D. H., Warner, K., & Rackis, J. J. (1976). Toasting and hexane:ethanol extraction of defatted soy flakes flavor of flours, concentrates and isolates. *Journal of Food Science*, 41, 642-646.

Honig, D. H., Warner, K. A., Selke, E., & Rackis, J. J. (1979). Effects of residual solvents and storage on flavor of hexane/ethanol azeotrope extracted soy products. *Journal of Agricultural and Food Chemistry*, 27, 1383–1386.

Hornostaj, A. R., & Robinson, D. S. (2000). Purification of hydroperoxide lyase from pea seeds. *Food Chemistry*, 71, 241-247.

Hostettmann, K., & Marston, A. (1995). *Saponins: Chemistry and pharmacology of natural products*. Cambridge: Cambridge University Press.

How, J. S. L., & Morr, C. V. (1982). Removal of phenolic compounds from soy protein extracts using activated carbon. *Journal of Food Science,* 47, 933-940.

Huang, A. S., Hsieh, O. A. L., & Chang, S. S. (1982). Characterization of the nonvolatile minor constituents responsible for the objectionable taste of defatted soybean flour. *Journal of Food Science*, 47, 19-23.

Itoh, A., & Vick, B. A. (1999). The purification and characterization of fatty acid hydroperoxide lyase in sunflower. *Biochimica et Biophysica Acta*, 1436, 531-540.

Iwuoha, C. I., & Umunnakwe, K. E. (1997). Chemical, physical and sensory characteristics of soymilk as affected by processing method, temperature and duration of storage. *Food Chemistry,* 59, 373-379.

Jakobsen, H. B., Hansen, M., Christensen, M. R., Brockhoff, P. B., & Olsen, C. E. (1998). Aroma volatiles of blanched green peas (*Pisum sativum* L.*). Journal of Agricultural and Food Chemistry*, 46, 3727–3734.

Januszkiewicz, J., Sabik, H., Azarnia, S., & Lee, B. (2008). Optimization of headspace solid-phase microextraction for the analysis of specific flavors in enzyme modified and natural Cheddar cheese using factorial design and response surface methodology. *Journal of Chromatography A*, 1195, 16-24.

Kalbrener, J. E., Warner, K., & Eldridge, A. C. (1974). Flavors derived from linoleic and linolenic acid hydroperoxides. *Cereal Chemistry*, 51, 406-416.

Kataoka, H., Lord, H. L., & Pawliszyn, J. (2000). Applications of solid-phase microextraction in food analysis. *Journal of Chromatography* A, 880, 35–62.

Kato, H., Doi, Y., Tsugita, T., Kosai, K., Kamiya, T., & Kurata, T. (1981). Changes in volatile flavour components of soybeans during roasting. *Food Chemistry*, 7, 87-94.

Kato, T., Ohta, H., Tanaka, K., & Shibata, D. (1992). Appearance of new lipoxygenases in Soybean cotyledons after germination and evidence for expression of a major new lipoxygenase gene. *Plant Physiology*, 98, 324-330.

Keppler, J. G. (1977). Twenty-five years of flavor research in a food industry. *Journal of the American Oil Chemists' Society,* 54, 474-477.

King, J. M., Svendsen, L. K., Fehr, W. R., Narvel, J. M., & White, P. J. (1998). Oxidative and flavor stability of oil from lipoxygenase-free soybeans. *Journal of the American Oil Chemists' Society*, 75, 1121-1126.

King, A. J., Readman, J. W., & Zhou, J. L. (2003). The application of solid-phase micro-extraction (SPME) to the analysis of polycyclic aromatic hydrocarbons (PAHs). *Environmental Geochemistry and Health*, 25, 69–75.

Kobayashi, A., Tsuda, Y., Hirata, N., Kubota, K., & Kitamura, K. (1995). Aroma constituents of soybean [*Glycine max* (L.) *Merril*] milk lacking lipoxygenase isoenzymes. *Journal of Agricultural and Food Chemistry*, 43, 2449-2452.

Koehler, P. E., Mason, M. E., & Newell, J. A. (1969). Formation of pyrazine compounds in sugar-amino acid model systems. *Journal of Agricultural and Food Chemistry*, 17, 393-396.

Li, S., & Weber, S. G. (1997). Determination of barbiturates by solid-phase microextraction and capillary electrophoresis. *Analytical Chemistry*, 69, 1217-1222.

Liu, K. (1997). Soybeans: Chemistry, technology, and utilization. New York: Chapman & Hall.

Lovegren, N. V., Fisher, G. S., Legendre, M. G., & Schuller, W. H. (1979). Volatile constituents of dried legumes. *Journal of Agricultural and Food Chemistry*, 27, 851-853.

MacLeod, G., & Seyyedain-Ardebili, M. (1981). Natural and simulated meat flavors (with particular reference to beef). *Critical Reviews in Food Science and Nutrition*, 14, 309-437.

MacLeod, G., Ames, J., & Betz, N. L. (1988). Soy flavor and its improvement. *Critical Reviews in Food Science and Nutrition,* 27, 219-400.

Maga, J. A. (1973). A review of flavor investigations associated with the soy products, raw soybeans, defatted flakes and flours, and isolates, *Journal of Agricultural and Food Chemistry,* 21, 864-868.

Maga, J. A., Sizer, C. E., & Myhre, D. V. (1973). Pyrazines in foods. *Critical Reviews in Food Science & Technology,* 4, 39-115.

Maga, J. A., & Katz, I. (1975). The role of sulfur compounds in food flavour. Part II: Thiophenes. *Critical Reviews in Food Science and Nutrition*, 6, 241-270.

Märk, J., Pollien, P., Lindinger, C., Blank, I., & Märk, T. (2006). Quantitation of furan and methylfuran formed in different precursor systems by proton transfer reaction mass spectrometry. *Journal of Agricultural and Food Chemistry*, 54, 2786-2793.

Marsili, R. T. (2007). Comparing sensory and analytical chemistry flavor analysis. In R. Marsili (Ed.), Sensory-directed flavor analysis (pp. 1-22). Boca Raton: Taylor & Francis.

Matoba, T., & Hata, T. (1972). Relationship between bitterness of peptides and their chemical structures. *Agricultural and Biological Chemistry,* 36, 1423-1431.

Matoba, T., Hidaka, H., Narita, H., Kitamura, K., Kaizuma, N., & Kito, M. (1985). Lipoxygenase-2 isozyme is responsible for generation of n-hexanal in soybean homogenate. *Journal of Agricultural and Food Chemistry*, 33, 852-855.

Messina, M. J. (1999). Legumes and soybeans: Overview of their nutritional profiles and health effects. *American Journal of Clinical Nutrition*, 70, 439S-450S.

Moharram, Y. G., Abou-Samaha, A. R., & EI-Mahady, A. R. (1986). Effect of cooking methods on the quality of lentils. *Zeitschrift für Lebensmittel-Untersuchung und - Forschung,* 182, 307-310.

Moll, C., Biermann, U., & Grosch, W. (1979). Occurrence and formation of bitter-tasting trihydroxy fatty acids in soybeans. *Journal of Agricultural and Food Chemistry*, 27, 239-243.

Morita, M., & Fujimaki, M. (1973). Minor peroxide components as catalysts and precursors to monocarbonyls in the autoxidation of methyl linoleate. *Journal of Agricultural and Food Chemistry*, 21, 860-863.

Mottram, D. S., Szauman-Szumski, C., & Dodson, A. (1996). Interaction of thiol and disulfide flavor compounds with food components. *Journal of Agricultural and Food Chemistry,* 44, 2349-2351.

Mtebe, K., & Gordon, M. H. (1987). Volatiles derived from lipoxygenase-catalysed reactions in winged beans (*Psophocarpus tetragonolobus*). *Food Chemistry*, 23, 175-182.

Müller, R., & Rappert, S. (2010). Pyrazines: Occurrence, formation and biodegradation. *Applied Microbiology and Biotechnology*, 85, 1315–1320.

Murray, K. E., Shipton, J., Whitfield, F. B., Kennett, B. H., & Stanley, G. (1968). Volatile flavour components from green peas (*Pisum sativum*): 1. Alcohols in unblanched frozen peas. *Journal of Food Science*, 33, 290-294.

Murray, K. E., Shipton, J., & Whitfield, F. B. (1970). 2-Methoxypyrazines and the flavour of green peas (*Pisum sativum*). *Chemistry and Industry*, 4, 897-898.

Murray, K. E., & Whitfield, F. B. (1975). The occurrence of 3-alkyl-2-methoxypyrazines in raw vegetables. *Journal of the Science of Food and Agriculture*, 26, 973-986.

Murray, K. E., Shipton, J., Whitfield, F. B., & Last, J. H. (1976). The volatiles of off-flavoured unblanched green peas (*Pisum sativum*). *Journal of the Science of Food and Agriculture,* 27, 1093-1107.

O'Connor, T. P., & O'Brien, N. M. (1991). Significance of lipoxygenase in fruits and vegetables. In P.F. Fox (Ed.), *Food enzymology* (pp. 337-372). Barking: Elsevier Applied Science.

Ohta, N., Kuwata, G., Akahori, H., & Watanabe, T. (1980). Isolation of a New Isoflavone Acetyl Glucoside, 6"-O-Acetyl Genistin, from Soybeans. *Agricultural and Biological Chemistry*, 44, 469-470.

Ohren, J. A. (1981). Process and product characteristics for soya concentrates and isolates. *Journal of the American Oil Chemists' Society*, 58, 333-335.

Okubo, K., Iijima, M., Kobayashi, Y., Yoshikoshi, M., Uchida, T., & Kudou, S. (1992). Components responsible for the undesirable taste of soybean seeds. *Bioscience Biotechnology and Biochemistry*, 56, 99-103.

Oomah, B. D., & Liang, L. S. Y. (2007). Volatile compounds of dry beans (*Phaseolus vulgaris* L.). *Plant Foods for Human Nutrition*, 62, 177-183.

Ouyang, G., & Pawliszyn, J. (2008). A critical review in calibration methods for solid-phase microextraction. *Analytica Chimica Acta*, 6, 184-197.

Pattee, H. E., Salunkhe, D. K., Sathe, S. K., & Reddy, N. R. (1982). Legume lipids. *Critical Reviews in Food Science and Nutrition*, 17, 97-139.

Pawliszyn, J. (1995). New directions in sample preparation for analysis of organic compounds. *Trends in Analytical Chemistry*, 14, 113-122.

Pawliszyn, J. (1997). *Solid phase microextraction: Theory and practice*. New York: Wiley-VCH Inc.

Penüalver, A., Pocurull, E., Borrull, F., & Marcé, R. M. (1999). Trends in solid-phase microextraction for determining organic pollutants in environmental samples. *Trends in Analytical Chemistry,* 18, 557-568.

Pérez, A. G., Sanz, C., Olías, R., & Olías, J. M. (1999). Lipoxygenase and hydroperoxide lyase activities in ripening strawberry fruits. *Journal of Agricultural and Food Chemistry,* 47, 249-253.

Perkins, E. G. (1989). Gas chromatography and gas chromatography-mass spectrometry of odor and flavor components in lipid foods. In D. B. Min, & T. H. Smouse (Eds.), *Flavor chemistry of lipid foods* (pp. 35-56). Champaign: American Oil Chemists' Society.

Piggott, J. R., (1984). *Sensory analysis of food*. London: Elsevier Applied Science Publishers.

Pinthong, R., Macrae, R., & Rothwell, J. (1980). The development of a soya-based yoghurt. Part II: sensory evaluation and analysis of volatiles. *Journal of Food Technology*, 15, 647-652.

Potter, S. M., Flores, R. J., & Pollack, J., Lone, T. A., & Berber-Jimenez, M. D. (1993). Protein-saponin interaction and its influence on blood lipids. *Journal of Agricultural and Food Chemistry,* 41, 1287-1291.

Price, K. R., Griffiths, N. M., Curl, C. L., & Fenwick, G. R. (1985). Undesirable sensory properties of the dried pea (*Pisum sativum*): The role of saponins. *Food Chemistry*, 17, 105-115.

Prosen, H., & Zupančič-Kralj, L. (1999). Solid-phase microextraction. *Trends in Analytical Chemistry*, 18, 272-282.

Rackis, J. J., McGhee, J. E., Honig, D. H., & Booth, A. N. (1975). Processing soybeans into foods: Selected aspects of nutrition. *Journal of the American Oil Chemists' Society*, 52, 249A-253A.

Rackis, J. J., Sessa, D. J., & Honig, D. H. (1979). Flavor problems of vegetable food proteins. *Journal of the American Oil Chemists' Society*, 56, 262-271.

Robinson, K. M., Klein, B. P., & Lee, S. Y. (2004). Utilizing the R-index measure for threshold testing of model soy isoflavone solutions. *Journal of Food Science*, 69, SNQ1-4.

Sathe, S. K., & Salunke, D. K. (1984). Technology of removal of unwanted components of dry beans. *Critical Reviews in Food Science and Nutrition*, 21, 263-287.

Sathe, S. K. (2002). Dry bean protein functionality. *Critical Reviews in Biotechnology*, 22, 175-223.

Saxby, M. J. (1993). *Food taints and off-flavours*. Glasgow: Blackie Academic & Professional.

Schaller, E., Bosset, J. O., & Escher, F. (1998). *Electronic noses and their application to food*. Lebensmittel-Wissenschaft und-Technologie, 31, 305-316.

Seals, R. G., & Hammond, E. G. (1970). Some carbonyl flavour compounds of oxidized soybean and linseed oils. *Journal of the American Oil Chemists' Society*, 47, 278-280.

Self, R., Casey, J. C., & Swain, T. (1963). The low-boiling volatiles in cooked foods. *Chemistry and Industry*, 4, 863-864.

Seo, W. H., Lee, H. G., & Baek, H. H. (2008). Evaluation of bitterness in enzymatic hydrolysates of soy protein isolate by taste dilution analysis. *Journal of Food Science*, 73, S41-S46.

Sessa, D. J., Warner, K., & Rackis, J. J. (1976). Oxidized phosphatidylcholines from defatted soybean flakes taste bitter. *Journal of Agricultural and Food Chemistry*, 24, 16-21.

Sessa, D. J., & Rackis, J. J. (1977). Lipid-derived flavors of legume protein products. *Journal of the American Oil Chemists' Society*, 54, 468-473.

Sessa, D. J., Gardner, H. W., Kleiman, R., & Weisleder, D. (1977). Oxygenated fatty acid constituents of soybean phosphatidylcholines. *Lipids*, 12, 613-619.

Sessa, D. J. (1979). Biochemical aspects of lipid-derived flavors in legumes. *Journal of Agricultural and Food Chemistry*, 27, 234-239.

Sessa, D. J. (1985). *Role of phospholipids in flavor problems*. In B. F. Szuhaj, & G. R. List (Eds.), Lecithins (pp. 347-374). Champaign: American Oil Chemists' Society.

Shankaranarayana, M. L., Raghavan, B., Abraham, K. O., Natarajan, C. P., & Brodnitz, H. H. (1974). Volatile sulfur compounds in food flavors. *Critical Reviews in Food Science and Nutrition,* 4, 395-435.

Shi, J., Arunasalam, K., Yeung, D., Kakuda, Y., Mittal, G., & Jiang, Y. (2004). Saponins from edible legumes: Chemistry, processing, and health benefits. *Journal of Medicinal Food,* 7, 67-78.

Shipton, J., & Last, J. H. (1968). Estimating volatile reducing substances and off-flavor in frozen green peas. *Food Technology*, 22, 917-920.

St-Angelo, A. J., & Ory, R. L. (1975). Effects of lipoxygenases on proteins in raw and processed peanuts. *Journal of Agricultural and Food Chemistry*, 23, 141-146.

Stephan, A., Bücking, M., & Steinhart, H. (2000). Novel analytical tools for food flavours. *Food Research International*, 33, 199-209.

Stone, H., & Sidel, J. L. (1993). *Sensory evaluation practices*. San Diego: Academic Press, INC.

Sugawara, E., Ito, T., Odagiri, S., Kubota, K., & Kobayashi, A. (1985). Comparison of compositions of odor components of natto and cooked soybeans. *Agricultural and Biological Chemistry*, 49, 311-317.

Tharanathan, R. N., & Mahadevamma, S. (2003). Grain legumes: A boon to human nutrition. *Trends in Food Science and Technology*, 14, 507-518.

Torres-Penaranda, A. V., & Reitmeier, C. A. (2001). Sensory descriptive analysis of soymilk. *Journal of Food Science*, 66, 352-356.

Tsukamoto, C., Shimada, S., Igita, K., Kudou, S., Kokubun, M., Okubo, K., & Kitamura, K. (1995). Factors affecting isoflavone content in soybean seeds: Changes in isoflavones, saponins, and composition of fatty acids at different temperatures during seed development. *Journal of Agricultural and Food Chemistry*, 43, 1184-1192.

Uzogara, S. G., & Ofuya, Z. M. (2007). Processing and utilization of cowpeas in developing countries: A review. *Journal of Food Processing and Preservation*, 16, 105-147.

van den Ouweland, G. A. M., Peer, H. G., & Tjan, S. B. (1978). Occurrence of amadori and heyns rearrangement products in processed foods and their role in flavor formation. In C. Charalambous, & G. E. Inglett (Eds.), *Flavor of foods and beverages* (pp. 131-143). New York: Academic Press.

van Ruth, S. M., Roozen, J. P., & Cozijnsen, J. L. (1995). Volatile compounds of rehydrated French beans, bell peppers and leeks: Part I. Flavour release in the mouth and in three mouth model systems. *Food Chemistry,* 53, 15-22.

van Ruth, S. M., & O'Connor, C. H. (2001). Evaluation of three gas chromatography-olfactometry methods: Comparison of odour intensity-concentration relationships of eight volatile compounds with sensory headspace data. *Food Chemistry*, 74, 341-347.

Vas, G., & Vékey, K. (2004). Solid-phase microextraction: A powerful sample preparation tool prior to mass spectrometric analysis. *Journal of Mass Spectrometry*, 39, 233-254.

Vernin, G., & Parkanyi, C. (1982). Mechanisms of formation of heterocyclic compounds in Maillard and pyrolysis reactions. In G. Vernin (Ed.), *Chemistry of heterocyclic compounds in flavours and aromas* (pp. 151-207). Chichester: Ellis Horwood.

Voragen, A. G. J. (1998). Technological aspects of functional food-related carbohydrates. *Trends in Food Science and Technology*, 9, 328-335.

Walker, A. F., & Kochhar, N. (1982). Effect of processing, including domestic cooking on nutritional quality of legumes. *Proceedings of the Nutrition Society*, 41, 41-51.

Wan, H. B., Chi, H., Wong, M. K., & Mok, C. Y. (1994). Solid phase microextraction using pencil lead as sorbent for analysis of organic pollutants. *Analytica Chimica Acta*, 298, 219-223.

Wang, Z. H., Dou, J., Macura, D., Durance, T. D., & Nakai, S. (1998). Solid phase extraction for GC analysis of beany flavours in soymilk. *Food Research International*, 30, 503-511.

Wang, B., Xiong, Y. L. & Wang, C. (2001). Physicochemical and sensory characteristics of flavoured soymilk during refrigeration storage. *Journal of Food Quality*, 24, 513-526.

Weber, F., & Grosch, W. (1976). Co-Oxidation of a carotenoid by the enzyme lipoxygenase: Influence on the formation of linoleic acid hydroperoxides. *Zeitschrift für Lebensmittel-Untersuchung und -Forschung*, 161, 223-230.

Werkhoff, P., Güntert, M., Krammer, G., Sommer, H., & Kaulen, J. (1998). Vacuum headspace method in aroma research: Flavor chemistry of yellow passion fruits. *Journal of Agricultural and Food Chemistry*, 46, 1076-1093.

Whitfield, F. B., & Shipton, J. (1966). Volatile carbonyls in stored unblanched frozen peas. *Journal of Food Science*, 33, 328-331.

Whitfield, F. B. (1992). Volatiles from interactions of Maillard reactions and lipids. Critical *Reviews in Food Science and Nutrition*, 31, 1-58.

Willet, J. (1987). *Gas chromatography*. Chichester: John Wiley & Sons.

Wilkens, W. F., Mattick, L. R., & Hand, D. B. (1967). Effect of processing method on oxidative off-flavors of soybean milk. *Food Technology*, 21, 1630-1633.

Wilkens, W. F., & Lin, F. M. (1970). Gas chromatographic and mass spectral analyses of soybean milk volatiles. *Journal of Agricultural and Food Chemistry*, 18, 333-336.

Wolf, W. J., & Cowan, J. C. (1971). Soybeans as a food source. *Critical Reviews in Food Science and Nutrition,* 2, 81-158.

Wolf, W. J. (1975). Lipoxygenase and flavor of soybean protein products. *Journal of Agricultural and Food Chemistry,* 23, 136-141.

Xu, B. J., & Chang, S. K. C. (2007). A Comparative study on phenolic profiles and antioxidant activities of legumes as affected by extraction solvents. *Journal of Food Science,* 72, S159 - S166.

Yousif, A. M., Kato, J., & Deeth, H. C. (2007). Effect of storage on the biochemical structure and processing quality of Adzuki bean (*Vigna angularis*), *Food Reviews International,* 23, 1-33.

Zambonin, C. G. (2003). Coupling solid-phase microextraction to liquid chromatography: A review. *Analytical and Bioanalytical Chemistry*, 375, 73−80.

Zhang, Y. & Ho, C. T. (1989). Volatile compounds from thermal interaction of 2,4-decadienal with cysteine and glutathione, *Journal of Agricultural and Food Chemistry,* 37, 1016-1020.

In: Progress in Food Science and Technology, Volume 1 ISBN: 978-1-61122-314-9
Editor: Anthony J. Greco © 2012 Nova Science Publishers, Inc.

Chapter 2

MICRONUTRIENT DISORDERS AND VARIOUS MITIGATION OPTIONS IN INDIAN AGRICULTURE

M. P. S. Khurana, K. P. Patel, V. P. Ramani G. Mahajan and Kuldip-Singh

Department of Soils, PAU, Ludhiana, India

The deficiency of micronutrient in Indian soils is attributed to several factors which include: adoption of high-yielding crop cultivars during last two decades that removed large quantities of micronutrients in the harvested crop resulting in mining of nutrients from the soil, particularly those not applied in required quantities through fertilizers and manures, increased cropping intensity, increased irrigation facility, increased use of high analysis fertilizers, reduced recycling of micronutrient contained in crop residues and organic manures. Micronutrient deficiencies in India are widely reported across the different soil types and crop species. As much as 48, 12, 5, 4 and 33 per cent soils in India are afflicted with deficiency of zinc (Zn), iron (Fe), manganese (Mn), copper (Cu) and boron (B) respectively (Singh, 1999). This causes failure of normal crop growth thereby resulting in low crop yields. Ample evidences has become available on hand from hundreds of experiments that micronutrient fertilization of crops is essentially required to ensure balanced nutrition for producing optimum yield, enhancing fertilizer use efficiency of other nutrients besides improving quality of crops. It is further noticed that crop response were not related to total content but the size of available pool of particular micronutrient.

Deficiency of zinc has started showing decline trend in states like Punjab and Haryana because of extensive use of zinc sulfate fertilizer having considerable residual effect. Adoption of rice-wheat system on highly permeable coarse textured soils has not only caused deficiency of Fe in rice but also deficiency of Mn has now become a serious problem where wheat is followed by rice. This problem is now prevalent in north India particularly in parts of Punjab, Haryana, Utter Pradesh and Bihar where striking responses of wheat to applied manganese has been observed. The Mn deficiency in soils is presently limited to 4-5 percent of samples but in coming years as much as 25% of soils will be suspected to be afflicted with severe manganese deficiency in Punjab and Haryana. Copper deficiency in Indian soils is not a severe problem as only about 3-4 per cent of soils analyzed were found deficient. Thus, by and large most soils do not respond to copper fertilization except peat and mollisol soils having high organic matter contents. The incidence of boron deficiency was highest in the acid soils of West Bengal followed by the calcareous soils of Bihar. Molybdenum deficiencies are becoming more prevalent in acid soils of Uttar Pradesh. This article reviews the extent of micronutrient deficiencies and various management approaches to alleviate micronutrients deficiencies through application of external sources as foliar sprays /soil application, supplementation through organic sources and mobilization through cultivation of efficient crop cultivars.

DISTRIBUTION OF MICRONUTRIENTS IN INDIAN SOILS TOTAL

Total micronutrient content of Indian soils varies from less than 1 ppm Mo up 27% Fe. With the exception of Cl and Fe, generally remaining micronutrients range between 10 and 250 ppm. Parent materials from which soils are formed plays a dominant role in micronutrient distribution. According to Katyal and Friesen (1987), soils derived from shale, loess and fine-textured alluvium are typically high in total B, whereas those arising from acid igneous rocks and fresh sedimentary deposits contain the lowest amounts of B. Similarly, soils born out of basalt are richer in total Cu compared to those formed from granite. Apart from parent material, other soil forming processes like leaching which influences the retention of rock inherited micronutrients. Compared to a mean value of around 100 ppm Zn, some highly leached acid sands of tropical India contain as low as 10 ppm Zn. Except for some extreme values of micronutrients, their total contents hardly bear any relevance to plant availability. Soil characteristics like pH, organic matter content, texture, calcareousness and anoxic conditions play far bigger role in determining micronutrient availability.

Available Micronutrient Contents

Diethylene triamine penta acetic acid (DTPA) extraction method (Lindsay and Norvell 1978) has been the most widely adopted method of assessing availability of Zn, Cu, Mn and Fe. Analysis of more than 250 thousand soil samples revealed Zn deficiency in about one half of them which was further confirmed by plant analysis. Response studies at farmers' fields gave further credence to extensive Zn deficiency in Indian soils and crops (Katyal 1985). Deficiency of other nutrients is limited in extent and occurs under very specific conditions.

Hosts of factors which promote incidence of Zn deficiency are: coarse texture, low organic matter, alkaline pH favored by calcareousness and sodicity are among the prominent ones (Katyal and Rattan 1993; Katyal and Sharma 1991; Rattan et al. 1997; Rattan et al. 1999; Sharma et al. 2000 and Singh 1999). Same group of adversaries dominate availability of Cu, Mn, and Fe, albeit in varying extent and intensity. In addition to these soil characteristics, fluctuating pedogenic moisture regimes and seasonal temperatures assert a striking influence on the endemic distribution of micronutrient availability. Katyal and Sharma (1991) reported that as the soil moisture regime becomes drier (a shift from aquic through ustic to aridic), DTPA- extractable Zn, Cu, Mn and Fe contents become lower. This negative effect was led by a coinciding increase in pH and lime and decrease in organic matter. Compared to dry regions, leaching is a prominent phenomenon of humid regions. Leaching creates acidity in soil and robes effective soil profile of certain nutrient elements. Boron is among the nutrients that are susceptible to soil acidity and leaching. Accordingly, B deficiency is widespread in light textured flood-prone soils of Bihar (Sakal et al. 1996). In contrast, in non-leaching dry soils, B gets accumulated in toxic amounts. Soil acidic reaction, whether arising from parent material or climate-led leaching, makes Mo less available- which is a contrasting behavior to all other micronutrients.

Micronutrient Status of Indian Soils

Large experimental evidences indicated that the total content of micronutrients *per se* is a poor predictor of their supplying power to the plants. It is the soil available micronutrient pool that represents the native level of plant usable forms; it is more often the basis to decide on the occurrence of deficiency or sufficiency status (Table 1).

Table 1. Total and available micronutrient contents of the benchmark soils of India *vis-a- vis* soils of the world

Micronutrients	Total content (mg kg^{-1})			Available content (mg kg^{-1})	
	India		World	India	
	Range	Mean	Range	Range	Mean
Zn	20-97	55	17-125	0.12-2.80	0.54
Fe	13000-80000	33000	-	3.40-68.1	20.5
Mn	38-1941	537	60-1300	4.00-102.0	26.0
Cu	11-141	41	6-80	0.15-5.33	1.7
B	2.8-630	-	9-85	0.04-7.40	1.7
Mo	Traces-12.3	-	1-3	Tr.-2.80	-

Source: Rattan et al. (1999)

Under the aegis of the "All India Coordinated Research Project on Secondary, Micronutrients and Pollutant Elements in Soils and Plants", analysis of 250,000 soil samples from 20 states of the country revealed that 49% of the soil samples were deficient in Zn (< 0.6 ppm DTPA-extractable content). Hundreds of field experiments also proved incidence of Zn deficiency, which in extent and severity varied across soil types and agro-ecological zones. Coarse texture, high pH, calcareousness and diminishing organic carbon often accentuated the Zn deficiency (Katyal and Rattan 1993). Irrespective of these soil properties, irrigated crops whose productivity is two to three times higher than rain fed crops suffered more from Zn deficiency. It is pertinent to note that lowland rice is invariably affected by Zn deficiency regardless of soil type and eco-climatic conditions.

On the basis of severity; extent of B deficiency is next to Zn in order. Thirty-three per cent of the 36825 soil samples analyzed were found to be deficient in available B. Among different states of the country, extent of B-deficient samples varied from 2% in Gujarat to a maximum of 68% in West Bengal (Singh 1999). In general, its deficiency is most widespread (39-68%) in red lateritic soils of Karnataka, leached and acidic soils of West Bengal, Orissa & Maharashtra and highly calcareous and old alluvium of Bihar. The deficiency of Fe, Mo and Mn reported in agricultural soils of India is 13, 7 and 4% respectively. Iron deficiency is most common in upland crops particularly grown on calcareous/ alkaline soils of arid region. The adoption of rice-wheat cropping system in place of maize-wheat or groundnut-wheat in non-traditional rice growing areas on highly permeable coarse-textured soils has been responsible for occurrence of Mn deficiency. The deficiency of Mo is common in acid soils of humid region. Deficiency of chloride and Nickel has not been reported so far in the Indian soils.

PROGRESS MADE IN DIAGNOSING THE MICRONUTRIENT DEFICIENCIES

Over the years, soil and plant analyses have emerged as a useful tool for monitoring micronutrient status in soils and plants. The different extractants like dithiozone for Zn, NH_4OAc for Fe, Mn and Cu; $NH_4H_2PO_4$ and dilute H_3PO_4 for Mn have been widely used with variable success up to late 1970's. Since early 1980's, the use of these extractants have been vanishing barring a few instances, as these have been replaced by the DTPA soil test of Lindsay and Norvell (1978) and this is now the soil test for available micronutrient cations (Rattan et al. 2008). Although researchers has confirmed the validity and general applicability of these soil test for Zn in a number of pot and field trials across the country. On the other side, its usefulness as predictor of iron availability is still uncertain (Katyal and Sharma, 1980). There are a few successful cases of assessing available Fe in soils with 1M NH_4OAc (Banerjee and Das 1978). The predication of micronutrient deficiencies based on total content in plant tissues has been reasonably successful for all micronutrients except Fe because chlorotic plants generally have as much or more Fe than the green ones. In this regard, Fe^{2+} content in leaves proved to be more useful than total Fe as an indicator of its nutritional status in plant (Pal et al. 2008). Since reducible Mn in soils were reported to correlate very well with Mn uptake by plants in the wheat-growing areas of certain pockets of Punjab (percolating coarse soils), Takkar and Randhawa (1978) advocated the use of this form of Mn as an index of its availability in soils. Subsequent work on this aspect indicated that extractants like 0.005 M DTPA, 2% hydroquinone + 1 N NH_4OAc, 0.02 M sodium pyrophosphate, 0.1 N H_3PO_4 and double acid (0.05 N HCl + 0.025 N H_3PO_4) could be used to predict the availability of Mn in coarse textured soils of Punjab (Nayyar et al. 1990). However, greenhouse and field trials all over the world consistently demonstrated that the most successful approach to identify Mn-deficient soils is the use of an availability index incorporating extractable Mn and soil pH. We feel that introspection is required on the extensive use of DTPA soil test in relation to initial soil pH. The DTPA soil test is a good test which identify very well in near neutral and calcareous soils with insufficient available micronutrients (Zn, Cu, Fe and Mn). Whenever, one strays from the original design of this soil test, one should be aware of the possible consequences and pass awareness to others (O' Connor 1988).

Hot water extraction for B estimation (Berger and Truog 1939) is a commonly used method for obtaining an index of plant available B in soil. Extracting the soil with boiling water poses problems due to colour impurities from organic matter and turbidity from suspended clay particles in the colorimetric estimation of B. Apart from these problems, this method is difficult to adopt for the analysis of large number of samples on routine basis. Over the years various modifications have been suggested including the use of alternate extractants such as hot 0.01 M $CaCl_2$, NH_4OAc, mannitol-$CaCl_2$, tartaric acid and salicylic acid. Considering the accountability and advantageous features of 0.05 M mannitol-0.01 M $CaCl_2$ (Cartwright et al. 1983) for alkaline/ calcareous soils and 0.1 M salicylic extraction (Datta et al. 1998) for acid soils, these method is simple as compared to hot water extraction. At the same time, suitability of these two extractants may be tested across the different soils, crops and agro-ecological regions for delineation work, when a large number of samples are to be analyzed. Under AICRP micronutrients project, numbers of soil samples analyzed for available Mo are far less as compared to other micronutrients due to the tedious method of estimation.

Critical Levels of Deficiency (CLD)

Soil scientists and plant nutritionists have set the limits of micronutrient availability in soils and their contents in plants to define occurrence of deficiency. These limits are known as critical limits distinguishing deficiency from sufficiency which are generally employed to advice on need for micronutrient fertilization. Critical limits of deficiency of various micronutrients in soils and plants used in the country for delineation purpose are presented in table 2.

Table 2. Critical levels of deficiency (CLD) of micronutrients in soils and plants used for delineation purposes

Element	CLD in soil		CLD in mature plant tissue (mg kg^{-1} dry matter)
	Extractant used	Value (mg kg^{-1})	
Zinc	DTPA*	0.6	10-20
Iron	DTPA	2.5-4.5	50
Manganese	DTPA	2.0-3.5	15-25
Copper	DTPA	0.2	2-5
Boron	Hot water	0.5	5-30
Molybdenum	Ammonium Oxalate**	0.2	0.03-0.15

Source: Katyal and Rattan (2003)
* DTPA refers to 0.05 M DTPA-0.01 M $CaCl_2$-0.1 M Triethanolamine (pH 7.3) of Lindsay and Norvell (1978).
** 0.1 M $(NH_4)_2$ C_2O_4 + 0.175 $H_2C_2O_4$ (pH 3.25)

The values listed against each micronutrient are only indicative. For confident predictions on possible deficiencies, these limits must be defined and refined with reference to growing environment, certain soil characteristics and pre-defined plant parts of specific crops. For instance, lowland rice is more vulnerable than wheat to Zn deficiency. By reason of that rice is likely to suffer from Zn deficiency at higher levels of availability compared to wheat. In fact, the respective critical limits set by DTPA method for rice and wheat are 0.8 and 0.5 ppm. Employing DTPA for acid soils when it was designed basically to serve the cause of alkaline calcareous soils (O' Connor 1988) defies all the logics of chemical principles and competitive metal equilibria operating in the soils. Searching and developing alternative soil tests will need specific attention. This will be a step towards building confidence in soil test driven recommendations on micronutrient application.

Identification of Fe deficiency by soil testing needs specific mention. Neither plant analysis nor total Fe leads to clear advisories on incidence of its deficiency. In order to solve this enigmatic problem, Katyal and Sharma (1980) developed a plant test for active Fe (Fe^{2+}). Using that method, with reasonable degree of confidence, values less than 50 ppm Fe^{2+} in plants indicate incidence of Fe deficiency.

Review of literature indicated that CLD of DTPA extractable Zn for rice (0.74$\pm$ 0.18 ppm) and maize (0.72 $\pm$ 0.10 ppm) were higher compared to wheat (0.61 $\pm$ 0.01 ppm) and other crops (0.59 $\pm$ 0.13 ppm) which includes pearl millet, mustard, chickpea, green gram, pigeon pea, groundnut and gram.various crops irrespective of soils and agro-ecological regions. Similarly, exercise for hot water extractable B for crops, *viz.* rice, maize, soybean, black gram and cauliflower irrespective of the soil types indicated that CLD concentrated around the mean value of 0.48 ppm with standard deviation of $\pm$ 0.04. This essentially means that CLD of hot water extractable B have been revolving around 0.5 ppm only, which was established more than six decades ago by Berger and Truog (1939). Lowland rice is more vulnerable to Zn deficiency than wheat. By this reasoning, rice is likely to suffer from Zn deficiency at higher levels of availability as compared to wheat. In fact, the respective general critical limits set by DTPA method for rice and wheat are 0.8 and 0.5 ppm (Katyal and Sharma 1980). Then these limits change with changing soil organic matter and pH. Negative effect of unusually high values of these parameters on Zn uptake by crops necessitates fixing higher critical limits than those under normal conditions.

When a nutrient (any nutrient) becomes deficient, for whatever reason, essential plant processes which depend on it are slowed down or disrupted. The deficiency of a nutrient results in a setback to its essential functions. For example, since boron is essential for cell-division and formation of different organs (differentiation) which takes place at the growing tips, B- deficiency damages these tips. As Mo is essential for biological nitrogen fixation (BNF), Mo-deficient legumes will neither nodulate well nor fix the normal amounts of N. Chlorotic leaves cannot carry out normal photosynthesis for which it requires healthy green leaves.

Plants have their own way of showing that they are not getting a particular micronutrient in adequate amount. When the nutrient is marginally deficient, the plant suffers silently and does not produce any visible signs of hunger but its productivity goes down. This is known as the hidden hunger stage. A farmer must ensure that their crop is not suffering from hidden hunger because otherwise lower yield will obtained than expected. This can be ensured by getting plant analysis done in a reliable laboratory and then by following its recommendations. Severe nutrient deficiencies often produce typical signs of hunger or symptoms which can be diagnosed by knowledge, training analysis and experience.

All crops are not equally sensitive to the deficiency of a nutrient, even under similar growing conditions. Such differences can even be large. In fact the same holds true even for varieties of a given crop. This can be due to genetic differences, differences in nutrient requirement, in the efficiencies of absorption from the soil and in the efficiency of utilization of the absorbed nutrients. Crops which are considered to be relatively more sensitive to the deficiency of an element are listed in Table 3.

Micronutrient Disorders and Various Mitigation Options in Indian Agriculture   41

Table 3. Crops considered being relatively more susceptible to the deficiency of particular micronutrients

Crop	B	Cu	Fe	Mn	Mo	Zn
Alfalfa	B	Cu			Mo	
Apple	B			Mn		Zn
Barley		Cu			Mo	
Beans				Mn	Mo	Zn
Brocolli	B				Mo	
Cauliflower	B				Mo	
Carrot	B	Cu				
Citrus		Cu	Fe	Mn		Zn
Clover					Mo	
Coffee	B					Zn
Cotton	B					
Field beans			Fe			Zn
Flax			Fe			Zn
Grapes	B		Fe	Mn		
Groundnut	B		Fe			
Lettuce		Cu		Mn	Mo	
Mint			Fe			
Maize						Zn
Mustards/ Rape	B					
Oats		Cu		Mn		
Oilpalm*	B					
Onion		Cu				
Ornamentals			Fe			
Pea				Mn	Mo	
Peach				Mn		Zn
Pear						Zn
Potato				Mn		
Raddish				Mn		
Rice		Cu				Zn
Sorghum			Fe	Mn		Zn
Soybean			Fe	Mn	Mo	Zn
Spinach		Cu		Mn	Mo	
Strawberry				Mn		
Sudan grass			Fe	Mn		Zn
Sugar beet	B			Mn		
Sunflower	B					
Tobacco		Cu				
Turnip	B					
Wheat		Cu		Mn		

* Oilplam and coconut are also sensitive to chloride deficiency.
Source: Shorrocks and Alloway (1988)

This means that under conditions of inadequate supply, a sensitive crop will suffer more and earlier than a tolerant crop. Same thing is true for different varieties of a given crop. No amount of theoretical knowledge can substitute for a good understanding of crop physiology, the soil characteristics of the field in question, cropping pattern, past history of fertilizer application and changes induced by weather (Bell and Dell 2008).

Some knowledge of the relative susceptibility or tolerance of a particular crop (or variety) to a nutrient deficiency can be of practical use. For example, in a deficient soil, a susceptible variety should receive the nutrient application on a priority basis as compared to a tolerant one. Likewise, if the required micronutrient fertiliser is not available, a farmer should plant a tolerant variety and not one which is sensitive to that particular deficiency even if the seeds of the tolerant variety cost a little extra. If a particular area is known to be deficient in certain nutrient, then sowing of crop which is relatively tolerant to that deficiency in that area can increase not only agricultural production but also nutrient use efficiency. Such practice need to be adopted.

The fact that some crops are more sensitive than others to the deficiency of a micronutrient implies that the critical concentration or critical level (CL) of a nutrient is not the same for all crops. It is not even same for a given crop under different soil-climatic conditions. That is why the susceptibility or tolerance rating of crops to nutrient deficiencies in the literature shows considerable variation as discussed later. This may primarily be due to wide hereditary variability within a crop species affecting plant response to nutrient stress, but can cause confusion and even mistakes at the field level. Further, terms such as low, moderate or high are very relative, as they do not mean much and should be used with caution.

Bronzing	:development of bronze/copper colour on the plant tissue
Chlorosis	:loss of the chlorophyll pigment, resulting in loss of green colour leading to paleness and yellow tissue.
Decline	: onset of general weakness as indicated by loss of vigour, poor growth andlow productivity (yield).
Die-back	:collapse of the growing tip, affecting the youngest leaves and twigs at the top.
Firing	:burning appearance of leaf tissue with dark brown or reddish brown colour.
Lesion	: a localised wound of the tissue (leaf, stem) accompanied by loss of normal colour.
Necrosis	:death of tissue (such tissue is called necrotic).
Scorching	:burning of the tissue accompanied by light brown colour (This can also result from foliar spray, salt injury etc).

Technical literature on nutrient deficiency symptoms contains a number of descriptive terms. All workers at the field level, extension, and training and in the trade may not be fully familiar with such technical jargon. Therefore, some of these are explained in the accompanying box.

Relative mobility of a nutrient influences the site of appearance of its deficiency symptoms. In a relatively mobile element, deficiency symptoms generally appear first on the older (lower) leaves because the plant is able to move that nutrient from older to younger leaves. It thus nourishes younger leaves at the expense of older leaves to the extent possible. This is similar to a mother sharing her food with a hungry child at her own expense. In contrast, deficiency symptoms of an immobile nutrient usually appear on the younger leaves because the plant is not able to move that nutrient from older to the younger leaves after absorption.

Micronutrient Disorders and Various Mitigation Options in Indian Agriculture 43

Table 4. Forms of nutrients absorbed by roots, their mean concentration and their mobility in the plant

Element	Form absorbed* by roots	Mean concentration in plant dry matter	Mobility in plant with age
Boron	$H_2BO_3^-$, H_3BO_3	20 ppm	relatively immobile
Chlorine	Cl^-	100 ppm	mobile
Copper	Cu^+, Cu^{2+}	6 ppm	relatively immobile but mobile under sufficiency condition
Iron	Fe^{2+}	100 ppm	relatively immobile
Manganese	Mn^{2+}	50 ppm	relatively immobile
Molybdenum	MoO_4^{2-}	0.1 ppm	moderately immobile
Nickel	Ni^{2+}	0.1 ppm	relatively immobile
Zinc	Zn^{2+}	20 ppm	low mobility

* Chelated forms can also be absorbed, though there is disagreement about the relevance of this process.

Quite often, nutrient deficiency symptoms in the field are not as straight forward or clear cut as shown in books. Most of are obtained pictures under very controlled "studio-like" conditions. In the field, a number of factors can complicate the scene which can even lead to wrong diagnosis. In addition, several nutrients perform similar or related roles in the plant with the result that their deficiency symptoms can also be confused.

Many of these confusing symptoms can be sorted out with the help of a good plant analysis laboratory (Singh, 1982).

In some cases, nutrient deficiencies can occur due to antagonisms (negative interaction) between two nutrients. An example is P induced Zn deficiency. There is however a need to distinguish between (i) antagonism where the supply of one element reduces the absorption or translocation of another and (ii) negative interactions where an inverse relation is observed between the concentration of two elements without their being any direct relationship between them.

Fe deficiency can be confused with	Mn deficiency
Leaf stripe disease of oat can be confused with	Mn deficiency
Effect of viruses, little leaf etc. can be confused with	Zn or B deficiency
Brown streak disease of rice can be confused with	Zn deficiency
Mo deficiency can be confused with	N deficiency
Ca deficiency can be confused with	B deficiency

Some commonly-known antagonisms are between P and Zn, K and Fe, S and Mo, Zn and Mn, Zn and Fe, Mn and Fe, Cu and Mo; and between Cu and Fe (Ishizuka 1971). A well balanced crop nutrition program will ensure that in most cases, antagonisms are kept to the minimum. Thus the application of one nutrient should take into account its possible effect on others so as to maintain optimum ratios and concentrations in the plants.

MANAGEMENT OF MICRONUTRIENTS IN CROPS AND CROPPING SYSTEMS

A commendable progress has been made in India in micronutrient research related to optimization of sources, rates and methods of micronutrient application in crops and cropping systems. Such information is not only important in maximizing the productivity of the country but has also been proving pivotal for other countries having more or less similar agro-ecological conditions.

ZINC

Zinc is an essential component of various enzyme systems for energy production, protein synthesis, and growth regulation. It is involved in the bio synthesis of a plant harmone Indole acetic acid (IAA) and is component of various enzymes such as carbonic anhydrase, alcholic dehydrogenase etc. The metabolic effects of zinc deficiency are suggestive of its role in photosynthesis and metabolic reaction in plants. Zinc deficient plants also exhibit delayed maturity. Zinc is not mobile in plants so zinc-deficiency symptoms occur mainly in new growth. Poor mobility in plants suggests the need for a constant supply of available zinc for optimum growth. The most visible zinc deficiency symptoms are short internodes and a decrease in leaf size. Delayed maturity also is a symptom of zinc-deficient plants. Uptake of zinc also is adversely affected by high levels of available phosphorus and iron in soils. On an overall basis, response studies confirmed that in terms of extents; Zn deficiency comes next only to N in lowland rice. While it ranks after N and P in case of upland crops.

Carriers, Mode, Rate and Method of Zinc Application

Of the several inorganic Zn carriers namely $ZnSO_4$ $7H_2O$, ZnO, $ZnCO_3$, Zn_3 $(PO4)_2$ and Zinc frits, $ZnSO_4$ $7H_2O$ has proved to be most efficient in correcting the Zn deficiency irrespective of the soil and crop situation (Nayyar et al., 1994). Majority of the field experiments involved studies on responses to Zn application through $ZnSO_4$ $7H_2O$. These studies were conducted both on farmers' fields (Katyal 1985) and research stations (Rattan et al. 1997). Critical analysis of the findings obtained from Farmers' field experiments indicated that Zn application enhanced yield by more than 200 kg ha^{-1} in at least half of the experiments. Between rice and wheat the former produced more often and greater yield advantage attributable to Zn application (Rattan et al. 1997).

Water insoluble ZnO and Zn frits were at par with $ZnSO_4$ $7H_2O$ in their performance in fine textured soils (Takkar and Nayyar 1986) but were inferior in coarse textured and sodic soils for rice and wheat (Gupta et al. 1986).

Soil application of Zn to annual crops is a preferred method over less efficient foliar sprays. Bi-weekly foliar sprays with 0.5% $ZnSO_4$ + 0.25% lime suspension are recommended using 500 liters of water per hectare on crops exhibiting Zn deficiency symptoms; spraying continues until the disappearance of the deficiency symptoms. Zinc sprays are almost exclusively used to alleviate Zn deficiency in trees and the Zn-sources are more effective if the sprays are made before the spring flush of the growth. As per Katyal and Rattan (1990), rates of Zn application mostly fall in the range of 5 to 20 kg Zn ha[-1]. Lower rates of Zn application are recommended for upland crops grown on coarse textured than those on heavy textured soils (5 kg Zn ha[-1] vs. 10 kg Zn ha[-1]). Broadcasting of 5-10 kg Zn ha[-1] as $ZnSO_4$. $7H_2O$ before the last plowing followed by mixing has been found to be efficient in enhancing grain yields of wheat; incorporation of the broadcasted zinc into the soil provides the mixing conditions usually necessary for improving the Zn utilization by different crops. Compilation by Katyal and Randhawa (1983) put the rate of soil application of Zn as zinc sulfate and zinc oxide through broadcast and band placement at 5-20 and 3-5 kg Zn ha[-1] respectively; similar rates for band placement were found to be 0.5-1.0 and 0.5-4.0 kg Zn ha[-1] as chelated-Zn ($Na_2ZnEDTA$; NaZn-NTA; Na_2Zn-HEDA) and zinc polyflavonoids, respectively.

Efforts were also made to develop the micronutrient-fertilization schedule for the whole cropping system rather than the individual crops. Most of the Zn, unutilized by the direct rice becomes potentially available to the upland wheat, grown immediately after rice. On the Zn-deficient Ustifluvents, whereas 5 kg Zn ha[-1] rate was the most optimum for rice but for the rice-wheat system as a whole 10 kg Zn ha[-1] exhibited superiority because of its superior residual effect on the subsequent wheat crop.

The yield maximizing rates of soil application of zinc sulfate for rice-wheat system were different under different agroclimatic conditions even under same type of soil as 5.5 kg Zn ha[-1] to every 3rd crop of rice on Haplustepts from Karnal; 10 kg Zn ha[-1] to every 5th crop on Haplustepts of north Bihar; and 11 kg Zn ha[-1] to every 7th crop on Typic Haplustepts of Punjab (Singh, 1999). On semi-reclaimed sodic soils, application of 20 kg $ZnSO_4$ ha[-1] to both the crops proved adequate to ensure optimum yields of the system during initial years of reclamation (Singh and Abrol 1986); zinc application could be withdrawn to the succeeding crops if 20 kg $ZnSO_4$ ha[-1] annum[-1] was applied continuously up to 4 to 5 years. Thus for rice-wheat cropping system, consensus is to fertilize rice annually with 5 kg Zn ha[-1] as zinc sulfate and subsequently raise wheat crop on residual Zn fertility. On red loam Zn-deficient soil of Tamil Nadu, the productivity of rice-rice system was optimized by application of 15 kg Zn ha[-1] annum[-1]. On Zn-deficient Calciorthents of Bihar, 10 kg Zn ha[-1] emerged as the best rate in maximizing grain yield of and Zn uptake by maize-rice system (Das et al. 1993); zinc sulfate was superior to chelamin (chelated-Zn). On alluvial soils of Punjab, 10 kg Zn ha[-1] proved to be adequate for 4 cropping years of maize-wheat-cow-pea fodder system on Haplustepts, as its application not only maintained the initial available Zn status but also improved it (Nambiar 1995). Water insoluble ZnO and Zn frits were at par with $ZnSO_4$ $7H_2O$ in their performance in fine textured soils (Takkar and Nayyar 1986) but were inferior in coarse textured and sodic soils for rice and wheat (Gupta et al. 1986). Zn-frits could be an excellent source for broadcast application and mixed uniformly in the soil (Singh 1999; Rattan et al. 2008). Five kg Zn ha[-1] rate adequately met the Zn requirements of rice-rice system on Vertisols (Deb et al. 1986). On these soils, $ZnSO_4$ proved superior to ZnO, Zn-frits and $ZnCO_3$ at 10 kg equivalent Zn ha[-1] in enhancing the productivity of rice-rice system (Takkar et al. 1989). Chelated- Zn (Zn-EDTA), free-flowing crystalline powder salt, typically guarantees a minimum of 12% Zn. Agronomical efficiency of chelated-Zn is 4-5 times higher than zinc sulphate but former is 15-20 times more costlier than latter. Nayyar et al. (1994) compared the efficiency of Zn –EDTA with $ZnSO_4$ $7H_2O$ at equivalent dose of zinc showed Zn-EDTA to be superior but was not economically feasible (Table 5). Fortified or customized zinc fertiliser such as zincated urea /zincated super and multi-nutrient micronutrient mixture have been found to be inferior to conventional inorganic fertilizers.

Table 5. Effect of sources and rates of zinc application on yield and zinc uptake by rice

Zinc source	Zinc rates (kg ha^{-1})	Grain Yeld (t ha^{-1})	Zinc uptake (g ha^{-1})
$ZnSO_4$ $7H_2O$	2.8	6.10	225
	5.6	6.65	283
	11.2	7.10	351
Zn -EDTA	2.8	6.25	240
	5.6	6.90	304
	11.2	7.60	378
Control		5.30	123
LSD (0.05)		0.54	43

Source: Nayyar et al. (1994)

Enrichment of organic manures (5t FYM) with Zn (5kg Zn ha^{-1}) gave as much yield as 10 kg Zn ha^{-1} alone with a similar residual effect on the subsequent wheat crop (Rathore et al. 1995). On Calciorthents soils application of 5.6 kg Zn ha^{-1} $annum^{-1}$ to direct wheat through $ZnSO_4$ along with poultry manure, biogas slurry, compost, sewage sludge and press mud maximized the yields of wheat-rice system. Prasad et al. 1989 found that zinc use efficiency of organic sources was superior (2.0-8.1%) to zinc sulfate (0.32-1.5%).

IRON

Being a structural component of chytochrome and ferrodoxin, it is regarded as necessary element for the synthesis and maintenance of chlorophyll in plants. It is essential component of various diverse metalloenzymes and carriers such as catalase, peroxidase and several cytochromes. The iron performs essential role in oxidation and reduction reactions in plants and nucleic acid metabolism. Cytochromes operates the respiratory mechanism of living cells. Its role in photosynthesis and nitrogen assimilation underlines the vital function it performs in overall plant metabolism. Iron is associated with sulphur in plants to form compounds that catalyze other reactions. Iron deficiencies are mainly manifested by yellow leaves due to low levels of chlorophyll. Leaf yellowing first appears on the younger upper leaves in interveinal tissues. Severe iron deficiencies cause leaves to turn completely yellow or almost white and then brown as leaves.

The problem related to iron deficiency is more due to the size of its available pool in the soil rather than its total content. The extent of Fe deficiency is approximately one fourth of Zn deficiency in the country and is largely prevalent in vast areas of alkaline to calcareous soil tracts. Iron deficiency seems to be only second in importance to Zn deficiency in the states of Karnataka, Haryana and Punjab.

Although there are many reports on responses of crops to this micronutrient, they are largely localized in nature. For example, on an average, responses of rice, wheat, sorghum, chickpea, lentil, black gram to Fe application on calcareous soils of Bihar are reported as 5.7, 2.4, 5.8, 9.4, 3.8, 3.1 and 15.3 q ha^{-1} (Sakal et al. 1996).

Enriching the organic manure with Fe further improved the yield. Gupta et al. (1994) reported that both soil application and foliar spray of FeSO$_4$ have been found to be equally effective for wheat in soils of Haryana. Rapid conversion of added Fe to unavailable forms is the major factor and therefore uneconomical as compared to foliar spray. Generally Fe deficiency in rice occurs in upland and highly percolating coarse textured soils because of solubilization and reduction of Fe^{3+} to Fe^{2+} does not occur to desired degree. Puddling of such soils have been shown to create conditions conducive for reduction of Fe^{3+} and help combating Fe chlorosis as evidenced by significant increase in grain yield of rice (Nayyar and Takkar, 1989). Maji and Bandopadhyay (1992) have also reached to the same conclusion that submergence in all the thirteen soils of West Bengal enhanced appreciably the availability of Fe. Similarly, the findings of Srivastava and Srivastava (1994) have shown 3.9 fold increases in the available Fe content when sodic soils were subjected to water logging in comparison to field capacity conditions. Nayyar and Takkar (1989) compared the efficiency of green manure, soil applied Fe coupled with foliar application and green manure + foliar applied Fe for three years. Highest grain yield was obtained with combined application of green manure and foliar spray of 1 % ferrous sulphate solution followed by green manure or foliar sprays. Soil application of Fe proved least effective. Soil application of ferrous sulfate is distinctly inferior to foliar sprays; for soil application to be effective, few quintals of ferrous sulfate will suffice. On Inceptisols of Delhi, foliar application of Fe (3% FeSO$_4$.7H$_2$O) solution, thrice at 40, 60 and 75 days after sowing of rice *i.e.* 45 kg FeSO$_4$.7H$_2$O ha^{-1} was most effective and economical in correcting Fe deficiency in aerobic rice, followed by soil application of 150 kg FeSO$_4$.7H$_2$O ha^{-1} + 10 t FYM ha^{-1} and 305 kg FeSO$_4$.7H$_2$O ha^{-1} (Pal et al. 2008).

Carriers, Mode, Rate and Method of Iron-Application

Iron deficiency is the most difficult nutritional disorder to correct. A number of sources of Fe are available to combat its deficiency. These include inorganic salts, frits, natural and synthetic chelates. Among the inorganic salts, ferrous sulphate and ferric sulphate are more commonly used. They are the cheapest Fe sources, although not very effective when applied to the soil. As a foliar spray they perform well. Iron frits are used only under specific situations. One important instance is high percolating acid sands. Chelates such as Fe- EDTA, Fe-EDDHA and multi-nutrient mixtures containing variable amount of Fe are available. Organic manures like FYM and Piggery manure are also good source of Fe. Arora et al. (1975) reported that manure such as poultry, piggery and FYM contained 1075, 1600 and 1465 mg kg^{-1} of total iron respectively.

Sakal et al. (1996) reported still higher content of total Fe in sewage sludge, FYM, press mud, municipal waste and poultry manure which were as 82, 2528, 3910, 5375 and 3400 µg Fe g^{-1} respectively. Sakal et al. (1996) compared the effect of Fe on rice yield on calcareous soils of Bihar applied as soil application of FeSO$_4$, compost, pyrite, FeSO$_4$ with compost, pyrite with compost (Table 6).

They observed that the yield produced by 50 kg FeSO$_4$ ha^{-1} were at par with the yield obtained with the application of 1t pyrite ha^{-1} and 10 t compost ha^{-1}. The response of rice to FeSO$_4$ or pyrite improved appreciably when applied in combination with compost. Prasad et al. (1989) concluded that application of 10 t ha^{-1} of different organic manures significantly increased the rice yield over soil application of 10-20 kg Fe ha^{-1} as ferrous sulphate.

Enriching the organic manure with Fe further improved the yield. They are known to enhance the availability of native Fe and increase the efficiency of added Fe.

Table 6. Effect of iron carriers on yield and iron uptake by rice

Treatment	Rice yield (t ha^{-1})		Fe uptake (kg ha^{-1})
	Grain	Straw	
Control	3.40	5.93	1.25
50kg FeSO$_4$ ha^{-1}	4.27	7.27	1.25
10t compost ha^{-1}	4.00	6.98	1.19
1t pyrite ha^{-1}	3.67	6.53	1.53
50 kg FeSO$_4$ ha^{-1} + 10t compost ha^{-1}	4.80	7.83	1.73
1 t pyrite +10t compost ha^{-1}	4.47	7.40	2.07
	0.84	1.16	0.28

Sakal et al. (1996)

Gupta et al. (1994) reported that both soil application and foliar spray of FeSO$_4$ have been found to be equally effective for wheat (Table 7). Rapid conversion of added Fe to unavailable forms is the major factor and therefore uneconomical as compared to foliar spray.

Table 7. Grain yield of wheat as influenced by soil and foliar application of iron

Treatment	Grain yield (t ha^{-1})	
	Experiment -1	Experiment -2
Control	3.8	3.8
Soil application		
10 kg ha^{-1}	4.1	4.4
20 kg ha^{-1}	4.7	4.7
Foliar application (Through FeSO$_4$)		
0.5 %	-	4.5
2.0 %	4.4	-
LSD(0.05)	0.5	0.3

Source: Gupta et al. (1994)

Iron chelates are highly effective as soil treatment, about 0.5 -1.0 kg Fe ha^{-1} through Fe-EDDHA should overcome Fe deficiency. For better results, Fe chelates are placed in bands than broadcast. Since Fe chlorosis affects even young seedling, chelates must be applied to soil before sowing or placing.

Effect of green manuring on iron availability: incorporation of green manure under submergence conditions substantially increased the Fe and Mn concentration and partial pressure of carbon dioxide and decreased the pH and EC. The increase in Fe and Mn concentration in soil solution could be attributed to the formation of complexes of Fe^{2+} with organic acids produced during anaerobic decomposition of green manure and also due to sharp decrease in pH and EC.

Iron efficient cultivars: Marked genotypic differences among cereals for Fe use efficiency have been reported. Screening of hyper accumulator or supra cultivars having high micronutrient density in seed has been found of more importance in mobilization of micronutrients from seed to seed (soil to root, root to shoot, shoot to seed) by solubilizing soil native Fe to overcome malnutrition (Welch and Graham, 2004). Solubilization of Fe^{3+} depends upon acidity of the rhizoshere and amount of reluctant present in root exudates. Thus, iron efficient crop varieties for mobilizing higher amounts of Fe needs to be developed. On the basis the degree of iron chlorosis, Singh et al. (1985) sorted out rice genotypes into four classes viz. highly susceptible, susceptible, moderately tolerant and tolerant. In addition to Fe chlorosis, they explained that variations among genotypes occurred due to variable shoot yield, chlorophyll content, active Fe and total Fe. All these parameters showed increasing trend with the increase in tolerance or decrease in Fe chlorosis.

MANGANESE

The role of manganese in photosynthesis has been clearly established through its absolute necessity for water splitting in photosynthesis. It is also known to be important in carbon dioxide assimilation and in nitrogen metabolism. An inadequate supply of this element results in poor growth, development and obstruct the normal yield of the crops. Interveinal chlorosis is a characteristic manganese-deficiency symptom. In very severe manganese deficiency cases, brown necrotic spots appear on leaves, resulting in premature leaf drop. Delayed maturity is another deficiency symptom in some species. White/gray spots on leaves of some cereal crops is also a sign of manganese deficiency

The adoption of rice –wheat rotation in non conventional rice soils has not only caused Fe deficiency but also resulted in appearance of Mn deficiency in wheat. Its deficiency causes a significant decline in crop field and in extreme cases results in its failure. First of all its deficiency at large scale was observed on a large area of relatively coarse textured soils under rice- wheat rotation in 1979-80 in many villages of Ferozepur district (Takkar and Nayyar 1981). At some places, the impact of the deficiency was so severe that the farmers had to plough their wheat crops because of its extremely poor stand. Since then its deficiency had been increasing on such soils. This has now become a serious problem of highly permeable coarse textured soils where wheat is followed by rice. This is happening due to leaching losses of manganese during rice cultivation. A soil is considered deficient if it contains less than 3.5 mg kg^{-1} available manganese. In several crop plants Mn contents below 20 $\mu g\ g^{-1}$ dry matter are suggestive of its deficiency. In both the cases manganese application is required. This problem is now prevalent in many parts of India where striking responses of wheat to applied manganese has been observed. Its deficiency has been noticed in calcareous soils of Bihar and coarse textured soils of Haryana. Significant decline in wheat yield due to manganese deficiency have been reported in many parts of the state (Bansal et al. 1991 & Khurana et al. 2005). The extent of yield loss, however varied with the magnitude of manganese deficiency in the soils and variety sown (Nayyar et al. 1985 and Bansal and Nayyar, 2000). The presence of Mn in higher oxidation states as very insoluble hydroxides and oxides accounts for low Mn availability in well drained alkaline and calcareous soils.

This has necessitated the studies to evolve technology for efficient management of this problem by conducting research on Mn carriers, time of application and screening of Mn efficient crop cultivars. Manganese deficiency for field crops in the country is mainly at localized sites where rice-wheat crop rotation is practiced in coarse-textured soils.

Carriers, Mode, Rate and Method of Manganese Application

Various sources of manganese such as manganese sulphate ($MnSO_4$ H_2O), manganese oxide (MnO) and manganese frits as well as micronutrient mixtures has been evaluated for their efficacy to correct Mn deficiency. Manganese sulphate has been rated as the most efficient carrier followed by Mn frits and MnO.

Table 8. Response by wheat to foliar application of Mn at different sites in Punjab

Name of the village		Grain Yield (q ha^{-1})		Response (q ha^{-1})
	District	Control	Sprays	
Moorkarima	Ludhiana	35.3	48.5	13.3
Batha Dhua	Ludhiana	33.0	50.0	17.0
Basti -Nanaksar	Faridkot	27.3	40.8	13.5
Beehlewala	Faridkot	30.8	42.3	11.5
Passan	Jalandhar	43.0	52.0	9.0
Bheelan	Jalandhar	34.5	45.0	10.5
	Mean	35.7	45.7	11.5
	SD±	5.7	4.6	1.8

Source: Khurana et al. (2005)

Comparison of some multi-nutrients mixtures containing $MnSO_4$ as carriers of Mn under field conditions has shown to be distinctly inferior to $MnSO_4$ (Sadana *et al.*, 1989 & Bansal et al. 2005). Teprosyn-Mn, a formulation meant for seed treatment to combat Mn deficiency, was much inferior to $MnSO_4$ in meeting Mn requirement of wheat crop. Mn chelates, by and large, perform poorly as soil application (Table 9). In some cases they aggravate Mn deficiency rather than ameliorating it. It is due to substitution of Fe for Mn in chelate source, thereby enhancing Fe availability. A high Fe availability in soils discourages Mn uptake by plants. Mn chelate (Mn- EDTA) have been successfully used as foliar spray.

Table 9. Yield and Mn uptake by wheat as influenced by different sources of Mn

Treatment	Mode	Rate	Yield (qha^{-1}) Grain Straw		Mn content (μg g^{-1}) Grain Straw		Mn uptake (g ha^{-1})
$MnSO_4.H_2O$	Foliar	0.5 %	39.0	62.2	11.5	6.5	85.3
		1.0%	42.8	66.1	14.0	8.2	114.1
Micronutrient mixture (15% Mn +4%Zn)		0.5 %	36.7	60.7	8.5	5.8	66.4
		1.0%	39.8	62.6	10.3	7.3	86.7
Teprosyn -Mn	Seed	5ml kg^{-1} seed	33.4	56.9	8.0	4.3	51.2
$MnSO_4.H_2O$		75g kg^{-1} seed	34.0	60.2	8.5	5.0	56.4
Control			30.5	54.0	7.1	4.0	43.2
LSD(p=0.05)			3.8	6.4	3.0	1.5	25.0

Bansal and Khurana (2007)

While both soil and the foliar application cause a significant increase, yet the foliage applied Mn proves strikingly superior to soil application as was evident from 1.92 to 2.47 t ha[-1] increases in wheat grain yield with foliar application against 1.13 to 1.19 t ha[-1] with soil application (Takkar et al. 1986). In an other study, soil application of manganese at the rate of 20-40 kg Mn ha[-1] at seeding produced yield comparable to that obtained with three sprays with 0.5 percent $MnSO_4$ solution (Nayyar et al. 1985), thereby indicating non economic viability of soil application. Since its deficiency in wheat appears after first irrigation to crop, its first spray should be imitated before first irrigation (Table 10). A supplementary study involving two field experiments was conducted by Takkar et al. (1986) to evaluate the efficiency of rate, mode and time of application of manganese application to wheat on a manganese deficient soil. A significant increase in manganese uptake by wheat and marked increases in wheat grain and straw yields were recorded due to both soil (5, 10 and 20 kg Mn ha[-1]) and foliar (0.5, 1.0 and 2.0% $MnSO_4$ solution) application of manganese, but foliar application was more effective. A series of sprays initiated before first irrigation to the wheat (26 days after sowing); gave significantly better results rather than foliar application of Mn after the first irrigation (32 days after sowing); four sprays of 0.5% $MnSO_4$ solution produced the highest grain yield, which was followed by three sprays of 1% solution.

Table 10. Effect of concentration of $MnSO_4$ and the time of its sprays on wheat

Solution Concentration (%)	Number of Sprays		Grain Yield (t ha[-1])	Mn content ($\mu g\ g^{-1}$)
	Before Ist irrigation	After Ist irrigation		
0.5	1	2	4.75	17.4
	0	3	4.00	13.5
1.0	1	3	5.20	18.7
	0	4	4.20	18.8
Control	-	-	2.67	12.3
LSD (p=0.05)			0.31	1.7

(Takkar et al.1986)

Singh and Nayyar (1994) studied the response of winter crops (Lentil, Gobhi, sarson, oat, lucerene, yellow sweet clover and winter maize) to soil applied (10, 20 and 40 mg kg Mn soil[-1]) and foliar applied Mn (1% $MnSO_4$ solution) on Mn deficient loamy sand soil (1.8 mg kg[-1] soil) in a pot culture experiment (Table 11).

Though the dry matter yield of all the crops at 60 days growth continued to increase with Mn application, the significant differences were noted only in lentil, oat and annual yellow sweet clover. It indicates that higher rates of Mn are needed for these crops for obtaining their best yields. Foliar application of Mn also significantly increased the dry matter yield of all the crops compared with control and increase in yield was almost equal to the yield obtained with the highest rate of Mn applied to soil. It suggests that on soils very low in available Mn, 40 mg Mn kg[-1] soil or 4 foliar sprays of 1% $MnSO_4$ H_2O solution can efficiently cure Mn deficiency in crops. Nevertheless, soil application of Mn is uneconomical as large amount of $MnSO_4.H_2O$ is required as compared to the small quantity applied through foliar spray to obtain similar yield.

Table 11. Dry matter weight (g pot^{-1}) of different crop species as influenced by Mn application

Crop	Control (Mn$_0$)	Mn application				CD (p=0.05)
		Soil (mg Mn kg^{-1} soil)			Foliar (1% MnSO$_4$ Solution)	
		Mn$_{10}$	Mn$_{20}$	Mn$_{40}$		
Lentil	1.8	2.9	3.8	5.1	4.4	1.1
Gobhi Sarson	9.2	12.2	15.9	16.9	15.8	1.9
Oat	1.4	2.4	3.2	4.8	5.1	1.0
Lucerene	2.5	4.0	4.6	4.9	5.1	1.6
Yellow sweet clover	8.9	10.6	10.7	12.7	12.7	1.0
Winter maize	13.4	14.8	15.8	16.9	16.4	2.0

Singh and Nayyar (1994)

Table 12. Classification of wheat cultivars into different groups of Mn efficiency based on absolute yield and rating of Mn deficiency symptoms

Group No	Grain yield at 0 Mn application (t ha^{-1})	No of cultivars	Percent cultivars in a category showing Mn deficiency symptoms	Rating for Mn efficiency
	1-12	(PBW 205*), (PDW 206*), (PDW 208*), (DWL 5023*), (PDW34*), (DWR 39*), (NI 8625*), (N 59*), (Bijaga Yellow*), (NI 5439*) (MACS 1967*), (HS 223*)	100	Inefficient
	13-15	CPAN 1990*, CPAN 1992, HUW 1045*	100	Inefficient
	16-18	C 306*, HD 2278*, (HD4502*)	100	Inefficient
	19-27	HD2285*, PBW 175, HD 2385, HUW 283*, PBW175, HD 2385, HUW283*, CPAN***, HD 2400, HD 1135< NI 8841***, HD2189*	67	Slightly efficient
	28-39	IWP72, K 8434, CPAN1905*, (HD2780), DL254-4(Raj 6276*** NI 8796, NI1077 NI5439***, NI8858**, Sonalika*	67	Slightly efficient
	40-50	HD2449, PBW186, PBW189, HD2461**, HUW280, Kalyan Sona, Swati** , HD2469**, CPAN1990, RAJ 3077, HRD43	27	Moderately Efficient
	51-56	PBW212, WL410, DL330-1 HD237, HUW284, NI8838	0	Efficient
	57-61	HD2412, PBW188, J478, RAJ 3163, CPAN 1922	0	Efficient

Source: Bansal et al 1991

* and ** indicate cultivars developed severe and mild deficiency symptoms, respectively. Cultivars within the bracts are *Triticum durum* while the others are *Triticum aestivum*

From the above studies, it is observed that the common method of correcting Mn deficiency is the foliar application of manganese in the form of three spray of 0.5 per cent manganese sulphate solution which appears to be a temporary remedy. This means repeated sprays of manganese sulphate solution are required every year for correcting Mn deficiency.

Manganese efficient crop cultivars: Durum varieties of wheat have been found to be more susceptible than aestivum varieties with reference to Mn deficiency. Wide variation exists among varieties in respect of their response to Mn sprays have been observed. Sixty one cultivars were tested in a Mn-deficient field containing 3.0 mg kg^{-1}-soil available Mn. The cultivars were divided into eight categories according to grain yield produced without Mn application. Based on absolute yields and Mn deficiency symptoms, these cultivars were classified in to four groups of Mn efficiency (Table12). Eighteen cultivars which produced grain yield of less than 20.t ha^{-1} and also developed severe Mn deficiency symptoms were rated as inefficient. Eleven cultivars which produce more than 305 t ha^{-1} grain and did not show any visible symptoms were classified as efficient. It is interesting to note that of 18 cultivars rated as Mn inefficient, 12 belong to the *triticum durum* group, indicating that durum cultivars are more prone to Mn deficiency. (Bansal et al. 1991)

BORON

A primary function of boron is related to cell wall formation, so boron-deficient plants may be stunted. Sugar transport in plants, flower retention; pollen formation and germination are also affected by boron deficiency. Seed and grain production are reduced with low boron supply.

Deficiency of boron is wide spread in India which is next to zinc in extent. Its deficiency is found in nearly 30% soils of the country which are highly calcareous, leached and sandy in nature. Its deficiency is more prevalent in soils of *terai* region of West Bengal (Mondel et al. 1991) and calcareous soils of north Bihar (Sakal et al. 1991) and Madhya Pradesh (Takkar et al. 1989). It is also more commonly observed in light textured acidic Entisols and Inceptisols receiving high precipitation (Mandal and De, 1993). It has now started appearing in alluvium derived coarse textured, low organic matter, alkaline and calcareous soils of Punjab. According to various sources, about 5-10 percent soils are deficient in available boron in Punjab (Bansal et al. 2003; Singh and Nayyar 1999 & Sharma and Nayyar 2004) which has obstructed to attain the full yield potential particularly of legume crops and thus expected to respond to its application. As suggested critical levels of boron in soils below which deficiency is expected have a B content less than 0.5 mg/kg of soil (topsoil). Critical levels in reference leaves are usually in the range 10-20 ppm for most crops. However, lack of boron nutrient can cause serious physiological damage, thus retarding plant growth and reducing the crop yield substantially.

Boron deficiency symptoms first appear at the growing points. This results in a stunted appearance (rosetting), barren ears due to poor pollination, hollow stems and fruit (hollow heart), brittle discolored leaves. Boron deficiencies are mainly found in acid soils, on sandy soils in regions of high rainfall and soils with low organic matter content. Borate ions are mobile in soil and can be leached from the root zone. Boron deficiencies are more pronounced during drought periods when root activity is restricted. Crops that are susceptible to B deficiency are alfalfa, sugar beets, clovers and some vegetable crops. The incidence of boron deficiency was highest in the acid soils of West Bengal followed by the calcareous soils of Bihar. Alluvial soils of Assam are more deficient (44%) in B as compared to lateritic soils (34%) based on the critical level of 0.5 mg kg^{-1} HWS-B (Borkakati and Takkar 2000). The deficiency of B was highest under fruit crop ecosystem (56-65%) followed by field crop (42-47%), pasture (37-45%) and forest ecosystem (5-17%). In acid soils of Meghalaya under rice cultivation, 35 per cent of the 136 soil samples indicated deficiency of B (Nongkynrih et al.1996). Similarly, in acid alluvial soils of Coochbehar, Jalpaiguri and west Dinajpur districts of Bengal, 70 out of 88 surface soil samples were deficient in available B (Mondal et al. 1991). In young alluvial soils of Bihar, the incidence of B deficiency was higher (47%) in calcareous as compared to non-calcareous soils (Sakal and Singh 1999).

Carriers, Mode, Rate and Method of Boron-Application

Among the various boron carriers, borax ($Na_2B_4O_7.10H_2O$ with 11% B), sodium tetraborate ($Na_2B_4O_2.5H_2O$ with 14% B) and sodium borate ($Na_2B_4O_7$ with 20% B) have been commonly used for soil application; whereas boric acid (H_3BO_3 with 17% B) has been popularly employed for foliar application of B. Solubor ($Na_2B_4O_7.5H_2O+Na_2B_{10}O_{18}.10H_2O$) is used for both soil and foliar application because of its higher solubility Colemanite ($Ca_2B_6O_{11}.5H_2O$ with 10% B) and B-frits containing 2-6% B are considered to be more promising on highly leached sandy soils as well as for long duration field crops and fruit plants owing to their low solubility and slow release of B. Boronated superphosphate has also been tried for correcting B deficiency and is found promising on preliminary basis (Patil et al. 1987).

Field studies involving B have been limited in number and geographical spread. Most of the studies have been located in eastern India where soils are relatively light in texture, remain largely flooded and are prone to excessive leaching. For example, extensive on-farm trials conducted in Assam, Bihar, Orissa and West Bengal (all states of eastern India) indicated that rainy-season rice, wheat and mustard responded positively (response>200 kg ha^{-1}) to applied B in 69, 70, 79 and 71% of the experiments (Ali 1992). Findings from other studies conducted in flood-prone regions (mostly Calcifluvents) of Bihar confirmed significant yield advantage attributable to B application (Sakal et al. 1996; Takkar et al.1997). Even in conventionally non-B deficient states like Punjab and Haryana, response to B application in B- loving crops like groundnut was obtained on the light textured soils. Positive response of cereals, pulses, oilseeds and cash crops to B application (0.5 to 2.5 kg B ha^{-1}) have largely been reported from Bihar, Orissa, West Bengal, Assam and Punjab (Takkar et al. 1997). Significant increase in yield of wheat was reported due to addition of 20 kg borax ha^{-1} on some silty loam acid soils of North Bengal where additional grain yield of 18.9 q ha^{-1} over control was recorded (Mitra and Jana 1991); application of B @ 1.135 kg ha^{-1}, half in soil and other half through foliar spray in two equal splits, at late jointing and just before flowering was superior in increasing the grain yield (9.9 q ha^{-1}) of soybean over control to that with full quantity either through foliar spray or in the soil at the time of sowing (Roy and Pradhan 1994). On acid Inceptisols of Ranichauri, 20 kg sodium tetraborate ha^{-1} applied directly to soybean and potato was superior to two foliar sprays, while soil application left significantly higher residual effect on wheat grown after soybean (Dwivedi and Dwivedi 1992). On B-deficient (hot water-B 0.4 mg ka^{-1}) calcareous soil of Bihar, yield response to 1.5 kg B ha^{-1} as borax for groundnut, maize, onion, yambean and sweet potato in *kharif* were 2.4, 5.2 44.8, 16.7 and 70 q ha^{-1}, respectively, whereas in rabi, the magnitude of yield response were 3.1, 8.7, 3.2, 28.3 and 2.4 q ha^{-1} for mustard, maize, sunflower, onion and lentil, respectively (Sinha et al. 1991). Boron deficiency is invariably corrected by its soil application depending upon soil type (Table 14). In calcareous soils of Bihar, Sakal et al. (1988) and Sinha et al. (1991) observed the optimum rate varying between 1 to 2.5 kg B ha^{-1} for different crops. Due to their higher boron requirement, oilseeds and pulses responded more to B application as compared to cereals and response was found to be quadratic. Singh et al (1988) reported optimum level of B to be 2.08 kg ha^{-1} for chick pea and 1.66 kg ha^{-1} for maize. Similarly in blakgram , the optimum rate of B has been reported to be 1.96 kg ha^{-1} with a response of 0.23 t ha^{-1}(Singh and Singh)

On Calciorthents of North Bihar, application of 8 kg borax ha^{-1} to each crop or application of 16 kg borax ha^{-1} to alternate crops of rice-wheat and maize-mustard cropping systems sustained higher crop yields, ensured higher B uptake and maintained higher levels of the hot water soluble B levels in the soil (Sakal et al. 1996). On B-deficient coarse textured soils of Tamil Nadu, initial application of 2 kg B ha^{-1}followed by application of 0.5 kg B ha^{-1} to alternate crops of groundnut-maize cropping systems sustained highest system's productivity and total B uptake (Singh, 2000).Soil application of B @ 20 kg sodium tetraborate ha^{-1} as well as two foliar sprays with 0.2% solution of this salt proved equally effective in increasing soybean grain yield and the residual effect of soil applied B on subsequent wheat significantly out yielded the direct effect of foliage sprays (Table 13). Addition of the same rate of B after liming the soil produced significantly higher yields of soybean and wheat as compared to its without lime (Dwivedi et al. 1990) Since B undergoes less leaching in heavy textured soils, its one application may produce residual effect. In view of very sharp and narrow difference between optimum and the toxic levels of B, great precaution is required in its repeat application, particularly in medium to heavy textured soils.

Table 13. Effect of modes of B application in soybean-wheat system

Treatment	Grain yield (t ha^{-1})			
	Soybean		Wheat	
	-Lime	+Lime	-Lime	+Lime
Soil Application (20 kg sodium tetraborate ha^{-1})	1.41	2.33	1.46	2.65
Foliar application solution) (0.2% sodium tetraborate soil.	1.33	-	1.35	-
Control	0.89	1.48	0.66	1.63
LSD (p=0.05)	0.1		0.11	

Dwivedi et al. 1990

However, in recent study, Starker et al. (2007) conducted experiments for increasing use efficiency of boron fertilisers by rescheduling the time and methods of application for crops in light textured acidic *Entisols* and *Inceptisols* receiving high precipitation. Crop use efficiency of fertilizer B is also low under such B leaching environments. For these high leaching regimes, the optimum timing and methods of B application would vary with the crop sensitivity to B deficiency and periods of peak demand for boron. They suggested that foliar application, in general had higher economic benefits than soil application. Split application of B either to soil (as basal and 25 days after sowing) or foliar sprays (at 25, 40 and 25 and 45 days after sowing for mustard and potato, respectively) had an edge over a single application. For wheat, a single late application of B (at 45 or 60 days after sowing through soil and foliar sprays respectively) was more effective than the early or split application in increasing yields. Better use efficiency of B can thus be achieved if it is applied late for wheat but in splits over a longer period for mustard and potato with higher economic benefits.

Table 14. Response dose as affected by soil type in ground nut and wheat

Crop	State of India	Type of soil Soil Type	Response dose (Soil application)	Reference
Groundnut	Punjab	Coarse textured alkaline soil	0.5kg B ha[-1]	Arora et al. 1985
Groundnut	Bihar	Calcareous soils	2kg B ha[-1]	Sakal et al. 1980
Wheat	Bihar	Calcareous soils	1 kg B ha[-1]	Sakal et al. 1980
Wheat	Assam	Acid soils	1 kg B ha[-1]	Ali and Monoranjan 1989

COPPER

Copper is known to be associated with a number of metalloproteins such as diamine oxidase, ascorbate oxidase, O–diphenol oxidase, Cyt C oxidase and superoxide dismutase. Thus Cu deficiency is expected to cause decreased activities of these enzymes. As much as 70 percent of the Cu in plant is concentrated in the chlorophyll and its important function can be seen in assimilation. Copper deficiency leads to early aging of the chlorophyll with the result of a noticed decline in plants performance. Copper is also required for lignin synthesis which is needed for cell wall strength and prevention of wilting. Deficiency symptoms of copper are dieback of stems and twigs, yellowing of leaves, stunted growth and pale green leaves that wither easily. Copper deficiencies are mainly reported on sandy soils which are low in organic matter. Copper uptake decreases as soil pH increases. Increased phosphorus and iron availability in soils decreases copper uptake by plants. The incidence of Cu deficiency in the indo gangetic plains (IGP) of India is less than 3 percent. Its deficiency has been reported to be higher (31 to 40%) in the states of Kerala and Gujarat (Singh1999). The visual symptoms of Cu deficiency in rice and wheat have not been recorded under field condition in such soils.

However symptoms of Cu deficiency have been developed and characterized under sand culture (Sharma at al. 1996; Karim and Vlamis 1962) Deficiency of copper causes distortion and chlorosis in leaves followed by narcosis of the tips of young leaves that proceeds along the margins of the leaf. In detail copper deficient plants show bluish-green leaves which later on turn yellowish-white near the tips.. Thereafter, dark brown necrotic lesions appear at the tips and then spread downwards along the midrib. The newly emerging leaves fail to unroll and they maintain a needle like appearance usually of the entire leaf or occasionally of the top-half leaf, leaving the basal end to develop normally. Effective tillering is severely depressed resulting in considerable loss in grain yield. The panicles are small with numerous small sterile grains. In wheat, the youngest leaves show pale green coloration and rolled and twisted appearance with their tips bleached which later spreads along the leaf margins towards the base of the leaves. Ears and stems may become noticeably darker on maturation.

Source, Rate and Method of Copper Application

Very few response studies have been conducted in India because of low magnitude of its deficiency. While summarizing results of 110 experiments on rice and 34 on wheat in soils of Bihar, Takkar et al. (1989) reported an average response on 0.46 and 0.38t ha[-1], respectively to 25 kg $CuSO_4$ ha[-1]. Earlier, Singh et al. (1979) observed average response up to 1.13 t ha[-1] in rice to foliar application of Cu at Varanasi, Nandyal and Chiplima after summarizing the results of 124 field experiments. Similarly, results of 87 experiments on wheat have shown response to the foliar application of Cu to the order of 0.25 to 0.98 t ha[-1]

The deficiency of Cu is generally corrected by copper sulphate ($CuSO_4.5H_2O$) which can be applied for both as soil and foliar application. Fungicide and germicide containing copper as an active ingredients and/or copper sulfate is frequently applied as fungicide and germicide in vegetables and fruit trees for controlling diseases because Cu^{2+} ion is toxic to lower forms of life. Thus, the deficiency of copper is taken care of by their foliar application

Its soil application has been found to help more efficient utilization of this element compared to foliar application in wheat on alkaline soils of Punjab and soybean in acid soils of Uttar Pradesh as shown in Table 15. In a loamy sand soil of Ludhiana, soil application of 5 and 10 kg Cu ha[-1] was proved superior to foliar application of 0.2 and 0.4% $CuSO_4$ solution in case of wheat grain yield. The increase over control with soil application of 5 and 10 kg Cu ha[-1] was 0.31 and 0.49 t ha[-1] respectively, compared to 0.11 and 0.09 t ha[-1] with foliar application of 0.2 and 0.4% $CuSO_4$ solution (Lal et.al., 1971). Similarly, Agrawal and Gupta (1994) on similar type of soil reported significant increase in grain yields of rice grown on at Varanasi with 10 and 20 kg of Cu as copper sulfate ha[-1] while the higher rate of 40 kg Cu ha[-1] resulted in decline in grain yield. It may be concluded from these studies that crop appears to be suffering from hidden hunger of this element as evident from crop responses but soil and plant analysis do not suggest its deficiency.

Table 15. Effect of soil and foliar application of Cu on crop yield

Mode of application	Cu rate	Grain yield (t ha[-1])		
		[1]Wheat	[2]Soybean	
			I	II
Soil	5 kg ha[-1]	5.21	-	-
	10 kg ha[-1]	5.39	1.63	1.54
Foliar	0.2% $CuSO_4$	5.01	1.20	1.15
	0.4% $CuSO_4$	4.99	-	-
Control		4.90	0.89	0.80
LSD (p=0.05)		1.5	1.0	1.3

[1]Lal et al. (1971); 2. Dwivedi et al. (1990)

An increse in wheat grain yield (0.91 t ha[-1]) over control yield (3.80 t ha[-1]) by an application of 20 kg $CuSO_4$ ha[-1] on clay loam soil was observed by Sharma et al. (1992).

In vegetables and food crops, however, the Cu deficiency is taken care of by its foliar application as Bordeaux mixture generally used for control of diseases. Since Cu is immobile in the plant, multiple sprays are required to achieve maximum production.

Molybdenum: The role of Mo in normal assimilation of N by plants is well known, because it is a component of nitrogen fixing enzymes such as nitrate reductase and nitrogenase. Thus it plays a key role in nitrogen fixation of legumes. Molybedenum is required in the synthesis of ascorbic acid and is implicated in making Fe physiologically available with in the plant. Molybdenum concentration is highly variable. Plants may contain from less than 0.1 ppm to greater than 300 ppm. Typically, plant concentration ranges between 0.1 and 2.0 ppm. A soil is classified deficient if it tests less than 0.2ppm Mo by Ammonium Oxlate (pH3.3). It is highly available in alkaline soils-a behaviour opposite to other essential micronutrients and is much less available in acid soils liming alleviate Mo deficiency in acid soils

The early symptoms of Mo deficiency resemble nitrogen deficiency as Mo is involved in N metabolism of plants and is particularly true for legumes. In wheat, young leaves become chlorotic along the apex and apical margins which later intensifies and extends to the apical half, turning dry and necrotic. Severely affected leaves turn light brown. Deficiency of Mo also reduces the yield of rice, mustard, chickpea, greengram, blackgram, pigeonpea, cotton, sugarbeet and cauliflower.

Sodium molybdate, ammonium molybdate and molybodenum trioxide are the common sources of Mo, but sodium molybdate is most commonly used. In view of the limited deficiency of Mo, very little work has been done on the management of its deficiency.

In acid hill soil of Uttar Pradesh, the increase in soybean yield with application of 0.5 kh ha^{-1} of sodium molybdate to a limed soil far exceeded the yield obtained with the same rate of application to an unlimed soil. The residual effect of this treatment was also apparent on wheat. Two foliar sprays of 0.05 percent solution of sodium molybdate proved significantly inferior to its soil application (Dwivedi et al 1990).

REFERENCES

Aggarwal H.P. and M.L. Gupta (1994) Effect of copper and zinc on copper nutrition of rice. *Annals of Agricultural Research,* 15(2):162-166.

Ali, M. H. (1992) *Proceedings of Workshop on Micronutrients,* Bhubaneswar, IBFEP-HFC, Calcutta, India. pp. 313.

Ali, S. J. and Monoranjan, R. (1989) Effect of nitrogen, phosphorus, potassium and micronutrients on controlling sterility of wheat. *Fertiliser News*, 34: 35-36.

Arora, C. L., Nayyar, V. K. and Randhawa, N. S. (1975) Note on secondary and micro-element contents of fertilisers and manures. *Indian Journal of Agricultural Science*, 45:80-85.

Arora, C.L., Singh, B. and Takkar, P. N. (1985) Secondary and micronutrient deficiency in crops. *Progressive Farmiing*, 21: 13-22.

Banerjee, S. K. and Das, D. (1978) Studies on available iron status and ferrous ferric transformation in soils of West Bengal under wet and dry conditions. *Fertiliser Technology*, 15: 137-141.

Bansal , R.L, Nayyar, V.K. and Brar, J.S (2003) Available boron status of soils in central plain region of Punjab. Journal of Research Punjab agricultural University ,40 : 172-176.

Bansal, R. L. and Khurana, M. P. S. (2007) Effectiveness of manganese carriers for the correction of its deficiency in wheat (*Triticum aestivum*). *Indian Journal of Ecology*, 34: 58-59.

Bansal, R. L. and Nayyar, V. K. (1989) Effect of zinc fertilisers on rice grown on *Typic Ustochrepts*. *IRRI Newsletter*, 14: 24-25.

Bansal, R. L. and Nayyar, V. K. (2000) Effect of foliar and seed treatments with manganese for correcting manganese deficiency in wheat. *Indian Journal for Sustainable use of chemicals in agriculture*, 1: 8-11

Bansal, R. L., Nayyar, V. K. and Takkar, P. N. (1991) Field screening of wheat cultivars for manganese efficiency. *Field Crops Reserach*, 29: 107-112.

Bansal, R.L.,V.K. Nayyar and M.P.S. Khurana. (2005) Efficiency of Mn carriesr for soybean (*Glycine max.*) and moong (*Vigana aureus*). *Indian Journal of Ecology,* 32 : 200-203.

Borkakati, K. and Takkar, P.N. 2000. *Forms of boron in acid alluvial and lateritic soilsin relation to ecosystem and rainfall distribution. In* International Conference on Managing Resources for Sustainable Agricultural Production in the 21st Century. Vol 2:127-128.

Bell, R. W. and Dell, B. (2008) Micronutrients for Sustained Food, Feed, Fibre and Bioenergy Production. IFA, Paris. pp.175.

Berger, K. C. and Troug, E. (1939) Boron determination in soils and plants. *Ind. Eng. Chem. Anal. Ed.*, 11, 540-544.

Cartwright, B., Tiller, B. A., Zarcinas, B. A. and Spouncer, L. R. (1983) The chemical assessment of boron status of soils. *Australian Journal of Soil Research*, 21:321-332.

Das, D. K., Singh, A. P. and Sakal, R. (1993) Relative performance of some zinc carriers in maize-rice sequence under calcareous soil. *Annals of Agricultural Research*, 14, 84-89.

Datta, S. P., Bhadoria, P.B.S. and Kar, S. (1998) Availability of extractable boron in some acid soils of West Bengal, India. *Communications in Soil Science and Plant Analysis,* 29:2285-2306.

Deb, D. L., Gupta, G. N., Meisheri, M. B., Rattan, R. K. And Sarkar, A. K. (1986) Radioisotope aided micronutrient research for increasing fertilizer use efficiency. *Fertilizer News*, 31 (2): 21-29.

Dwivedi, G. K. and Dwivedi, M. (1992) Efficacy of different modes of application of copper, zinc and boron to potato. *Annals of Agricultural Research*, 13: 1-6.

Dwivedi, G.K., Dwivedi, M. and Pal, S.S. (1990) Modes of application of micronutrients in acid soil in soybean- wheat crop sequence. *Journal of the Indian Society of Soil Science,* 38: 458-463.

Dwivedi, G. K., Dwivedi, M. and Pal, S. S. (1990) Mode of application of micronutrients in soybean –wheat crop sequence. *Journal of the Indian Society of Soil Science,* 38: 458-463.

Gupta, V. K., Gupta, S. P., Ram Kala, Potalia, B. S. and Kaushik, R. B. (1994) Twenty five years of micronutrients research in soils and crops of Haryana. Department Of Soil Science, CCS Haryana Agricultural University, Hissar. Pp: 1-99.

Gupta, V. K., Potalia, B. S. and Katyal, J. C. (1986) Response of wheat to different zinc carriers in a loamy sand (*Typic torripsamments*) soil. *Journal of Indian Society of Soil Science*, 34: 631-32.

Ishizuka, Y. (1971) *Nutrient Deficiencies of Crops.* FFTC, Taiwan, pp. 112.

Karim A. Q. M. B. and Vlamis.J (1962) Comparative study of the effects of ammonium and nitrate nitrogen in the nutrition of rice , *Plant and Soil,*16:32-41.

Katyal, J. C. (1985) Research Achievements of All India Co-ordinated Schemes on Micronutrients in Soils and Plants. *Fertilizer News*, 30 (4): 67-80.

Katyal, J. C. and Randhawa, N. S. (1983) Micronutrients, *FAO Fertilizer and Bull.* 7. FAO, Rome, Italy.

Katyal, J. C. and Rattan, R. K. (1993) Distribution of zinc in Indian soils. *Fertiliser News*, 38 (3): 15-26.

Katyal, J. C. and Sharma, B. D. (1980) A new technique for plant analysis to resolve iron chlorosis. *Plant and Soil*, 55: 105-119.

Katyal, J.C. and Friesen, D. K. (1987) *In Wheat: Production Constraints in Tropial Environments* (Klatt, A.R., ed.). UNDP/ CIMMYT. pp.99-127.

Katyal, J.C. and Rattan, R. K. (1990) Micronutrient use in the 90's. *In Soil Fertility and Fertiliser Use*, Vol. IV *Nutrient Management and supply system for sustaining agriculture in 1990's,* IFFCO, New Delhi, pp. 119-135.

Katyal, J.C. and Rattan, R. K. (2003) Secondary and micronutrients: Research gaps and future needs. *Fertilizer News*, 48 (4), 9-14 & 17-20.

Katyal, J.C. and Sharma, B.D. (1991) DTPA-extractable and total Zn, Cu, Mn and Fe in Indian soils and their association with some soil properties. *Geoderma*, 49: 165-179.

Khurana, M. P. S., Baddesha, H. S. and Gill, M. S. (2005) Management of manganese deficiency in wheat. *Indian Farmers' Digest* (December) 31-32.

Lal, C., Grewal, J. S. and Randhawa, N. S. (1971) Response of maize and wheat crops to soil and spray application of copper in Ludhiana soils. *Journalof Research Punjab agricultural Univiversity*, 8: 52-56.

Lindsay, W.L. and Norvell, W.A. (1978) Development of DTPA soil test for zinc, iron, manganese and copper. *Soil Science Society of America Journal*, 42: 421-428.

Maji, B. and Bandyopadhyay, B. K. (1992) Effect of submergence on Fe, Mn and Cu contents in some coastal rice soils of West Bengal. *Journal of Indian Society of Soil Science*, 40: 390-392.

Mandal B, De DK (1993) Depthwise distribution of extractable boron in some acidic Inceptisols of India. Soil Science, 155:256–262.

Mitra, A. K. and Jana, P. K. (1991) Effect of doses and method of boron application on wheat in acid Terai soil of North Bengal. *Indian Journal of Agronomy* 36, 72-74.

Mondal, A. K., Pal, S., Mandal, B. and Mandal, L. N. (1991) Available boron and molybdenum content in alluvial acidic soils of north Bengal. Indian Journal of Agricultural Sciences, 61: 502-504.

Nambiar, K. K. M. (1995) In *Agricultural Sustainability: Economic, Environmental and Statistical Considerations* (Barnett, V., Payne, R. and Stiener, R. Eds.) pp. 134-171.

Nayyar, V. K. and Takkar, P. N. (1989) Controlling iron deficiency in rice grown in sandy soils of Punjab. International Symposium on Managing Sandy Soils. CAZRI, Jaipur 1:379- 384.

Nayyar, V. K., Bansal, R. L. and Singh, and Khurana, M. P. S. (1994) Annual Report of AICRP of Micro- and Secondary nutrients and Pollutant Elements in Soils and Plants (1998-1999). Punjab Agricultural University, Ludhiana, India

Nayyar, V. K., Sadana, U. S. and Takkar, P. N. (1985) Methods and rates of application of Mn and its critical levels for wheat following rice on coarse textured soils. *Fertilizer Research*, 8: 173-178.

Nayyar, V. K., Takkar, P. N., Bansal, R. L., Singh, S. P., Kaur, N. P. and Sadana, U. S. (1990) *Micronutrients in soils and crops of Punjab*, Research bulletin, Department of Soils, PAU, Ludhiana.

Nongkynrih, P., Dkhar, P. S. and Khathing, D. T. (1996) Micronutrient elements in acid alfisols of Meghalaya under rice cultivation. *Journal of the Indian Society of Soil Science*, 44: 455-457.

O' Connor, G. A. (1988) Use and misuse of the DTPA soil test. *Journal of Environmental Quality*, 17: 715-718.

Pal, S., Datta, S. P., Rattan, R. K. and Singh, A. K. (2008) Diagnosis and amelioration of iron deficiency under aerobic rice. *Journal of Plant Nutrition*, 31: 919-940.

Pathak, A. N., Shanker, H. and Misra, R. V. (1968) Molybdenum status of some Uttar Pradesh soils. *J. Indian Soc. Soil Sci.*, 16: 400-404.

Patil, G. D., Patil, M. D., Patil, N. D. and Adsule, R. N. (1987) Effect of boronated superphosphate, single superphosphate and borax on yield and quality of groundnut. *J. Maharashtra Agric.University*, 12: 168-170.

Prasad, B., Kumar, M. and Prasad, J. (1989) Direct and residual effect of organic wastes and zinc sulphate on zinc availability under cropping sequence of wheat – rice in calcareous soil. *Indian Journal of Agricultural Sciences*, 59: 300-305.

Rathore G.S., Khamparia, R.S., Gupta, S.B., Dubey, S.B., Sharma, B.L.& Tomer, V.S. (1995) Twenty five years of micronutrient research in soils and crops of Madhya Pradesh. Research Bulletin, Department of Soil Science and Agricultural Chemistry. JNKVV, Jabalpur, Mad

Rattan, R. K., Datta, S. P. and Katyal, J. C. (2008) Micronutrient management – research achievements and future challenges. *Indian Journal of Fertilisers*, 4: 93-100, 103-106, 109-112 & 115-118.

Rattan, R. K., Datta, S. P., Saharan Neelam and Katyal, J.C. (1997) Zinc in Indian agriculture – A look forward. *Fertiliser News*, 42 (12), 75-89.

Rattan, R. K., Saharan, Neelam and Datta, S. P. (1999) Micronutrient depletion in Indian soil – extent, causes and remedies. *Fertiliser News*, 44 (2): 35-50.

Roy, S. K. and Pradhan, A. C. (1994) Effect of method and time of boron application on wheat production in Terai region of West Bengal. *Indian Journal of Agronomy*, 39: 643-645.

Sadana U. S., Nayyar V. K. and Kaur N. P. (1989) Response of wheat grown on manganese deficient soil to foliar application of different micronutrient formulations. *Micronutrient News*, 2: 1-2.

Sakal R., Sinha R. B., Singh, A. P. and Bhogal, N .S. (1991) Relative susceptibility of some important varieties of sesamum and mustard to boron deficiency in calcareous soil. *Fert ilizer News,* 36 (3) 43-49.

Sakal, R. and Singh, A.P. 1999. Available zinc and boron status of Bihar soils and response of oilseeds and pulses to zinc and boron application. In *national Sympoisum on Zinc Fertiliser Industry Whither To*(ed. Ramendra Singh and Abhay Kumar) Session III.

Sakal, R., Singh, A. P., Singh R. B. and Bhogal, N.S. (1996) Research Bulletin on Twenty five years of Research on Micronutrients in Soils And Crops of Bihar (1967-1992), Department of Soil Science, RAU, Bihar, India.

Sakal, R., Singh, A.P., Sinha R.B. and Bhogal N.S. (1980 , 1988). Annual progress Reports, ICAR All India Coordinated Scheme of Micro-and Secondary Nutrients in Soils and crops of Bihar, Res. Bull. Dept of Soil Science, RAU., Samastipur, Bihar

Sarkar, D., Biswapati, Mandal And Kundu, M. K. (2007) Increasing use efficiency of boron fertilisers by rescheduling the time and methods of application for crops. *Plant and Soil*, 301: 77-85.

Sharma , P.K and Nayyar, V.K (2004) Diagnosing micronutrient related constraints to productivity in Muktsar, Patiala, Hoshiarpur and Ludhiana districts. In information technology for sustainable agriculture in Punjab (It –SAP) UNDP-TIFAC sponsored project, pp 1-18, Punjab Remote Sensing centre.

Sharma , S.K.Swami, B.N. and Singh R. K. (1992) Response of wheat to micronutrients at two fertility levels in black soil. *Indian Journal of Agronomy* ,37:255-257.

Sharma, B. D., Mukhopadhyay, S. S., Sidhu, P. S. and Katyal, J. C. (2000) Pedospheric attributes in distribution of total and DTPA-extractable Zn, Cu, Mn and Fe in Indo-Gangetic Plains. *Geoderma*, 96: 131-151.

Sharma, C.P., C. Chatterjee, P.N. Sharma, B.D. Nautiyal, N. Nautiyal, N. Khurana, and P.Sinha. (1996). Micronutrient deficiency symptom in cereals. pp. 11-42. In: C.P. Sharma (ed.), Deficiency Symptoms and Critical Concentration of Micronutrients in Crop Plants. Botany Department, Lucknow University, Lucknow, India.

Shorrocks, V. M. and Alloway, B. J. (1988) Copper in Plant, Animal and Human Nutrition. CDA, UK. pp. 84.

Singh S P and Nayyar V K (1999) Available boron status of some alluvium derived arid and semi arid soils of Punjab. *Journal of the Indian Society of Soil Science* 47: 801-802.

Singh, B. P., Singh, R. A., Sinha, M. K. and Singh, B. N. (1985) evaluation of techniques for screening Fe efficient genotypes of rice in calcareous soil. *Journal of Agricultural Science*, Cambridge, 105: 193-197.

Singh, D. Leelawathi, C.R.,Krishana, K.S. and sarup, S. (1979) Monograph on crop responses to micronutrients . Indian Agricultural statistics research Institute (ICAR), New Delhi.

Singh, M. (1982) *Nutritional Disorders in Crops and Their Remedy*. HAU, Hisar. pp. 37.

Singh, M. V. (1999) Micronutrient deficiency delineation and soil fertility mapping. In *National Symposium on "Zinc Fertiliser Industry – Whither To"* (ed. Ramendra Singh and Abhay Kumar) Session II.

Singh, M. V. (2000) *Micro-, Secondary-Nutrients and Pollutant Elements Research in India.* All India Coordinated Research Project of Micro-, Secondary- Nutrients and Pollutant Elements in Soils and Plants, IISS, Bhopal.

Singh, M. V. and Abrol, I. P. (1986) Split application of zinc better than a single dose, *Indian Farming* (July Issue):35-37.

Singh, M. V., (1999) Current status of micro- and secondary nutrients deficiency and crop responses in different agro- ecological regions- Experiences of All India Coordinated Schemes of Micro- and Secondary Nutrients and Pollutant Elements in Soils and Plants. *Fertiliser News*, 44 (4): 63-82.

Singh, S. P. and Nayyar, V. K. (1994) Response of winter crops to manganese application on loamy sand soil. *Indian Journal of Agricultural Sciences*, 64: 627-629.

Sinha, R. B., Sakal, R., Singh, A. P. and Bhogal, N. S. (1991) Response of some field rops to boron application in calcareous soils. *Journal of the Indian Society of Soil Science.*, 39: 342-345.

Srivastava, A. K. and Srivastava, O. P. (1994) Response to iron and manganese application as affected by their availability in a *Typic Natraquaff. Arid Soil Research Rehabiltation*, 8: 301-305.

Takkar, P. N. and Randhava, N. S. (1978) Micronutrients in Indian agriculture. *Fertiliser News*, 23 (8): 3-26.

Takkar, P. N., Chibba I. M. and Mehta S. K. (1989) Twenty years of Co-ordinated Research on Micronutrient in soils and Plants. Bull. Indian Inst. Soil Sciience., Bhopal. Pp. 314.

Takkar, P. N., Singh, M. V. and Ganeshamurthy, A. N. (1997) In *Plant nutrient needs, supply, efficiency and policy issues: 2000-2025,* (Kanwar, J.S. and Katyal, J.C. Eds.), NAAS, New Delhi.

Takkar, P.N and Nayyar, V.K (1981) Preliminary field observation of manganese deficiency in wheat and berseem . *Fertilizer News*, 26 : 22-23.

Takkar, P.N., Nayyar V.K. and Sadana U.S. (1986) Response of wheat on coarse textured soils to mode and time of manganese application. Expl. Agriculture, 22 : 149-152.

Takkar, P.N., Nayyar, V.K., (1986). *Integrated approach to combat micronutrient deficiency.* Paper presented at the seminar on Gowth and modernisation of the fertilizer industry, 15-17,December 1986, FAI, New Delhi, India.

Welch, R.M. and Graham, D.G. (2004) Breeding for micronutrients in staple food crops from a human nutrition perspective. *Journal of Experimental Botany.* 55: 353-364.

In: Progress in Food Science and Technology, Volume 1   ISBN: 978-1-61122-314-9
Editor: Anthony J. Greco   © 2012 Nova Science Publishers, Inc.

Chapter 3

LEGUMES: PROPERTIES, NUTRITION, CONSUMPTION AND HEALTH

Tzi Bun Ng and Randy Chi Fai Cheung*
School of Biomedical Sciences, Faculty of Medicine, the
Chinese University of Hong Kong

INTRODUCTION

A legume is a plant or a fruit belonging to family Fabaceae (or Leguminosae). A legume fruit or pod is a simple dry dehiscent fruit. Alfalfa, beans, carob, clover, peas, lentils, lupins, mesquite, peanuts, locust trees (Gleditsia or Robinia), wisteria, and the Kentucky coffeetree (*Gymnocladus dioicus*) are examples of legumes. Legumes made their appearance early in America, Europe and Asia around 6,000 BC. They became a staple food, serving as in important protein supplement when there was an inadequate supply of meat.

PROPERTIES

Leguminous plants associate symbiotically relationship with nitrogen fixing bacteria (rhizobia) found in their root nodules. The nitrogen fixation reduces the use of fertilizers. It also enables legumes to be employed in rotation with other crops to replenish the nitrogen content of soil. The nitrogen fixing capacity of legumes is augmented by soil calcium and curtailed in the presence of abundant nitrogen. Legume seeds and leaves have richer protein content than plants belonging to other families owing to nitrogen-fixation. The high protein content accounts for the importance of leguminous plants in agriculture. Legumes contain many biologically active constituents including soy isoflavones, soy protein, saponins, proteases inhibitors, hemagglutinins, antifungal proteins and ribosome inactivating proteins (Lam et al., 1998; Wong and Ng, 2005)

---

* Corresponding author (Tel: 852-26098031, Fax: 852-26035123, Email: b021770@mailserv.cuhk.edu.hk

India and Pakistan are the biggest producer as well as consumer of legumes. Canada, Australia and the United States are large exporters. Cultivated legumes are divided into many agricultural classes, including fallow/green manure, bloom, forage, grain, pharmaceutical/industrial, and timber species. Forage legumes including alfalfa, clover, Arachis, Stylosanthes, or Vicia, are grown in pasture and eaten by farm animals. Other forage legumes such as Leucaena or Albizia are woody shrubs or trees which are either disintegrated by farm animals or chopped by humans to furnish feed to farm animals. Grain legumes including beans, lentils, lupins, peas, and peanuts are farmed to yield seeds for human and animal consumption or for the production of industrial oils. Lupins are cultivated commercially for their flowers. Indigofera and Acacia species are grown for dye and natural gum production, respectively. Fallow/green manure legume including Leucaena, Cyamopsis, and Sesbania species are cultivated and added back to the soil in order to exploit the high levels of captured atmospheric nitrogen by nitrogen fixation found in the roots of most legumes. Various legume species including numerous Acacia species and *Castanospermum australe*, are cultivated for timber production.

NUTRITION

Legumes include beans, lentils, peas, peanuts, and other food plants with pods. Legumes are important in the diet of people in many countries. Legumes are low in fat and are rich in complex carbohydrates, dietary fiber, protein, essential amino acids, iron, folate and many phytochemicals and micronutrients. Soluble fiber can reduce blood cholesterol level. Legumes have a protein content of 20 to 25% protein by weight, which is two to three times that of wheat and rice. Bean consumption is higher in people with a lower income. Vegetarians have a higher legume intake than non-vegetarians. Soybean has been studied intensively during the past decade because it represents a unique dietary source of isoflavones which have a variety of biological effects and lower the risk of some chronic diseases. Soybeans are unique among the legumes because they are abundant in isoflavones which have been claimed to lower the risk of cancer, cardiovascular disease, and osteoporosis, and also alleviate menopausal symptoms.

The protein content of beans is generally between 20% and 30% of energy. A serving of beans (<90 g or 1/2 cup cooked beans) provides <7–8 g protein or <15% of the recommended dietary allowance (RDA) for protein for a 70-kg adult (National Research Council, 1989).

The relatively low content of sulphur amino acid (SAA) in beans offers an advantage in terms of calcium retention because hydrogen ions produced from the metabolism of SAAs bring about bone demineralization and urinary excretion of calcium (Chan, 1974; Remer and Manz, 1994). Studies involving human subjects disclosed that soy protein consumption has a more pronounced hypercalciuric effect compared with consumption of whey protein (Anderson et al., 1987) or a mixture of animal proteins (Breslau et al., 1988). Legume protein, even among people preferring a plant-based diet, accounts for only a small percentage of total protein intake. Nevertheless, the hypocalciuric effect of bean proteins may be substantial for people who replace animal protein with soy protein for its hypocholesterolemic effect and athletes on soy protein supplements (Anderson et al., 1999).

Most beans have an extremely low fat content. Exceptions are chickpeas and soybeans, which have <15% and 47% fat, respectively. Linoleic acid is the main fatty acid in beans, α-linolenic acid is another fatty acid (US Department of Agriculture, 1988). The dietary contribution of most beans to α-linolenic acid intake is small due to the low fat content. By comparison, consumption of full-fat soyfoods has a significant contribution to α-linolenic acid intake because of the high fat content. The linoleic to α-linolenic acid ratio in soybeans is < 7.5 : 1 (α-linolenic acid accounts for < 7–8 % of the total fat) (US Department of Agriculture, 1988). n-3 Polyunsaturated fatty acids, especially eicosapentaenoic acid (EPA) and docosahexaenoic acid (DHA), have health promoting effects (Nair et al., 1997; Stone, 1997; Caygill et al., 1996; Simopoulos, 1999). DHA is important to infants (Oski, 1997). α-Linolenic acid can be converted into EPA and then into DHA, although efficiency of conversion of α-linolenic acid into EPA is relatively low, at 5–10% (Emken et al., 1994; Indu, 1992), and it is inhibited by linoleic acid (Emken et al., 1994).

Beans are rich in dietary fiber (Marlett, 1992). High fiber, high-bean diets reduce circulating cholesterol level in hypercholesterolemic patients (Anderson et al., 1984). Beans have very low glycemic indexes (Jenkins et al., 1980; Foster-Powell and Miller, 1995) owing to their fiber (Thorne et al., 1983), tannin (Thompson et al., 1984), and phytic acid contents (Yoon et al., 1983). Women consuming a high-glycemic-index diets were < 40 % more likely to develop diabetes than those consuming low-glycemic-index diets, even after controlling for several diabetes risk factors (Salmerón et al., 1997). Thus, beans may be a beneficial dietary component for diabetics and people prone to develop diabetes.

Beans are an excellent source of folate, which lowers the risk of neural tube defects (Daly et al., 1995). Beans are also high in iron, but iron bioavailability from legumes is not high, thus reducing their value as an iron source (Lynch et al., 1984). In contrast to iron bioavailability, zinc bioavailability from legumes is higher (Sandström et al., 1989). Many beans are good sources of calcium, although calcium bioavailability from beans is lower than that from milk and green leafy vegetables, it is still relatively good (Weaver et al., 1993). Calcium bioavailability from soybeans and soyfoods is quite good—essentially equivalent to calcium bioavailability from milk—despite the fact that soybeans are high in phytate and oxalate (Weaver et al., 1994).

Lectins, oligosaccharides, phytate (inositol hexaphosphate), protease inhibitors and saponins found in beans are considered to be antinutrients, More recent information indicate, however, that the antinutrient label may be an oversimplification, particularly in the case of oligosaccharides and saponins. Protease inhibitors in beans can reduce protein digestion, and induce pancreatomegaly and enhance chemically induced pancreatic tumors in some of animals (Grant, 1989). However, boiling can inactivate the protease inhibitor (Duarte-Rayas et al., 1992). The quantity of protease inhibitors ingested would not have any adverse consequences in humans (Liener, 1994). Lectins have been suggested as agents to prevent gastrointestinal atrophy during total parenteral nutrition, either alone or in conjunction, as different lectins act in different parts of the gastrointestinal tract. In rats, intraluminally administered PHA evokes proliferation in the gastric fundus and in the small intestine, peanut agglutinin exerts effects in the large intestine (Jordinson et al., 1999). In obese rats, inclusion of raw kidney bean in the diet, reduces lipid accumulation probably due to a decline of insulin levels brought about by lectins. However, there is no reduction in body or muscle protein, even at high doses, as with normal rats, suggesting that lectins may be useful as anti-obesity agents (Pusztai et al., 1998). As noted above, phytate is thought to contribute to the poor mineral bioavailability of beans. The phytate concentration in beans lies between 1% and 2% (Oberleas and Harland, 1981; Mage, 1982). Although phytate reduces mineral bioavailability in plant foods, it may play a role in reducing cancer risk because of its antioxidant effects (Graf and Eaton, 1990). It has been proposed that phytic acid may reduce the risk of colon cancer (Harland and Morris, 1995) and probably breast cancer (Vucenik et al., 1997). More than four decades ago, diets containing legumes were first demonstrated to increase flatulence (Steggerda and Dimmick, 1966). In 1970, it was shown that bean oligosaccharides caused gas production (Rackis et al., 1970). The oligosaccharide content of dry beans is < 25–50 mg/g (Carlsson et al., 1992; Kuo et al., 1988). Since there is no α-galactosidase present in the human intestinal mucosa to break the α-(1–6) galactose linkage in galactoside-containing oligosaccharides such as raffinose and stachyose, these oligosaccharides enter the large intestine and are metabolized by bacteria to produce large amounts of carbon dioxide, hydrogen, and sometimes methane. Due to the discomfort and social embarrassment brought about by flatulence, some people avoid beans completely. However, the oligosaccharides, by virtue of their growth-promoting effect toward bifidobacteria, have been hypothesized to enhance colonic health, promote longevity, and reduce colon cancer risk (Mitsuoka, 1982; Benno et al., 1989; Koo and Rao, 1991). Saponins are glycosides composed of a lipid-soluble aglycone consisting of either a sterol or, more often, a triterpenoid structure linked to water-soluble sugar residues that vary in their type and quantity. Legumes are the major sources of dietary saponins. The same bean can have different types of saponins. Saponins are very poorly absorbed. Most saponins form insoluble complexes with 3-β-hydroxysteroids and interact with and form large, mixed micelles with bile acids and cholesterol.

CONSUMPTION AND HEALTH

Isoflavones represent another group of phytochemicals present in legumes, but for practical purposes the soybean is the only nutritionally relevant source of these compounds (Coward et al., 1993; Wang and Murphy, 1994). Soybeans are unique among legumes since they represent a concentrated source of isoflavones which have weak estrogenic activity. The isoflavone genistein also affects signal transduction. Soyfoods and isoflavones have drawn a great deal of attention because of their potential role in preventing and treating cancer and osteoporosis. The low breast cancer mortality rates in Asian countries and the putative antiestrogenic effects of isoflavones have led to the proposal that soyfood intake lowers breast cancer risk that soy or isoflavones may decrease the risk of prostate cancer is more encouraging. The weak estrogenic activity of isoflavones and the structural similarity between soybean isoflavones and the synthetic isoflavone ipriflavone, which was shown to increase bone mineral density in postmenopausal women, suggest that soy or isoflavones may reduce the risk of osteoporosis (Messina, 1999).

Diabetes

Legume therapy in diabetic individuals depends on the type of diabetes and other factors such as lifestyle and metabolic needs of the patients. Legume protein has an effect on diabetes by virtue of its content of glycine and arginine, which tend to reduce blood insulin levels. Legume fiber may be useful due to its insulin-moderated effect. A legume diet may be a good option in type 2 diabetes patients due to its effect on hypertension, hypercholesterolemia, atherosclerosis and obesity, which are common in these patients (Holt et al., 1996). In addition, substituting animal protein for legume or other vegetable protein may also reduce renal hyperfiltration, proteinuria, and renal acid load and thus lowers the risk of renal disease in type 2 diabetes (Jenkins et al., 2003). A high fibre diet, particularly soluble fiber, is of value in glycemic concentration in diabetics (Messina, 1999; Rubio, 2002). Dietary fiber decreases or slows carbohydrates absorption (Rubio, 2002). The antiangiogenic effects of isoflavones could be of value in inhibiting the development of diabetic retinopathy (Messina, 1994). Investigators conducted in diabetic patients with legume diets show several potential advantages (Farriol et al., 2006).

Cancer

Isoflavones are a subclass of the more ubiquitous flavonoids. Isoflavones may elicit a physiologic response because serum concentrations of isoflavones are several orders of magnitude higher than those of physiologic estrogens in people who consume soyfoods. Isoflavones produce antiestrogenic actions in a high-estrogen environment, like that in premenopausal women, but estrogenic actions in a low-estrogen environment such as that in postmenopausal women (Mäkela et al., 1995).

Other than isoflavones, there are phytochemicals in soybeans that exhibit anticarcinogenic activity. These comprise phenolic acids, phytosterols, phytates, protease inhibitors, and saponins (Messina and Barnes, 1991). However, the bulk of the data indicate that isoflavones account for the anticancer effects of soy.

Daidzein, a primary isoflavone in soybeans, manifests anticancer activity. It inhibits the growth of HL-60 cells implanted in mouse subrenal capsules (Jing et al., 1993). Nevertheless, genistein has captured much of the attention. There is voluminous literature showing that genistein inhibits the proliferation of a diversity of both hormone-dependent and hormone-independent cancer cells *in vitro* with an IC_{50} between 5 and 40 mM (2–10 mg/mL), including breast (Peterson and Barnes, 1991; 1996; Pagliacci et al., 1994; Peterson et al., 1996; So et al., 1996; Clark et al., 1996; Zava and Duwe, 1997), prostate (Peterson and Barnes, 1993; Naik et al., 1994; Kyle et al., 1997), colon (Kuo et al., 1997; Kuo, 1996), and skin (Rauth et al., 1997) cancer cells (Adlercreutz and Mazur, 1997; Akiyama and Ogawara, 1991; Constantinou and Huberman 1995). Genistein suppresses the metastatic activity of breast (Scholar and Toewa, 1994) and prostate (Santibáñez et al., 1997) cancer cells *in vitro* independent of the effects on cell growth. The anticancer effects of genistein observed *in vitro* (Wei et al., 1993) are due to its inhibitory actions on enzymes involved in signal transduction, including ribosomal S6 kinase (Linassier et al., 1990), MAP kinase (Thorburn and Thorburn, 1994), and tyrosine protein kinases (Akiyama et al., 1987). Genistein also inhibits the activity of DNA topoisomerase II (Constantinou et al., 1990) and increases the *in vitro* concentrations of transforming growth factor β (TGF β) (Peterson et al., 1998) which may inhibit the growth of cancer cells (Benson and Colletta, 1995; Benson et al., 1996; Markowitz and Roberts, 1997). Genistein has an important role as a potent inhibitor of angiogenesis *in vitro* (Messina, 1999).

Research on the reduction of cancer risk by soy intake initially focused on breast cancer. Interest was spurred by relatively low breast cancer mortality rates in Asian countries with frequent soyfoods consumption. In Japan, the breast cancer mortality rate is only approximately 25 % of that of the United States (American Cancer Society, 1994). In addition to the low breast cancer mortality rates in Asia, results of other studies furnished a basis for the hypothesis that soy consumption reduces breast cancer risk: *1*) the potential antiestrogenic effects of soybean isoflavones, and *2*) decrease in number of 7, 12-dimethylbenz(a)anthracene–induced mammary tumors in rats after dietary intake of soy (Barnes et al.,1990). Since this hypothesis was initially proposed, several epidemiologic studies have examined the relation between soy intake and breast cancer risk.

In 1991, a case-control study investigation performed by Lee et al, (1991) in Singapore disclosed that regular soyfood consumption was linked to a drastic decline in breast cancer risk in premenopausal women but not in postmenopausal women. A large-scale case-control study conducted in Japan revealed that tofu (bean curd) intake was associated with reduction of breast cancer risk in premenopausal women, but not in postmenopausal breast cancer (Hirose et al., 1995). Relatively little epidemiologic evidence is available to suggest that soy intake is related to a lower risk of postmenopausal breast cancer. However, there are some limited data, albeit inconsistent, suggesting that soy intake is associated with a decreased risk of premenopausal breast cancer.

Genistein suppresses the growth of both estrogen-dependent and estrogen-independent breast cancer cells *in vitro*, but it is not known whether cellular concentrations of genistein *in vivo* would attain the *in vitro* concentrations necessary for inhibiting breast cancer–cell growth. Genistein inhibits serum and epidermal growth factor–stimulated proliferation of normal human mammary epithelial cells with IC_{50} values much lower than those for transformed human breast epithelial cells (Peterson and Barnes, 1994) suggesting that soy intake may inhibit the initiation of cancer cells, rather than preventing the proliferation of existing cancer cells. Neither genistein nor daidzein (injected intraperitoneally) inhibited N-methyl-N-nitrosourea–induced mammary tumor incidence in rats (Constantinou et al., 1996). Because synergistic effects between genistein and daidzein have been observed *in vitro*, it would be of interest to examine their combined effects *in vivo* (Evans et al., 1995; Franke et al., 1995). Of course there is also the possibility that other components of soybeans, separately or in combination with isoflavones, account for the hypothesized anticancer effects of soyfoods. It is apparent from the human studies by Wrensch et al (1991), McMichael-Phillips et al (1998), and Cassidy et al (1994; 1995) that soy or isoflavones have the potential to exert physiologic effects theoretically related to breast cancer risk. In particular, Cassidy et al (1994) have observed that the consumption of soy, in particular isoflavone-rich soy (Cassidy et al., 1995), prolongs the follicular phase and reduces serum concentrations of gonadotropins. Brown and Lamartiniere (1995), Lamartiniere et al (1995), and Murrill et al (1996) furnished data suggesting that early consumption of soyfoods by young girls may inhibit breast cancer development later in life. Early exposure to genistein (during the neonatal or prepubertal period of life) by subcutaneous administration prevents the development of dimethylbenz(a)anthracene-induced mammary tumors in rodents and increases the latency period (Brown and Lamartiniere, 1995; Lamartiniere et al., 1995; Murrill et al 1996). These findings offer a potential explanation for the findings of Wu et al (1996). Perhaps Asia-born Asian women are exposed to tofu at an earlier age than Asians born in the West.

Rates of clinical prostate cancer display much greater variations than rates of latent prostate cancer. In US, the incidence of clinical prostate cancer among whites is 10–15-fold higher than that among Japanese, while the overall incidence of latent prostate cancer is only < 50 % higher (Yatani et al., 1989), indicating that in Japanese, prostate cancer growth is slower, the onset of prostate cancer occurs later in life, or both. Delaying the appearance of clinical prostate cancer by even several years could have a significant influence on mortality because prostate cancer usually occurs in older men. Soyfoods intake may contribute to the low mortality rate due to prostate cancer in Japan.

Genistein suppresses the proliferation of both androgen-dependent and androgen-independent prostate cancer cells *in vitro* (Peterson and Barnes, 1993; Naik et al., 1994). It also reduces the metastatic potential of prostate cancer cells independent of cell growth inhibition, accompanied by a reduction in the tyrosine phosphorylation of an unidentified molecular species (Santibáñez et al., 1997). Although the role of estrogen in prostate cancer has not been fully elucidated, the potential estrogenic effects of isoflavones may be protective because estrogens have been found effective in the treatment of metastatic prostate cancer (Pienta and Esper, 1993). Also, it has been shown that genistein inhibits 5-α-reductase in genital skin fibroblasts and benign hyperplastic prostate tissue (Evans et al., 1995). This enzyme converts testosterone into a more active form, dihydrotestosterone, which stimulates the growth of prostatic tissue. Ross et al (1992) showed that biomarkers of 5-α-reductase activity are higher in white and black men compared with Japanese men. The *in vitro* results of Evans et al (1995) agree with data from Lu et al (1996), showing that following 1 month of soymilk consumption (36 oz/d), the serum concentration of 3α, 17β-androstanediol glucuronide, a dihydrotestosterone metabolite, was significantly lowered. In mice fed a diet containing soy for 9 months, the incidence of prostatic dysplasia, considered to be a preneoplastic prostate lesion, was significantly diminished. At 12 months, however, difference between the 2 groups was much attenuated. These results are in line with the epidemiologic data noted above and also with the results of a study of MNU-induced prostate tumors in Lobund-Wistar rats (Pollard and Luckert, 1997). Rats receiving a diet containing soy with a small quantity of isoflavones had a shorter latency period than those fed a diet containing soy high in isoflavones (Pollard and Luckert, 1997). The effect of soy or genistein on tumor development in rats implanted with prostate cancer cells was investigated (Naik et al., 1994; Zhang et al., 1997; Schleicher et al., 1998). A diet containing soy flour (33% by weight) for 16 weeks inhibited tumor growth in rats implanted with Dunning R3327 PAP tumors (Zhang et al., 1997). Schleicher et al (1998) Genistein (50 mg/kg body weight) administered subcutaneously in the dorsal scapular area every 12 h commencing at the time of tumor cell transplantation retarded prostate tumor development in rats implanted with prostate carcinoma cells and completely inhibited development of lung metastases. Dalu et al (1998) noted that in Lobund-Wistar rats, dietary genistein (1 mg genistein/g diet) brought about a weight reduction of the dorsolateral and ventral prostates and down-regulated the expression of tyrosine-phosphorylated proteins. Geller et al (1998) found that genistein (at concentrations of 1–15 mg/mL) dose-dependently inhibited ^{3}H-thymidine incorporation in cultured benign prostatic hypertrophy tissue. However, Naik et al (1994) found that although genistein inhibited prostate cancer cell growth *in vitro*, when rats received by injection to the right flank, the metastatic MAT-Lylu prostate cancer line, genistein (0.07, 0.143, and 0.285 mg/d) administered orally did not inhibit the prostate tumor development. The doses employed more closely mimicked human dietary intake than those used in the studies of Schleicher et al

(1998) and Dalu et al (1998). Higher doses of genistein (0.143, 0.285, and 0.428 mg/kg) injected intraperitoneally had little effect on tumor growth (Naik et al., 1994). There are limited human data available for use in addressing the soy–prostate cancer hypothesis, although Severson et al (1989) noted that tofu consumption was associated with a markedly reduced risk of prostate cancer. However, this difference did not quite reach statistical significance ($P < 0.054$) and the number of men with tumors in each of the tertiles was small (Severson et al 1989). Of potential relevance to the effects of isoflavones on prostate cancer risk is the finding that isoflavones appear in the prostatic fluid, and that concentrations are highest in men from soyfood-consuming countries (Morton et al., 1997). Furthermore, relative to plasma concentrations, isoflavones are concentrated several-fold in the prostatic fluid.

Fotsis et al (1993) observed that at high concentrations (IC_{50}, 150 mmol), genistein inhibited the action of bovine microvascular cells to invade collagen gels and produce capillary-like structures when exposed to basic fibroblast growth factor. Development of antiangiogenesis agents is an area of cancer treatment because inhibiting the tumor-stimulated angiogenesis prevents tumor growth beyond a size of 1–2 mm and thus from becoming clinically insignificant (Folkman and Klagsbrun, (1987). A low genistein concentration is required to inhibit angiogenesis *in vitro*, (IC_{50}, 8 mmol) (Adlercreutz and Mazur, 1997).

Estrogen may inhibit prostate cancer growth. It can be associated with both benign prostatic hyperplasia and prostate cancer. Estrogen induces apoptosis of prostate cancer cells and inhibits enzymes associated with different processes in cancer development. Soybean foods may be a factor contributing to the diminution of prostate cancer mortality (Adlercreutz, 2002). Genistein reduces DNA synthesis in human prostate cells *in vitro* and inhibits the effect of testosterone in prostate cancer development in rats (Jenkins et al., 2003; Adlercreutz et al., 2000). However, a daily soybean intake enough to decrease LDL-cholesterol, does not reduce serum concentration of prostate specific antigen (PSA) (Adlercreutz, 2002). Antifungal proteins (Ma et al., 2009) lectins/hemagglutinins (Lin et al., 2008) and proteases inhibitors (Fang et al., 2010) demonstrate antiproliferative activity toward tumor cells *in vitro*. Some of these show pH stability and thermostability and thus their aforementioned activities may be retained *in vivo*.

Saponins may have anticancer properties. A saponin-containing diet (3 % by wt) inhibits in rodents the development of azoxymethane-induced preneoplastic colonic lesions (Koratkar and Rao, 1997). However, it is unclear whether these results in rodents are relevant to humans to a major extent (Ridout et al, 1988).

Protease inhibitors can prevent or inhibit carcinogen-induced transformation *in vitro* and carcinogenesis in animals and achieved Investigational New Drug status from the US FDA in 1992 (Kennedy, 1995; 1998a, b). Some investigations on human beings have been initiated (Wang et al., 1999; Armstrong et al., 2000). Protease inhibitors also exert an anti-inflammatory action, by inhibiting proteases liberated from inflammation-mediating cells and interfering with secretion of superoxide anion radicals from immunocytes. Protease inhibitors reduced ulcerative colitis in mice, and may have therapeutic action in humans (Kennedy, 1995; Ware et al., 1999). The potential beneficial effects depend on native proteins and therefore would be lost after cooking. Lectins are promising active substances, and their exploitation for possible medical use is attractive, in view of their ability to stimulate hyperplasia of the small intestine, induce changes in its bacterial flora, alter hormone secretion, and enter the systemic circulation (Pusztai, 1993; Pusztai and Bardocz, 1996).

Legumes: Properties, Nutrition, Consumption and Health 73

Evidence has accumulated on regarding their possible beneficial effects as biomedical agents. The addition of bean phytohemagglutinin to the diet of mice dose-dependently inhibited tumor growth due to competition between the gut epithelium undergoing hyperplasia and the developing tumor for nutrients (Pryme et al., 1998).

Osteoporosis

The structural similarity between estrogen, isoflavones and the synthetic isoflavone, 7-isopropoxyisoflavone (ipriflavone), which increases bone mass in postmenopausal women (Valente et al., 1994; Brandi, 1992) and possess weak estrogenic activity of isoflavones led to the speculation that isoflavones may be beneficial in promoting bone health. Speculation about the potential benefits of isoflavones was also fueled by the similarity in chemical structure between the soybean isoflavones and the ipriflavone requires metabolism to be maximally effective, and one of the metabolites is daidzein, soybean isoflavone (Stephens, 1997). The usual dose of ipriflavone is between 600 and 1200 mg/d. Daidzein makes up 10% of the metabolic products of ipriflavone (Brandi, 1992), which inhibits osteoclast activity in vitro (Tsuda et al., 1986).

The lower rate of hip fracture in Japanese women compared to American women (Ross et al., 1991; Fujita and Fukase, 1992) is often cited as lending credence to a protective effect of isoflavones, but the reasoning appears to be defective. The bone density of Japanese women is similar to or lower than that of American women with a twice-as-high hip fracture rate (Kin et al., 1993; Ross et al., 1995; Russell-Aulet et al., 1993). The vertebral fracture rate of Japanese women is substantially higher than that of American women (Ross et al., 1995). The low Japanese hip fracture rate is at least partially attributed to anatomical differences between American and Japanese women, including the shorter hip axis length of Japanese women (Nakamura et al., 1994), and also to other factors such as a lower propersity to fall (Davis et al., 1997). Prior to 1995, there were little direct data indicating that the soybean isoflavones affect bone density. In 1995, Anderson et al (1995) reported that genistein exerted a biphasic effect on bone in ovariectomized young growing rats and ovariectomized lactating rats, which were both fed low-calcium diets. Three different doses of genistein (1.0, 3.2, and 10 mg/day) were employed. After 2 weeks of treatment for the young growing rats and 5 weeks of treatment for the lactating rats, genistein at the lowest dose prevented ovariectomy-induced, bone-related changes with an efficacy analogous to conjugate equine estrogens (5 mg/d). Arjmandi et al (1996) examined the effects of soy protein on ovariectomy-induced bone loss. Sprague-Dawley rats were divided into 4 groups: *1*) sham operated, *2*) ovariectomized plus casein, *3*) ovariectomized plus soy, and *4*) ovariectomized plus estrogen. The bone density of the right femur was highest in the group receiving estrogen and lowest in the ovariectomized animals given casein in the diet. The bone density of the soy group was significantly reduced compared with the estrogen and sham rats, but significantly higher than that of the ovariectomized rats receiving dietary casein. Bone density of the fourth lumbar vertebra of the soy group was similar to that of the estrogen group and significantly higher than that of both the casein and sham groups. This suggests that soy is more protective on trabecular bone than on cortical bone. Similar observations were reported by Anderson et al (1995). In a follow-up study by Arjmandi et al (1998), in which a similar experimental model as described above (Arjmandi et al., 1996) was used, a soy product high in isoflavones but not a soy product low in isoflavones increased bone density indicating that the isoflavones are responsible for these beneficial effects of soy. Two other studies suggest that genistein in particular affects bone density in rats (Blair et al., 1996; Fanti et al., 1998). Blair et al (1996) found that the dry femoral mass of ovariectomized rats fed 30 mmol genistein/day for 4 weeks was 12% higher ($P < 0.05$) than that of the controls. In a study by Fanti et al (1998), after 21 days of subcutaneous injection of 5 and 25 mg genistein/g body wt, tibial bone mineral loss in ovariectomized rats was significantly reduced. Two studies on the influence of soy consumption on bone mineral loss in postmenopausal women have been reported (Potter et al., 1998; Dalais et al., 1998). The results of both studies indicated that soy had favorable effects on bone density or content; however, the results of these investigations should be considered preliminary. Potter et al (1998) reported that following half a year of treatment, lumbar spine bone mineral density was significantly elevated compared to baseline values in postmenopausal women who had an intake of 40 g soy protein containing 2.25 mg isoflavones/g protein daily, while bone density stayed unaltered in women who ingested the same amount of soy protein but containing only 60 % isoflavones (1.39 mg isoflavones/g protein). Dalais et al (1998) found that early postmenopausal women had a 5% rise in bone mineral content compared with baseline values after consuming soy flour for 3 months. The magnitude of this increase raised questions about these findings, but the control subjects, who were fed wheat protein, also exhibited a rise in bone mineral content which is unexpected since all the subjects were early postmenopausal women (Dalais et al., 1998). Some insight has been gained into the possible mechanism(s) regulating the effect of isoflavones on bone health in rats. There are data suggesting that isoflavones may both stimulate and inhibit bone formation. Fanti et al (1998) noted that genistein enhanced the number of osteoblasts and serum osteocalcin levels, but did not affect the number of osteoclasts. Conversely, Blair et al (1996) found that avian osteoclast protein synthesis *in vitro* was significantly inhibited by genistein, probably due to the inhibitory effects of genistein on tyrosine phosphorylation. Estrogen and tamoxifen, both of which inhibit bone resorption, induce apoptosis in osteoclasts, an effect which is antagonized *in vitro* by antibodies to TGF-β (Hughes et al., 1996). As noted previously, genistein enhanced TGF-β *in vitro*. It appears that the effect of genistein on bone resorption may be mediated by TGF-β.

The relation between isoflavones and bone health is provocative. No long-term human studies have examined the effects of either soy or isoflavones on bone density, markers of bone formation and resorption, and fracture risk, no firm conclusions have been reached at this time.

People of western countries compared with Asians have an increased risk of osteoporosis and a higher incidence of many menopausal symptoms. There is a significant positive correlation between calcium excretion and consumption of animal protein, but this correlation is not observed in case of intake of vegetable protein (Itoh et al., 1998). Studies have been performed to examine the potential effects of soybean products on bone density and osteoporosis risk. Asians with a low dietary intake of animal protein and calcium have a much lower fracture rate than people in western countries (Messina, 1999). Intake of soybean or its isoflavones may be required to produce a moderate increase in bone mass. These findings suggest that isoflavones and soybean protein are responsible for the effect (Itoh et al., 1998; Brouns, 2002). It has been found that genistein has a direct inhibitory effect on bone resorption (Messina, 1999; Yamaguchi and Hua, 1998), and daidzein shows an increase of bone mass in postmenopausal women (Messina, 1999). The mechanisms of isoflavones on bone require to be investigated.

Cardiovascular Disease

The Food and Drug Administration (FDA) in USA approved a health claim that soybean protein included in a diet low in saturated fat and cholesterol lowers the risk of coronary heart disease. Both soybean protein and soybean dietary fiber can reduce plasma cholesterol level in hyperlipidemic patients and protect against cardiovascular disease (Anderson et al., 1995, 1999; Kushi et al., 1999; Lukaczer et al., 2006; Reynolds et al., 2006). Although saponins reduce plasma cholesterol level in some animals, the hypocholesterolemic effects of saponins in humans are more speculative (Milgate and Roberts, 1995).

Regular consumption of legumes significantly reduces incidence of cardiovascular disease (CVD) (Anderson and Major, 2002). Replacement of dietary animal protein with soybean protein brings about the plasma a decrease in concentration of total and low-density lipoprotein (LDL)-cholesterol and triglycerides without affecting high-density lipoprotein (HDL)-cholesterol concentrations (Anderson et al., 1995). There are different hypotheses to explain the hypocholesterolemic effect. One hypothesis is that amino acid composition or distribution in soybean alters cholesterol metabolism, possibly because of alterations in insulin:glucagon ratio and thyroid hormone concentration (Potter, 1995), as well as an increase in plasma thyroxine level associated with a reduction in plasma cholesterol level (Forsythe, 1995). Another hypothesis is that non-protein components such as fiber, isoflavones, minerals, phytic acid and saponins associated with soybean protein change cholesterol metabolism. The metabolic changes observed when soybean is introduced in the diet comprise an increased cholesterol synthesis, an increased bile acid synthesis or a faecal bile acid excretion, increased apolipoprotein B or E receptor activity and reduced hepatic lipoprotein secretion and cholesterol level in blood (Potter, 1995).

Early epidemiologic investigations on cereal and legume dietary fiber revealed that intake of dietary fiber is inversely related to risk of coronary disease. Legumes represent a source of dietary fiber, relatively rich in soluble fiber, which may prevent heart disease (Kushi et al., 1999). The major effects of soybean soluble fibers on serum lipoproteins appear to be related with bile acid binding and with a reduction in bile acid reabsorption (Anderson and Major, 2002). Hence, cholesterol is used to synthesize bile acids (Mälkki, 2001). The colonic fermentation of soluble fibers generates short-chain fatty acids like propionic acid that decrease hepatic cholesterol (Delzenne and Kok, 2001; Anderson and Hanna, 2002). The decrease in hepatic cholesterol synthesis is caused by a decrease in serum insulin concentrations because insulin activates an enzyme involved in cholesterol synthesis and, on the other hand, it might be attributed to a change in the hepatic bile acid profile (Mälkki, 2001). The isoflavones in diet containing soybean may retard atherosclerotic progression on account of their antioxidant properties against LDL oxidation, which triggers a cascade of events producing atherosclerotic plaques. Isoflavones may exhibit hypocholesterolemic activity due to interaction of isoflavones with estrogenic receptors, due to structural similarity between isoflavones and their metabolites and estrogens. Serum cholesterol concentrations may decrease by a similar mechanism (Anderson et al., 1999). Different clinical investigations reveal that it is important to consume soybean protein with its natural isoflavones to produce a cholesterol-lowering effect (Lichtenstein, 1998; Farriol et al., 2006). The hydrolysis of fibers to be used as supplements may also alter their physiological effects.

Menopausal Syndrome

Menopausal symptoms usually involve vasomotor (e.g., hot flashes and nocturnal perspiration), psychological (e.g., depression, anxiety, sleeplessness), and other clinical manifestations (e.g. palpitations, loss of libido) associated with estrogen deficiency. Among the various symptoms, vasomotor symptoms are the most common and most intensely investigated symptoms of menopause.

The Asian diet is rich in soybean foods compared with the western diet (Craig, 1997). Addition of isolated soybean protein with isoflavones to the diet of postmenopausal women lowers the incidence of hot flashes (Albertazzi et al., 1998; Setchell and Cassidy, 1999). In postmenopausal women, isoflavones bind to free estrogenic receptor and produces a weak estrogenic action and thus could be employed as a dietary alternative or supplement to postmenopausal hormone replacement therapy (Setchell and Cassidy, 1999; Duffy et al., 2003). Improved cognitive function was detected in postmenopausal women following consumption of soybean extract containing isoflavones. Two types of estrogenic receptors (ER-α and ER-β) are found in the brain. Isoflavones exhibit binding preference to ER-β receptors, which are abundant in brain regions involved in cognition. ER-β receptors play a role in cognitive function (Duffy et al., 2003).

The soy isoflavone extract alleviates vasomotor symptoms in postmenopausal women furnishing a safe and effective alternative therapy for these subjects (Nahas et al., 2007). Soy foods and soy isoflavones have been used by some women an alternative to hormone replacement therapy. The isoflavones found in soybean are mainly genistein, daidzein and glycitein. These phytoestrogens are nonsteroidal compounds that resemble estrogen in structure and bind weakly to estrogenic receptors (< 1 % of estradiol binding affinity) (Mackey and Eden, 1998). Soy isoflavones preferentially bind to β-estrogen receptors located in the bones, central nervous system, urogenital tract and vascular walls. In contrast to estrogens, isoflavones have little affinity to α-receptors of breast and uterine tissues (Morito et al., 2001). Depending on the concentrations of estradiol, isoflavones display proestrogenic responses in some tissues but inhibit estrogenic action in other tissue (Baker et al., 2000; Lissin and Cooke, 2000). Isoflavones diminish both the severity and the frequency of menopause-related vasomotor symptoms (The North American Menopause Society (NAMS), 2004). The weak estrogen-like effects of isoflavones may explain the low incidence of hot flushes experienced by women in Japan who consume products containing isoflavones (Nagata et al., 2001).

Soybean is a source of proteins. It is low in saturated fat, but is rich in dietary fiber and isoflavones. Many investigations have been conducted on the health benefits of legumes including chickpeas, lentils, soy and beans. Soybean protein may prevent cardiovascular disease because it brings about a reduction in circulating cholesterol level. Isoflavones may be useful for prevention and treatment of chronic diseases. The health properties of soy dietary fiber have been demonstrated. It would be interesting to consider the substitution of animal based foods with soybean foods for obtaining nutritional benefits (Mateos-Aparicio et al., 2008).

A consensus view of soybean phytoestrogens in clinical interventions in post-menopausal women is presented based on results from the EU-funded project Phytohealth (Cassidy et al., 2006). The phytoestrogens, mainly genistein and daidzein, were administered as soybean-protein isolates, whole-soybean foods or extracts, supplements or pure compounds. A comprehensive literature search was carried out using well-defined inclusion or exclusion criteria. For areas in which there is a lot of research, only placebo-controlled double-blind randomised controlled trials (RCT) performed on healthy post-menopausal women were included. For emerging areas, the review covered all available investigations in post-menopausal women. To facilitate cross comparisons between studies, the doses of isoflavones were computered as aglycone equivalents. The results suggest but do no furnish conclusive evidence, that consumption of isoflavones is beneficial to bone health. Ingestion of whole-soybean foods and soybean-protein isolates has some beneficial effects as judged by lipid markers of cardiovascular risk. Intake of isolated isoflavones does not affect blood lipid levels or blood pressure, however, endothelial function may be improved. There is limited evidence that soybean-protein isolates, soybean foods or red-clover (*Trifolium pratense* L.) extract are effective in alleviating menopausal symptoms but soybean isoflavone extracts may be efficacious in reducing hot flushes. There are insufficient RCT studies to enable conclusions on the effects of isoflavones on breast cancer, colon cancer, diabetes or cognitive function. The health benefits of soyabean phytoestrogens in healthy post-menopausal women are subtle and even some well-designed studies do not show protective effects. It was suggested that future studies focus on high-risk post-menopausal women, especially in the areas of diabetes, cardiovascular disease, breast cancer and bone health.

Epidemiological investigations furnish evidence that consumption of soybeans may be linked to a reduced incidence of some chronic diseases. Clinical data indicate that intake of soy proteins minimizes the risk of cardiovascular disease. The food-labeling that claims prevention of coronary heart disease by soy proteins was approved by the US FDA in 1999. Similar health claims for soy proteins have subsequently been approved in other countries. When 22 randomized trials conducted since 1999 were evaluated by the Nutrition Committee of the American Heart Association, it was discovered that isolated soy protein with isoflavones brought about a small reduction of LDL-cholesterol but it was devoid of a similar suppressive action on effect on HDL-cholesterol, triglycerides, lipoprotein(a), and blood pressure. The soybean source and procedures employed for the production of soy protein or isoflavones are important because they may determine the content and integrity of bioactive proteins (Xiao CW, 2008).

Most of the severe food-related allergic reactions are caused by peanut allergy. Recent allergy is usually presented early in life. People with peanut allergy usually do not outgrow it. In highly sensitized individuals, small amounts of peanut can initiate an allergic reaction. The prevalence, clinical characteristics, diagnosis, natural history and management of peanut allergy has been reviewed (Al-Muhsen et al., 2003).

Soy protein is gaining importance in the human diet. Soy protein has a shortage of methionine which is an essential amino acid. Methionine supplementation benefits soy infant formulas, but apparently not food intended for adults with an adequate nitrogen intake. The content of lysine in soy protein, which is also an essential amino acid, is lower than that of the milk protein casein albeit higher than that of wheat proteins. Undesirable outcome after consumption of raw soybean meal has been ascribed to the presence of protease inhibitors and lectins and to poor digestibility. To enhance the nutritional quality of soy foods, protease inhibitors and lectins are heat inactivated or removed by fractionation during food processing. However, protease inhibitors are more thermostable than lectins. Most commercially heated meals have up to 20% residual Bowman-Birk (BBI) chymotrypsin and trypsin inhibitor and Kunitz trypsin inhibitor (KTI). In order to increase the importance of soybeans to nutrition and health, a better understanding of the factors that affect the nutritional and health-promoting quality of soy proteins is warranted. Possible beneficial effects of soy-containing diets comprise lowering of cholesterol, anticarcinogenic effects of BBI, and protection against obesity, diabetes, irritants of the digestive tract, bone, and kidney diseases. Undesirable effects include poor digestibility and allergy to soy proteins. Strategies to decrease the level of soybean inhibitors by rearrangement of protein disulfide bonds, immunoassays of inhibitors in processed soy foods and soybean germplasm, the roles of phytoestrogenic isoflavones and lectins, and research needs in all of these areas are also discussed (Friedman and Brandon, 2001).

Messina (1999) reviewed the nutritional profiles and health effects of legumes including soybeans. Legumes are important in the traditional diets of many regions throughout the world. Beans are low in fat and abundant in protein, dietary fiber, micronutrients and phytochemicals. Among the legumes, soybeans are unique since they are rich in isoflavones. Isoflavones are weakly estrogenic and the isoflavone genistein influences signal transduction. Soyfoods and isoflavones may be of value for prevention and therapy of cancer and osteoporosis. The low breast cancer mortality rates in Asian countries and antiestrogenic effects of isoflavones suggest that soyfood intake may reduces breast cancer risk. The available data also suggest that soy or isoflavones may decrease the risk of prostate cancer. The low estrogenic activity of isoflavones and structural resemblance between soybean isoflavones and the synthetic isoflavone ipriflavone, which increases bone mineral density in postmenopausal women, suggest that soy or isoflavones may reduce the risk of osteoporosis. Human and rodent studies support this hypothesis, given the nutrient profile and phytochemical contribution of beans, nutritionists should make a concerted effort to encourage the public to consume more beans in general and more soyfoods in particular.

The nutritional significance of soy products in the human diet has been reviewed by Erdman and Fordyce (1989). Results of clinical investigations and animal studies on the protein quality and growth-promoting effects of soy protein, allergies in children, cholesterol lowering effects of soy protein and soy fiber, effects of soy products on glucose tolerance, and the bioavailability of zinc and iron from soy foods are discussed.

The leguminous seeds constitute a valuable source of feed and dietary protein, however, they contain also many different substances that are recognised as antinutritive factors. This review deals with harmful substances and composition of leguminous seeds, their profitable and negative activity considering also the effect on physiology of digestion, intestinal functions and health of animals (Jamroz and Kubizna, 2008).

DISCUSSION

Some nutritional advantages could be obtained by replacing many animal-based foods for legume foods. Legumes represent an excellent source of high-quality protein with a low content of saturated fat and a great amount of dietary fibre. Therefore, the possible use of legumes in functional food design is very interesting, since the consumption of legume protein and dietary fiber seems to reduce the risk of cardiovascular diseases and to improve glycemic control. Furthermore, soybean isoflavones are associated with a potential role in the prevention and treatment of different diseases. Therefore, soybean could play an important role for the promotion of health.

Attempts to disclose beneficial activity of isolated dietary legume components in the prevention and treatment of various diseases do not always produce unequivocal results. The reasons are the complexity of the physiological processes involved: the multi-factorial nature of the diseases, the synergism between bio-active legume, different experimental approaches adopted and conditions employed, which prevent the standardisation and comparison of the results. Specific proteins, and therefrom peptides, have unequivocally been demonstrated to play a role in plasma lipid and glucose homeostasis, inhibition of hydrolytic enzymes crucial to some pathophysiological phenomena, blood pressure control and immuno-modulation. All these basic functions are more or less directly related to a number of diseases, encompassing cancer, cardio-vascular diseases, diabetes, obesity, hypertension and immunity-related diseases (Duranti, 2006).

Continuing research on legumes will hopefully disclose more about the mechanism of health promoting action of legumes, in particular soybean.

REFERENCES

Adlercreutz H, Mazur W, Bartels P, et al. (2000) Phytoestrogens and prostate disease. *J Nutr* 130: 658-659.

Adlercreutz H, Mazur W. (1997) Phyto-estrogens and western diseases. *Ann Med* 29: 95–120.

Adlercreutz H. (2002) Phyto-estrogens and cancer. *Lancet Oncol* 3: 364-373.

Akiyama T, Ishida J, Nakagawa S, et al. (1987) Genistein, a specific inhibitor of tyrosine-specific protein kinases. *J Biol Chem* 262: 5592–5.

Akiyama T, Ogawara H. (1991) Use and specificity of genistein as inhibitor of protein-tyrosine kinases. *Methods Enzymol* 201: 362–70.

Albertazzi P, Pansini F, Bonaccori G, Zanotti L, Forini E, De Aloysio D. (1998) The Effect of Dietary Soybean Supplementation on Hot Flushes. *Obstetrics & Gynecology* 91: 6-11.

Al-Muhsen S, Clarke AE, Kagan RS. (2003) Peanut allergy: an overview. CMAJ. 168(10): 1279-85. Erratum in: *CMAJ*. 168(12): 1529.

Anderson JJB, Thomsen K, Christiansen, C. (1987) High protein meals, insular hormones and urinary calcium excretion in human subjects. In: Christiansen C, Johansen JS, Riis BJ, eds. *Osteoporosis*. Viborg, Denmark: Nørhaven A/S, 240–5.

Anderson JW, Hanna TJ. (2002) Impact of non-digestible carbohydrates on serum lipoproteins and risk for cardiovascular disease. *J Nutr* 129: 1457-1466.

Anderson JW, Johnstone BM, Cook-Newell ME. (1995) Meta-analysis of the effects of soybean protein intake on serum lipids. *N Engl J Med* 333: 276-282.

Anderson JW, Major AW. (2002) Pulses and lipaemia, short- and long-term effect: potential in the prevention of cardiovascular disease. *Br J Nutr* 88: 263-271.

Anderson JW, Smith BM, Washnock CS. (1999) Cardiovascular and renal benefits of dry bean and soybean intake. *Am J Clin Nutr* 70: 464-474.

Anderson JW, Story L, Sieling B, Chen W-JL. (1984) Hypocholesterolemic effects of high-fibre diets rich in water-soluble plant fibres. *J Can Diet Assoc* 47: 140–8.

Arjmandi BH, Alekel L, Hollis BW, et al. (1996) Dietary soybean protein prevents bone loss in an ovariectomized rat model of osteoporosis. *J Nutr* 126: 161–7.

Arjmandi BH, Birnbaum R, Goyal NV, et al. (1998) Bone-sparing effect of soy protein in ovarian-hormone–deficient rats is related to its isoflavone content. *Am J Clin Nutr* 68(suppl): 1364S–8S.

Armstrong WB, Kennedy AR, Wan XS, Atiba J, McLaren E, Meyskens FL. (2000) Single-dose administration of Bowman-Birk inhibitor concentrate in patients with oral leukoplakia. Cancer Epidemiol. *Biomarkers* 9: 43-47.

Baker VL, Leitman D, Jaffe RB. (2000) Selective estrogen receptor modulators in reproductive medicine and biology. *Obstet Gynecol Surv* 55(Suppl. 2): S21–47.

Barnes S, Grubbs C, Setchell KDR, Carlson J. (1990) Soybeans inhibit mammary tumors in models of breast cancer. In: Pariza MW, Aeschbacher H-U, Felton JS, Sato S, eds. *Mutagens and carcinogens in the diet*. New York: Wiley Liss, 1990: 239–53.

Benno Y, Endo K, Mizutani T, Namba Y, Komori T, Mitsuoka T. (1989) Comparison of fecal microflora of elderly persons in rural and urban areas of Japan. *Appl Environ Microbiol* 55: 1100–5.

Benson JR, Baum M, Colletta AA. (1996) Role of TGFβ in the anti-estrogen response/resistance of human breast cancer. *J Mammary Gland Biol Neoplasia* 1: 381–9.

Benson JR, Colletta AA. (1995) Transforming growth factor β. Prospects for cancer prevention and treatment. *Clin Immunother* 4: 249–58.

Blair HC, Jordon SE, Peterson TG, Barnes S. (1996) Variable effects of tyrosine kinase inhibitors on avian osteoclastic activity and reduction of bone loss in ovariectomized rats. *J Cell Biochem* 61: 629–37.

Brandi ML. (1992) Flavonoids: biochemical effects and therapeutic applications. *Bone Miner* 19 (suppl): S3–64.

Breslau NA, Brinkley L, Hill KD, Pack CYC. (1988) Relationship of animal protein-rich diet to kidney stone formation and calcium metabolism. *J Clin Endocrinol Metab* 66; 140–6.

Brouns F. (2002) Soybean isoflavones: a new and promising ingredient for the health foods sector. *Food Res Int* 35: 187-193.

Brown NM, Lamartiniere CA. (1995) Xenoestrogens alter mammary gland differentiation and cell proliferation in the rat. *Environ Health Perspect* 103: 708–13.

Cancer facts and figures. (1994) Atlanta: American Cancer Society.

Carlsson NG, Karlsson H, Sandberg AS. (1992) Determination of oligosaccharides in foods, diets, and intestinal contents by hightemperature gas chromatography and gas chromatography/mass spectrometry. *J Agric Food Chem* 40; 2404–12.

Cassidy A, Albertazzi P, Lise Nielsen I, Hall W, Williamson G, Tetens I, Atkins S, Cross H, Manios Y, Wolk A, Steiner C, Branca F. (2006) Critical review of health effects of soyabean phyto-estrogens in post-menopausal women. *Proc Nutr Soc.* 65(1): 76-92.

Cassidy A, Bingham S, Setchell KD. (1994) Biological effects of a diet of soy protein rich in isoflavones on the menstrual cycle of premenopausal women. *Am J Clin Nutr* 60: 333–40.

Cassidy A, Bingham S, Setchell KD. (1995) Biological effects of isoflavones in young women: importance of the chemical composition of soybean products. *Br J Nutr* 74: 587–601.

Caygill CPJ, Charlett A, Hill MJ. (1996) Fat, fish, fish oil and cancer. *Br J Cancer* 74: 159–64.

Chan JCM. (1974) The influence of dietary intake on endogenous acid production. *Nutr Metab* 16: 1–9.

Clark JW, Santos-Moore A, Stevenson LE, Frackelton AR. (1996) Effects of tyrosine kinase inhibitors on the proliferation of human breast cancer lines and proteins important in the RAS signaling pathway. *Int J Cancer* 65: 186–91.

Constantinou A, Huberman E. (1995) Genistein as an inducer of tumor cell differentiation: possible mechanisms of action. *Proc Soc Exp Biol Med* 208: 109–15.

Constantinou A, Kiguchi K, Huberman E. (1990) Induction of differentiation and DNA strand breakage in human HL-60 and K-562 leukemia cells by genistein. *Cancer Res* 50: 2618–24.

Constantinou AL, Mehta RG, Vaughan A. (1996) Inhibition of N-methyl-N-nitrosourea–induced mammary tumors in rats by the soybean isoflavones. *Anticancer Res* 16: 3293–8.

Coward L, Barnes NC, Setchell KDR, Barnes S. (1993) Genistein, daidzein, and their β-glycoside conjugates: antitumor isoflavones in soybean foods from American and Asian diets. *J Agric Food Chem* 41: 1961–7.

Craig WJ. (1997) Phytochemicals: Guardians of our health. *J Am Diet Assoc* 97: 199-204.

Dalais FS, Rice GE, Bell RJ, et al. (1998) Dietary soy supplementation increases vaginal cytology maturation index and bone mineral content in postmenopausal women. *Am J Clin Nutr* 68(suppl): 1518S (abstr).

Dalu A, Haskell J, Lamartiniere CA. (1998) Dietary genistein inhibits protein tyrosine phosphorylation in the dorsolateral prostate of the rat. *Am J Clin Nutr* 68(suppl): 1524S(abstr).

Daly LE, Kirke PN, Molloy A, Weir DG, Scott JM. (1995) Folate levels and neural tube defects. *JAMA* 274: 1698–702.

Davis JW, Ross PD, Nevitt MC, Wasnich RD. (1997) Incidence rates of falls among Japanese men and women living in Hawaii. J Clin Epidemiol 50: 589–94.

Delzenne NM, Kok N. (2001) Effects of fructans-type prebiotics on lipid metabolism. *Am J Clin Nutr* 73:456-458.

Duarte-Rayas P, Bergeron D, Nielsen SS. (1992) Screening of heat-stable trypsin inhibitors in dry beans and their partial purification from great Northern beans (Phaseolus vulgaris) using anhydrotrypsin-sepharose affinity chromatography. *J Agric Food Chem* 40: 32–42.

Duffy R, Wiseman H, File SE. (2003) Improved cognitive function in postmenopausal women after 12 weeks of consumption of a soya extract containing isoflavones. *Pharmacol Biochem Behav* 75: 721-729.

Duranti M. (2006) Grain legume proteins and nutraceutical properties. *Fitoterapia.* 77: 67-82.

Emken EA, Adlof RO, Gulley RM. (1994) Dietary linoleic acid influences desaturation and acylation of deuterium-labeled linoleic and linolenic acids in young adult males. *Biochim Biophys Acta* 1213: 277–88.

Erdman JW Jr, Fordyce EJ. (1989) Soy products and the human diet. *Am J Clin Nutr.* 49(5): 725-37.

Evans BAJ, Griffiths K, Morton MS. (1995) Inhibition of 5"-reductase in genital skin fibroblasts and prostate tissue by dietary lignans and isoflavonoids. *J Endocrinol* 147: 295–302.

Fang EF, Wong JH, Ng TB. (2010) Thermostable Kunitz trypsin inhibitor with cytokine inducing, antitumor and HIV-1 reverse transcriptase inhibitory activities from Korean large black soybeans. *J Biosci Bioeng.* 109(3): 211-7

Fanti O, Faugere MC, Gang Z, Schmidt J, Cohen D, Malluche HH. (1998) Systemic administration of genistein partially prevents bone loss in ovariectomized rats in a nonestrogen-like mechanism. *Am J Clin Nutr* 68(suppl): 1517S (abstr).

Farriol M, Jordà M, Delgado G. (2006) Past and current trends supplementation: a bibliographic study. *Nutr Hosp* 21(4): 448-51.

Folkman J, Klagsbrun M. (1987) Angiogenic factors. *Science* 235: 442–7.

Forsythe WA. (1995) Soy protein, thyroid regulation and cholesterol metabolism. *J Nutr* 125:619-623.

Foster-Powell K, Miller JB. (1995) International tables of glycemic index. *Am J Clin Nutr* (suppl); 62: 871S–90S.

Fotsis T, Pepper M, Adlercreutz H, et al. (1993) Genistein, a dietary-derived inhibitor of in vitro angiogenesis. *Proc Natl Acad Sci USA* 90: 2690–4.

Franke AA, Mordan LJ, Conney RV, et al. (1995) Dietary phenolic agents inhibit neoplastic transformation and trap toxic NO. *Proc Am Assoc Cancer Res* 36: 125 (abstr).

Friedman M, Brandon DL. (2001) Nutritional and health benefits of soy proteins. *J Agric Food Chem.* 49(3): 1069-86.

Fujita T, Fukase M. (1992) Comparison of osteoporosis and calcium intake between Japan and the United States. *Proc Soc Exp Biol Med* 200: 149–52.

Geller J, Sionit L, Partido C, et al. (1998) Genistein inhibits the growth of human-patient BPH and prostate cancer in histoculture. *Prostate* 34: 75–9.

Graf E, Eaton JW. (1990) Antioxidant functions of phytic acid. *Free Radic Biol Med* 8: 61–9.

Grant G. (1989) Anti-nutritional factors of soybean: a review. *Prog Food Nutr Sci* 13: 317–48.

Harland BF, Morris ER. (1995) Phytate: a good or a bad food component? *Nutr Res* 15: 733–54.

Hirose K, Tajima K, Hamajima N, et al. (1995) A large-scale, hospital based case-control study of risk factors of breast cancers according to menopausal status. *Jpn J Cancer Res* 86: 146–54.

Holt S, Muntyan I, Likyer L. (1996) Soybean-based diets for Diabetes Mellitus. *Alternative & Complementary Therapies*

Hughes DE, Dai A, Tiffee JC, Li HH, Mundy GR, Boyce BF. (1996) Estrogen promotes apoptosis of murine osteoclasts mediated by TGF-β. *Nat Med* 2: 1132–6.

Indu M, (1992) Ghafoorunissa. N-3 fatty acids in Indian diets—comparison of the effects of precursor (alpha-linolenic acid) vs product (long chain n-3 polyunsaturated fatty acids). *Nutr Res* 12: 569–82.

Itoh R, Nishiyama N, Suyama Y. (1998) Dietary protein intake and urinary excretion of calcium: a cross- sectional study in a healthy Japanese population. *Am J Clin Nutr* 67: 438-444.

Jamroz D, Kubizna J. (2008) Harmful substances in legume seeds-their negative and beneficial properties. *Pol J Vet Sci.* 11: 389-404.

Jenkins DJ, Kendall CW, Marchie A, Jenkins AL, Augustin LS, Ludwig DS, Barnard ND, Anderson JW. (2003) Type 2 diabetes and vegetarian diet. *Am J Clin Nutr* 78: 610-616.

Jenkins DJA, Kendall CWC, D'Costa MA, et al. (2003) Soybean consumption and phytoestrogens: effect on serum prostate specific antigen when blood lipids and oxidized low-density lipoprotein are reduced in hyperlipidemic men. *J Urol* 169: 507-511.

Jenkins DJA, Wolever TMS, Taylor RH, Barker HM, Fielden H. (1980) Exceptionally low blood glucose response to dried beans: comparison with other carbohydrate rich foods. *Br Med J* 281: 578–80.

Jing Y, Nakaya K, Han R. (1993) Differentiation of promyelocytic leukemia cells HL-60 induced by daidzein in vitro and in vivo. *Anticancer Res* 13: 1049–54.

Jordinson M, Goodlad RA, Brynes A, Bliss P, Ghatei MA, Bloom SR, Fitzgerald A, Grant G, Bardocz S, Pustzai A, Pignatelli M, Calam J (1999) Gastrointestinal responses to a panel of lectins in rats maintained on total parenteral nutrition. *Am. J. Physiol.-Gastr. Liver Physiol.* 39: G1235-G1242.

Kennedy AR (1995) The evidence for soybean products as cancer preventive agents. *J. Nutr.* 125: 733S-743S.

Kennedy AR (1998a) Chemopreventive agents: protease inhbibitors. *Pharmacol. Ther.* 78: 167-209.

Kennedy AR (1998b) The Bowman-Birk inhibitor from soybeans as an anticarcinogenic agent. *Am. J. Clin. Nutr.* 68: 1406S-1412S.

Kin K, Lee JH, Kushida K, et al. (1993) Bone density and body composition on the Pacific Rim: a comparison between Japan-born and U.S.-born Japanese-American women. *J Bone Miner Res* 8: 861–9.

Koo M, Rao AV. (1991) Long term effect of bifidobactaria and neosugar on precursor lesions. *Nutr Cancer* 16: 249–57.

Koratkar R, Rao AV. (1997) Effect of soya bean saponins on azoxymethane induced preneoplastic lesions in the colon of mice. *Nutr Cancer* 27: 206–9.

Kuo SM, Morehouse HF Jr, Lin CP. (1997) Effect of antiproliferative flavonoids on ascorbic acid accumulation in human colon adenocarcinoma cells. *Cancer Lett* 116: 131–7.

Kuo SM. (1996) Antiproliferative potency of structurally distinct dietary flavonoids on human colon cancer cells. *Cancer Lett* 110: 41–8.

Kuo TM, Van Middlesworth JF, Wolf WJ. (1988) Content of raffinose oligosaccharides and sucrose in various plant seeds. *J Agric Food Chem* 36: 32–6.

Kushi LH, Meyer KM, Jacobs DR. (1999) Cereals, legumes, and chronic disease risk reduction: evidence from epidemiologic studies. *Am J Clin Nutr* 70: 451-458.

Kyle E, Neckers L, Takimoto C, Curt G, Bergan R. (1997) Genistein induced apoptosis of prostate cancer cells is preceded by a specific decrease in focal adhesion kinase activity. *Mol Pharmacol* 51: 193–200.

Lam SS, Wang H, Ng TB. (1998) Purification and characterization of novel ribosome inactivating proteins, alpha- and beta-pisavins, from seeds of the garden pea Pisum sativum. *Biochem Biophys Res Commun.* 253(1): 135-42.

Lamartiniere CA, Moore JB, Brown NM, Thompson R, Hardin MJ, Barnes S. (1995) Genistein suppresses mammary cancer in rats. *Carcinogenesis* 16: 2833–40.

Lee HP, Gourley L, Duffy SW, Esteve J, Day NE. (1991) Dietary effects on breast cancer risk in Singapore. *Lancet* 337: 1197–200.

Lichtenstein AH. (1998) Soybean protein, isoflavones and cardiovascular disease risk. *J Nutr* 128: 1589-1592.

Liener IE. (1994) Implications of antinutritional components in soybean foods. *Crit Rev Food Sci Nutr* 34: 31–67.

Lin P, Ye X, Ng T. (2008) Purification of melibiose-binding lectins from two cultivars of Chinese black soybeans. *Acta Biochim Biophys Sin* (Shanghai). 40(12): 1029-38.

Linassier C, Pierre M, Le Peco JB, Pierre J. (1990) Mechanism of action in NIH-3T3 cells of genistein, an inhibitor of EGF receptor tyrosine kinase activity. *Biochem Pharmacol* 39: 187–93.

Lissin LW, Cooke JP. (2000) Phytoestrogens and cardiovascular health. *J Am Coll Cardiol* 35: 1403–10.

Lu LJ, Anderson KE, Nagamani M. (1996) Effects of one month soya consumption on circulating steroids in men. *Proc Am Assoc Cancer Res* 37: 270 (abstr).

Lukaczer D, Liska DJ, Lerman RH et al. (2006) Effect of a low glycemic index diet with soybean protein and phytosterols on CVD risk factors in postmenopausal women. *Nutrition* 22: 104-113.

Lynch SR, Beard JL, Dassenko SA, Cook JD. (1984) Iron absorption from legumes in humans. *Am J Clin Nutr* 40: 42–7.

Ma DZ, Wang HX, Ng TB. (2009) A peptide with potent antifungal and antiproliferative activities from Nepalese large red beans. *Peptides*. 30(12): 2089-94.

Mackey R, Eden J. (1998) Phytoestrogens and the menopause. *Climateric* 1: 302–8.

Mage JA. (1982) Phytate: its chemistry, occurrence, food interactions, nutritional significance, and methods of analysis. *J Agric Food Chem* 30: 1–9.

Mäkela SI, Pylkkänen LH, Santti RSS, Adlercreutz H. (1995) Dietary soybean may be antiestrogenic in male mice. *J Nutr* 125: 437–45.

Mälkki Y. (2001) Physical properties of dietary fiber as keys to physiological functions. *Cereal Foods World* 46: 196-199.

Markowitz SD, Roberts AB. (1997) Tumor suppressor activity of the TGFβ pathway in human cancers. *Cytokine Growth Factor Rev* 7: 93–102.

Marlett JA. (1992) Content and composition of dietary fiber in 117 frequently consumed foods. *J Am Diet Assoc* 92: 175–86.

Mateos-Aparicio I, Redondo Cuenca A, Villanueva-Suárez MJ, Zapata-Revilla MA. (2008) Soybean, a promising health source. *Nutr Hosp.* 23(4):305-12.

McMichael-Phillips DF, Harding C, Morton M, et al. (1998) Effects of soyprotein supplementation on epithelial proliferation in histologically normal human breasts. *Am J Clin Nutr* 68(suppl):1431S–6S.

Messina MJ, Barnes S. (1991) The role of soy products in reducing risk of cancer. *J Natl Cancer Inst* 83: 541–6.

Messina MJ. (1994) *The Simple Soybean and Your Health*. Avery Publishing Group, New York, pp. 150-151.

Messina MJ. (1999) Legumes and soybeans: overview of their nutritional profiles and health effects. *Am J Clin Nutr* 70: 439-450.

Milgate J, Roberts DCK. (1995) The nutritional and biological significance of saponins. *Nutr Res* 15: 1223–49.

Mitsuoka T. (1982) Recent trends in research on intestinal flora. Bifidobacteria Microflora 1: 3–24.

Morito K, Hirose T, Kinjo J, et al. (2001)Interaction of phytoestrogens with estrogen receptors α and β. *Biol Pharm Bull* 24: 351–6.

Morton MS, Matos-Ferreira A, Abranches-Monteiro L, et al. (1997) Measurement and metabolism of isoflavonoids and lignans in human male. *Cancer Lett* 114: 145–51.

Murrill WB, Brown NM, Zhang JX, et al. (1996) Prepubertal genistein exposure suppresses mammary cancer and enhances gland differentiation in rats. *Carcinogenesis* 17: 1451–7.

Nagata C, Takatsuka N, Kawakami N, Shimizu H. (2001) Soy product intake and hot flashes in Japanese women: results from community-based prospective study. *Am J Epidemiol* 153: 790–3.

Nahas EA, Nahas-Neto J, Orsatti FL, Carvalho EP, Oliveira ML, Dias R. (2007) Efficacy and safety of a soy isoflavone extract in postmenopausal women: a randomized, double-blind, and placebo-controlled study. *Maturitas*. 58: 249-58.

Naik HR, Lehr JE, Pienta KJ. (1994) An *in vitro* and *in vivo* study of antitumor effects of genistein on hormone refractory prostate cancer. *Anticancer Res* 14: 2617–20.

Nair SSD, Leitch JW, Falconer J, Garg ML. (1997) Prevention of cardiac arrhythmia by dietary (n23) polyunsaturated fatty acids and their mechanism of action. *J Nutr* 127: 383–93.

Nakamura T, Turner CH, Yoshikawa T, et al. (1994) Do variations in hip geometry explain differences between Japanese and white Americans? *J Bone Miner Res* 9: 1071–6.

National Research Council (1989) *Recommended dietary allowances*. 10th ed. Washington, DC: National Academy Press

Oberleas D, Harland BE. (1981) Phytate content of foods: effect on dietary zinc bioavailability. *J Am Diet Assoc* 79: 433–6.

Oski FA. (1997) What we eat may determine who we can be. *Nutrition* 13: 220–1.

Pagliacci MC, Smacchia M, Migliorati G, Grignana F, Riccardi C, Nicoletti I. (1994) Growth-inhibitory effects of the natural phytoestrogen genistein in MCF-7 human breast cancer cells. *Eur J Cancer* 30A: 1675–82.

Peterson G, Barnes S. (1991) Genistein inhibition of the growth of human breast cancer cells: independence from estrogen receptors and the multi-drug resistance gene. *Biochem Biophys Res Commun* 179: 661–7.

Peterson G, Barnes S. (1993) Genistein and biochanin A inhibit the growth of human prostate cancer cells but not epidermal growth factor receptor autophosphorylation. *Prostate* 22: 335–45.

Peterson G, Barnes S. (1994) Genistein potently inhibits the growth of human primary breast epithelial cells: correlation with lack of genistein metabolism. *Mol Biol Cell* 5: 384a (abstr).

Peterson G, Barnes S. (1996) Genistein inhibits both estrogen and growth factor-stimulated proliferation of human breast cancer cells. *Cell Growth Differ* 7: 1345–51.

Peterson G, Coward L, Kirk M, Falany C, Barnes S. (1996) The role of metabolism in mammary epithelial growth inhibition by the isoflavones genistein and biochanin A. *Carcinogenesis* 17: 1861–9.

Peterson TG, Kim H, Barnes S. (1998) Genistein may inhibit the growth of human mammary epithelial (HME) cells by augmenting transforming growth factor beta (TGFβ) signaling. *Am J Clin Nutr* 68(suppl) : 1527S (abstr).

Pienta KJ, Esper PS. (1993) Risk factors for prostate cancer. *Ann Intern Med* 118: 793–803.

Pollard M, Luckert PH. (1997) Influence of isoflavones in soy protein isolates on development of induced prostate-related cancers in L-W rats. *Nutr Cancer* 28: 41–5.

Potter SM (1995) Overview of proposed mechanism for the hypocholesterolemic effect of soybean. *J Nutr* 125: 606-611.

Potter SM, Baum JA, Teng H, Stillman RJ, Shay NF, Erdman JW Jr. (1998) Soy protein and isoflavones: their effects on blood lipids and bone density in postmenopausal women. *Am J Clin Nutr* 68(suppl): 1375S–9S.

Pryme IF, Pustzai A, Bardocz S, Ewen SWB (1998) The induction of gut hyperplasia by phytohaemagglutinin in the diet and limitation of tumour growth. *Histol. Histopathol.* 13: 575-583.

Pusztai A (1993) Dietary lectins are metabolic signals for the gut and modulate immune and hormone functions. *Eur. J. Clin. Nutr.* 47: 691-699.

Pusztai A, Bardocz S (1996) Biological effects of plant lectins on the gastrointestinal tract: metabolic consequences and applications. *Trends Glycosci. Glycotechnol.* 8: 149-165.

Pusztai A, Grant G, Buchan WC, Bardocz S, de Carvalho AFFU, Ewen SWB (1998) Lipid accumulation in obese Zucker rats is reduced by inclusion of raw kidney bean (Phaseolus vulgaris) in the diet. *Br. J. Nutr.* 79: 213-221.

Rackis JJ, Sessa DJ, Steggerda FR, Shimizu T, Anderson T, Pearl SL. (1970) Soybean factors relating to gas production by intestinal bacteria. *J Food Sci* 35: 634–9.

Rauth S, Kichina J, Green A. (1997) Inhibition of growth and induction of differentiation of metastatic melanoma cells in vitro by genistein: chemosensitivity is regulated by cellular p53. *Br J Cancer* 75: 1559–66.

Remer T, Manz F. (1994) Estimation of the renal net acid excretion by adults consuming diets containing variable amounts of protein. *Am J Clin Nutr* 59: 1356–61.

Reynolds K, Chin A, Lees KA, Nguyen A, Bujnowski D, He J. (2006) A Meta-Analysis of the Effect of Soybean Protein Supplementation on Serum Lipids. *Am J Cardiol* 98: 633-640.

Ridout CL, Wharf G, Price KR, Johnson LT, Fenwick GR. (1988) UK mean daily intakes of saponins—intestine-permeabilizing factors in legumes. *Food Sci Nutr* 42F: 111–6.

Ross PD, Fujiwara S, Huang C, et al. (1995) Vertebral fracture prevalence in women in Hiroshima compared to Caucasians or Japanese in the US. *Int J Epidemiol* 24: 1171–7.

Ross PD, Norimatsu H, Davis JW, et al. (1991) A comparison of hip fracture incidence among native Japanese, Japanese Americans, and American Caucasians. *Am J Epidemiol* 133: 801–9.

Ross RK, Bernstein LA, Lobo RA, et al. (1992) 5-Alpha-reductase activity and risk of prostate cancer among Japanese and US white and black males. *Lancet* 339: 887–9.

Rubio MA. (2002) Implicaciones de la fibra en distintas patologías. *Nutr Hosp* XVII(2): 17-29.

Russell-Aulet M, Wang J, Thornton JC, Colt EW, Pierson RN Jr. (1993) Bone mineral density and mass in a cross-sectional study of white and Asian women. *J Bone Miner Res* 8: 575–82.

Salmerón J, Manson JE, Stampfer MJ, Colditz GA, Wing AL, Willet WC. (1997) Dietary fiber, glycemic load, and risk of non-insulin dependent diabetes mellitus in women. *JAMA* 277: 472–7.

Sandström B, Almgren A, Kivistö B, Cederblad A. (1989) Effect of protein level and protein source on zinc absorption in humans. *J Nutr* 119: 48–53.

Santibáñez JF, Navarro A, Martinez J. (1997) Genistein inhibits proliferation and in vitro invasive potential of human prostatic cancer cell lines. *Anticancer Res* 17: 1199–1204.

Schleicher R, Zheng M, Zhang M, Lamartiniere CA. (1998) Genistein inhibition of prostate cancer cell growth and metastasis in vivo. *Am J Clin Nutr* 68(suppl): 1526S(abstr).

Scholar EM, Toewa ML. (1994) Inhibition of invasion of murine mammary carcinoma cells by the tyrosine kinase inhibitor genistein. *Cancer Lett* 87: 159–62.

Setchell KDR, Cassidy A. (1999) Dietary isoflavones: Biological effects and relevance to human health. *J Nutr* 129: 758-767.

Severson KJ, Nomura AMY, Grove JS, Stemmermann GN. (1989) A prospective study of demographics, diet, and prostate cancer among men of Japanese ancestry in Hawaii. *Cancer Res* 49: 1857–60.

Simopoulos AP. (1999) Essential fatty acids in health and chronic disease. *Am J Clin Nutr* 70(suppl): 560S–9S.

So FV, Guthrie N, Chambers AF, Moussa M, Carroll KK. (1996) Inhibition of human breast cell proliferation by flavonoids and citrus juice. *Nutr Cancer* 26: 167–81.

Steggerda FR, Dimmick JF. (1966) Effects of bean diets on concentration of carbon dioxide in flatus. *Am J Clin Nutr* 19: 120–4.

Stephens FO. (1997) Phytoestrogens and prostate cancer: possible preventive role. *Med J Aust* 167: 138–40.

Stone NJ. (1997) Fish consumption, fish oil, lipids, and coronary heart disease. *Am J Clin Nutr* 65: 1083–6.

The North American Menopause Society (NAMS). (2004) Treatment of menopause-associated vasomotor symptoms: position statement of NAMS. *Menopause* 11: 11–33.

Thompson LU, Yoon JH, Jenkins DJ, Wolever TM, Jenkins AL. (1984) Relationship between polyphenol intake and blood glucose response of normal and diabetic individuals. *Am J Clin Nutr* 39: 745–51.

Thorburn J, Thorburn T. (1994) The tyrosine kinase inhibitor, genistein, prevents a-adrenergic-induced cardiac muscle cell hypertrophy by inhibiting activation of the Ras-MAP kinase signaling pathway. *Biochem Biophys Res Commun* 202: 1586–91.

Thorne MJ, Thompson LU, Jenkins DJ. (1983) Factors affecting starch digestibility and the glycemic response with special reference to legumes. *Am J Clin Nutr* 38: 481–8.

Tsuda M, Kitazaki T, Ito T, Fujita T. (1986) The effect of ipriflavone (TC-80) on bone resorption in tissue culture. *J Bone Miner Res* 1: 207–11.

US Department of Agriculture, Nutrient Data Research Branch, Nutrition Monitoring Division. (1988) Provisional table on the content of omega-3 fatty acids and other fat components in selected foods. Hyattsville, MD: *Human Nutrition Information Service* (Publication HNIS/PT-103.)

Valente M, Bufalino L, Castiglione GN, et al. (1994) Effects of 1-year treatment with ipriflavone on bone in postmenopausal women with low bone mass. *Calcif Tissue Int* 54: 377–80.

Vucenik I, Yang GY, Shamsuddin AM. (1997) Comparison of pure inositol hexaphosphate and high-bran diet in the prevention of DMBA induced rat mammary carcinogenesis. *Nutr Cancer* 28: 7–13.

Wang HJ, Murphy PA. (1994) Isoflavone content in commercial soybean foods. *J Agric Food Chem* 42: 1666–73.

Wang XS, Meyskens FL, Armstrong WB, Taylor TH, Kennedy AR. (1999) Relationship between protease activity and neu oncogene expression in patients with oral leukoplakia treated with the Bowman-Birk inhibitor. *Cancer Epidemiol. Biomarkers* 8: 601-608.

Ware JH, Wan XS, Newberne P, Kennedy AR. (1999) Bowman-Birk inhibitor concentrate reduces colon inflammation in mice with dextran sulfate sodium-induced ulcerative colitis. *Digest. Dis. Sci.* 44: 986-990.

Weaver CM, Heaney RP, Proulz WR, Hinders SM, Packard PT. (1993) Absorbability of calcium from common beans. *J Food Sci* 58: 1401–3.

Weaver CM, Plawecki KL. (1994) Dietary calcium: adequacy of a vegetarian diet. *Am J Clin Nutr* 59(suppl): 1238S–41S.

Wei H, Wei L, Frenkel K, Bowen R, Barnes S. (1993) Inhibition of tumor promoter-induced hydrogen peroxide formation in vitro and in vivo by genistein. *Nutr Cancer* 20: 1–12.

Wong JH, Ng TB. (2005) Lunatusin, a trypsin-stable antimicrobial peptide from lima beans (Phaseolus lunatus L.). *Peptides.* 26(11): 2086-92.

Wrensch MR, Petrakis NL, King EB, et al. (1991) Breast cancer incidence in women with abnormal cytology in nipple aspirates of breast fluid. *Am J Epidemiol* 135: 130–41.

Wu AH, Ziegler RG, Horn-Ross PL, et al. (1996) Tofu and risk of breast cancer in Asian-Americans. *Cancer Epidemiol Biomarkers Prev* 5: 901–6.

Xiao CW. (2008) Health effects of soy protein and isoflavones in humans. *J Nutr.* 138(6):1244S-9S.

Yamaguchi M, Hua GY. (1998) Inhibitory effect of genistein on bone resorption in tissue culture. *Biochem Pharmacol* 55: 71-76.

Yatani R, Kusano I, Shiraishi T, Hayashi T, Stemmerman GN. (1989) Latent prostatic carcinoma: pathological and epidemiological aspects. *Jpn J Clin Oncol* 19: 319–26.

Yoon JH, Thompson LU, Jenkins DJ. (1983) The effect of phytic acid on in vitro rate of starch digestibility and blood glucose response. *Am J Clin Nutr* 38: 835–42.

Zava DT, Duwe G. (1997) Estrogenic and antiproliferative properties of genistein and other flavonoids in human breast cancer cells in vitro. *Nutr Cancer* 27: 31–40.

Zhang JX, Hallmans G, Landström M, et al. (1997) Soy and rye diets inhibit the development of Dunning R3327 prostatic adenocarcinoma in rats. *Cancer Lett* 114: 313–4.

In: Progress in Food Science and Technology, Volume 1
Editor: Anthony J. Greco

ISBN: 978-1-61122-314-9
© 2012 Nova Science Publishers, Inc.

Chapter 4

ROLE OF PROCESSING INDUSTRY FOR FOOD SECURITY IN INDIA

S. Mangaraj[*], M. K. Tiripathi and S. D. Kulkarni
Central Institute of Agricultural Engineering, Bhopal, India

INTRODUCTION

Agro-processing development could make a significant contribution to the vocation of agriculture and hence, in national and international development. Agro-processing at village/rural level can expand the local markets for primary agricultural products, add value by vertically integrating primary production and food processing systems and minimize post harvest losses. Being mindful of the pitfalls and obstacles to agro-processing related development, it may be instructive to understand the concept of establishing agro-processing center and its economic perspective associated to increase income generation and sustainable rural development. The country has achieved breakthrough in agricultural production. However, level of productivity is low compared to many neighboring countries, let alone developed world. As a result margin of profit to the farmers in most of the agricultural commodities is very low. Post harvest loss prevention, value addition, and entrepreneurship development are important for higher income and rural employment generation (Shukla, 1993). Agriculture contributes 25% to annual gross domestic products and provides livelihood to more than 76% of the people. The agricultural processing sector has immense potential for value addition and employment generation. Sun drying, winnowing, paddy hulling, pulse milling, oil expelling, wheat milling, pickle making, gur and khandsari, ghee & khoa etc. are the major processing activities undertaken by the farmers. The traditional processing equipment used by the farmers include supa, chalni, chakiya, janta, silbatta, Okhli, mathani, puffing pan, mini oil ghanis/kolhus, rice hullers and flour chakkis etc. The quality of the products made by the traditional methods has been accepted and popular in the rural markets but have little scope for marketing in urban areas due varied consumers preferences. High capacity modern machines introduced in urban and sub-urban areas for processing of

---

* Corresponding Author: sukhdev0108@gmail.com

agricultural produce have helped in increasing the income of the processors. But, many of the agro-processing activities earlier being undertaken in the rural areas are now gradually being shifted to urban areas as a result rural workers are deprived of their due share of modernized agriculture. Farmers and industries mutually depend on each other for inputs and raw materials for processing. A balance has to be maintained in the agricultural development so that the farmers equally share the fruits of the higher production by involving them in primary processing and value addition. Also agro industries lead to the creation of forward and back ward linkages on large scale by maximizing complementarities of agriculture and industries (Desai, 1986).

The Food Processing Industry sector in India is one of the largest in terms of production, consumption, export and growth prospects. The government has accorded it a high priority, with a number of fiscal reliefs and incentives, to encourage commercialization and value addition to agricultural produce; for minimizing pre/post harvest wastage, generating employment and export growth. To diversify agriculture for domestic and export, agro processing-industrial development and agri-business promotion are the major opportunities. Grading, regulated ripening, packaging, proper storage, controlled temperature containers and cold chain facility, controlled atmospheric and aseptic packaging, modern material handling equipment and services are some of the issues that need scientific and technical in puts. Preference of consumers for pre-processed ready to cook (RTC) and ready to eat (RTE) convenience and packed food, value added products from fruits and vegetables provide ample scope for domestic and export. The export earning from the agricultural sector has increased to Rs.2948.586 billion which is 14.19% of the total National export. Marine products, oil meals, rice, spices, cashew, tea, sugar etc are the major export items. Export of fresh and processed fruits and vegetables has increased to Rs 978.84 crores and Rs.708.7, respectively in 2001-2002 (Alam and Singh, 2003). Onion alone accounts for 70% export. Out of the total global processed foods business of Rs.16, 000 billion, the Indian share is Rs.1400 billion (8.75%).

AGRO PROCESSING INDUSTRY

Important sectors in agro processing industries are:- Fruit & Vegetable Processing, Grain Processing, Fish-processing, Milk Processing, Meat & Poultry Processing, Packaged/ Convenience Foods, Alcoholic beverages & Soft drinks etc.

Primary processing: Purification of raw material by removing foreign matter, immature grain and then making the raw material eligible for processing by grading in different lots or conversion of raw material into the form suitable for secondary processing.

Secondary processing: Processing of primarily processed raw material into product which is suitable for food uses or consumption after cooking, roasting, frying etc.

Tertiary processing: Conversion of secondary processed material into ready to eat form.

The Broad Area of Agro-Processing Industry

The rural sector of India is rich in resources, but they are in scattered position. The socio-economic condition of the people does not allow them to use the resources to their benefit. To organize this sector, an integrated approach is required. To provide direct benefit, the agro-based processing activity at village level can play a major role. Selection of activities under the APC depends on the availability of raw material and resources. Some of the agro-based activities, which can be established in rural areas, are Primary processing industries involving operations like cleaning, grading, shorting, drying, packaging, transportation, storing, etc. The Secondary processing industries would be for value addition, milling, dehydration, product development, cooking, oil expelling, etc.

Concept of Agro Processing Centre (APC)

The agro produce processing unit established to meet the need and become a source of income and employment for rural people is referred to as agro-processing center (Shukla, 1990 and 1993). It is an enterprise where the required facilities for primary and secondary processing of certain agro produce and also the processed product are made available on rental/payment basis to the rural people. They may be managed and operated by an individuals, co-operatives, community organization, and voluntary organizations like Self Help Groups (SHGs). The center would procure and use the excess produce of a village or a cluster of villages and provide income and employment to rural people (Shukla, 1993).

Principle of Establishing APC

The basic principle for establishing APC is based on surplus production in a defined area. In general, the following model is followed for the establishment of agro processing centre:

$$S = P_t - (Un + L)$$

Where, S = Surplus produce, P_t = Total production, Un = Total utilization of produce by the population and L = Total loss during various operations.

The above model helps to decide the capacity of processing center to be installed. It helps in the selection of equipment and related material; in deciding the capital investment and its recovery time and in search of market for the produce. The management structure for proper functioning of the center should be based on the recourse available in the rural area. In rural areas, the chances of success of establishing small enterprises are more compared to the big industries. Development of agro-processing activity in villages may also pose problem because of lack of education among the population. The rural people require some important knowledge to run such processing activities.

Guidelines to Establishment of APC

Starting of a business to earn profit is the main concern of the rural entrepreneur. There are several problems related to socio-economic, political, financial, administrative and technical nature for establishing agricultural enterprises for rural people. The entrepreneurs would be required to handle systematically the complex nature of the above problems and their interactions. The following are some basic steps to start an APC:

Steps to Start an APC

Planning

i. Setting up system boundary conditions with correct definition.
ii. Collection of reliable data.
iii. Designating in-house project team.
iv. Determination of functions to satisfy the objectives.
v. Decision making about the reliable equipment, instrument and machines.
vi. Correct estimation in renovation requirement and decision taking for future modernization.

Execution

i. Selection of location
ii. Creation of confidence among the beneficiary
iii. Satisfying the social factors
iv. Finance allocation/searching availability
v. Proper utilization of available space
vi. Overcoming the problem of licensing the plant/centre
vii. Creation of industrial/business environment in rural areas/villages
viii. Collection of required materials, machines, equipment etc.

Installation

i. Installation of the machines according to the plan and processed chart.
ii. Avoiding interference of instrument, cables, walls etc. in installation of machine.
iii. Getting connection of power (electricity) and water lines.

Production

i. Training the entrepreneur to operate the machine.
ii. Production of the processed products and overall management of the center.

iii. Correct planning and schedule for production and development of feasible control procedure.
iv. Deciding the need and quality of the product to be processed.
v. Proper testing and demonstration of all equipment to the prospective users.

Implementation

i. Testing of mechanical control.
ii. Attending to the repair and maintenance needs.
iii. Ensuring availability of raw materials.
iv. Transport.
v. Deciding the demand of the products.
vi. Storage of raw materials and processed products.
vii. Getting cooperation of financing agencies.
viii. Registration of products under food acts.
ix. Keeping correct amount of asset and proper distribution of profit.
x. Creation of management information system.
xi. Problem measurement efficiency.
xii. Collective cooperation of beneficiary and proper education.

Marketing

i. Creation of market.
ii. Meeting the requirement of marketing the product.
iii. Production of superior and inexpensive products.
iv. Gearing to face market competition.
v. Avoiding delay in marketing.

Economic Perspective of APC

Table 1. Assumptions

i	Working days in a year	300 (3h/day)/ 150 (6h/day)
ii	Interest rate / year (if known)	14%
iii	Rented (room/shed/machinery, if assumed)	---
iv	Depreciation rate	10% (life of the machine 10 year)
v	Any other	
	i) Marginal Money	25%

Agriculture is considered as a technology oriented industry in India. It feeds the nation and national economy is bases on it. Most of the rural youth are going away from agriculture in search of employment in urban areas. Establishment of APC in rural areas can provide income and employment to rural youth, both men and women. It requires dedicated work and involving all feasible resource for its success. For study on the MODEL APC, the processing equipment, namely, flour mill, *dal* mill, burr mill and multi purpose grain mill has been under taken for installation for the processing of wheat, pulses (pigeon pea, green gram, masor, chickpea) and spice (turmeric, coriander and chili) grinding. The overall cost-economics was calculated considering cost, capacity and time of operation etc., to ascertain the economic feasibility in establishing APCs in the rural areas of India. The details of the procedure adopted were as follows: The marginal money means the amount of own money used for established the agro processing centre, rest of the money is acquired through loans from the agency. Here 25% of the total cost for running APC is the marginal money.

FIXED COST

Table 2. Machinery / equipment list and details

Name of the machine	Motor hp	Capacity Kg/h	No. of units	Production per day kg	Unit cost of machine Rs.	Total cost Rs.	Final Product
Flour mill	3	80	1	240	18000	18000	Wheat flour
Multi purpose grain mill	2	30	1	90	8500	8500	Turmeric powder, Coriander powder, Chili Powder
Dal mill	2	100 (50 split *dal*)	for1	300	14500	14500	Pigeon pea *dal* Green gram *dal* Masoor *dal*
Bur mill	3	60 (30 split *dal*)	for1	180	10000	10000	Chick pea *dal*
Foot/power driven grain cleaner/grader	1	500	1	1500	8500	8500	Cleaned and graded food grains
Weighing balance	---	200 kg	1	---	5000	5000	---
Packaging and sealing machine		150	1	---	5000	5000	Packed product
Metallic bin		1000 kg	1	---	1000	4000	Storage of product
Total cost of the machine						73500	

* The desirable split *dal* is produce in two passes in the *dal* mill and burr mill

* Considering expected life of the above machines/equipments are 10 years

Installation cost @ 10% = Rs. 7350/-; Total cost of machine and installation = Rs. 80850/-.

VARIABLE COST

Table 3 a. Raw material

S.No	Raw material	Rate Rs/Kg	Quantity Kg/day	Kg/year	Total cost in Rs. Per month	Per year
1	Wheat	7	240	72000	42000	504000
2	Turmeric	36	30	9000	27000	324000
3	Coriander	36	30	9000	27000	324000
4	Chili	32	30	9000	24000	288000
5	Pigeon pea	18	50	15000	22500	270000
6	Green gram	24	50	15000	30000	360000
7	Lentil	20	50	15000	25000	300000
8	Chickpea	18	90	27000	40500	486000
Total cost of raw material					238000	2856000

Table 3 b. Employment

S.No	Labors*	Quantity	Rate Rs/month	Total cost Rs/year
2	Skilled man power	1	2875	34500
3	Unskilled worker	3	7200	86400
Total cost of employment			10075	120900

* Manager, skilled labour, casual Labour, watchman, peon etc.

Table 3 c. Utilities (Electrical power)

S.No	Item	Consumption Kwh/day	No. of units	Total Cost Rs./month	Rs./year
1	Motor	24.6	4	2460	29520
2	Light	2.0	4	200	2400
3	Fan	2.23	4	223	2676
Total cost of power				2883	34596

Table 3 d. Utilities (Water)

S.No	Water requirement 1	Water Charges Rs./month	Rs./year	Total Cost Rs./year
1	100 lt.	100	1200	1200

Table 3 e. Other Expenditures (As per requirement)

S.No	Name	Quantity required/day	Rate/unit	Monthly expenditure Rs.	Total Cost Rs./yr
1	Packaging material	1.10 kg	Rs. 80/kg	2200	26400
2	Repair and maintenance	---	6% of machinery cost	477	5730
3	Transportation cost	495 kg/day	0.10/kg	1237	14844
3	Miscellaneous	---	---	500	6000
Total other expenditure				4414	52968

- Quantity of final product produced per day 495 kg
- 450 kg of material can be packed in 1 kg of polythene

Working Capital required for 1 month = (a+b+c+d) = Rs. 255472.
Land purchase cost = Rs. 40,000.
Shed / building construction cost = Rs. 1,00000.
Fixed capital investment = (Land + building) + Machinery cost + Working capital = Rs. 476322

FIXED COST PER YEAR

i. Interest / yr= 0.75 (capital investment x 0.14) + 0.25 (capital Investment x 0.09) = Rs. 60730.

ii. Depreciation/yr = (Total cost of the machinery – Salvage value)/ Life = Rs. 6615

iii.   Insurance/yr =Cost of the machine x 1.5% = Rs. 1102

Total fixed cost per year = (i + ii + iii) = Rs. 68447.
Total variable cost per year = Rs. 3065664.
Total cost of operation per year = Total fixed cost per year + Total variable cost per year = Rs. 3134111

The different quantities of final product produced from various machines in the APC are as follows:

Table 4. Quantity of final products produced in the APC

Sl. No.	Final Product	Product produced Kg/year
1	Wheat	69120
2	Turmeric powder	8460
3	Coriander Powder	8460
4	Chili Powder	8460
5	Pigeon pea pulse	11250
6	Green gram pulse	11250
7	Masor pulse	11250
8	Chick pea pulse	20250
Total quantity of final product produced per year		148500
Total quantity of final product produced per month		12375
Total quantity of final product produced per day		495

Quantity of final product produced per year = 148500 kg

The details of the cost – economics of the various machines were calculated using the above mentioned format and are follows. The selling price of the processed product was fixed considering 20% profit as compared to the cost of production of that particular product.

Table 5. Cost-economics of Agro Processing Machineries

Name of the machine	Raw material used	Cost of raw material (Rs)	Final Product	Fixed Capital Investment (Rs)	Cost of production (Rs/kg)	Break Even Point (kg)	Return on Investment (%)	Pay Back Period (Yr/Month)
Flour mill	Wheat	7.00	Wheat flour	101925	8.38	7840	109	0.50 / 6
Multi purpose grain mill	Turmeric	36.00	Turmeric powder	129058	40.00	1927	142	0.24 / 3
	Coriander	36.00	Coriander Powder	129058	40.00	1927	142	0.24 / 3
	Chili	32.00	Chili Powder	120058	35.00	2314	117	0.31 / 4
Dal mill	Pigeon pea	18.00	Pigeon pea dal	122845	25.00	3436	107	0.40 / 5
	Green gram	24.00	Green gram dal	145345	32.50	3342	112	0.33 / 4
	Masoor	20.00	Masoor dal	130345	27.00	4545	83	0.50 / 6
Bur mill	Chick pea	18.00	Chick pea dal	89442	26.00	2075	91	0.58 / 7
Pedal-cum-power operated grain cleaner cum grader	Food grains (Calculation made for wheat)	7.00	Cleaned and graded food grains (Operated for 1 h/day)	134504	7.50	11004	165	0.2 / 2.4

ECONOMIC TOOLS/ECONOMIC INDICATORS OF APC

The details of the economic tools of the agro processing centre are calculated and are as follows:

$$\text{Break Even Point} = \frac{\text{Fixed cost per year (Rs/year)}}{\text{Selling price per unit (Rs/unit)} - \text{Variable cost per unit (Rs/unit)}}$$

The Break Even point of the APC would be achieved after the production of 7840 kg of wheat flour, 642 kg of turmeric powder, 642 kg of coriander powder, 771 kg of chili powder, 1145 kg of pigeon pea dal, 1114 kg of green gram dal, 1515 kg of masor dal and 2075 kg of chickpea *dal*.
Or

The Break Even point of the APC would be achieved after the operation/running of 98 h of flour mill, 21 h for turmeric, 21 h for coriander and 26 h for chili grinding in the multi purpose grain mill, 23 h for pigeon pea, 22 h of green gram and 30 h for masor milling in the *dal* mill, 69 h of chick pea milling in the burr mill.

$$\text{Return on Investment} = \frac{\text{Net Profit per year (Rs/year)}}{\text{Fixed Capital Investment (Rs/year)}} \times 100$$

Total Revenue = Total final product x Selling Price of the product
Total cost of operation = Fixed cost + Variable cost
Net Profit = Total Revenue − Total cost of operation

$$\text{Return on Investmentment of the APC} = \frac{\text{Net profit from all the machinaries}}{\text{Fixed capital investment of the APC}} \times 100$$

= (111353 + 61466 + 61466 + 47088 + 44192 + 54485 + 36373 + 81993) X 100/ 476322
= 498416 x 100 / 476322 = 104 %

$$\text{Pay Back Period} = \frac{\text{Fixed Capital Investment (Rs/year)- Working Capital (Rs/ period)}}{\text{Net Profit (Rs/year)}}$$

= (476322-255472) / 498416 = 0.44 Year = 5.3 Months

Status of Agro-Processing Industry

Food items are marketed in different forms as raw, primary processed, secondary processed and tertiary processed. The farmers in general prefer to sell their agricultural produce immediately after harvest leaving a part for own consumption and seed purposes. Indian Grain Storage and Management Research Institute, Hapur has estimated that the farmers retain 44% of the total wheat and 48 % of the paddy. Mandies and grain traders procure the balance for processing and/or for marketing. The agro produce at rural level is stored *in Kothi, Kuthla*, gunny bags, steel barrels etc to avoid losses. The country has buffer stock of more than 60 million tones. The Government agencies like Food Corporation of India and Central and State Warehousing Corporations store them in silos, godowns, and cap storage structures. The total storage facility available with FCI is 27 million tonnes and with CWC 18 million tonnes for all kind of goods. About 16-25 million tonnes of storage space is required for rice and wheat alone at a time to keep food grains buffer stock for public distribution system. Inadequate storage leads to huge losses due to damage by rain and rodents.

Table 6. Investment in processing food sector

Processing sector	Total industrial approval Nos.	Investment Rs. Crores	Foreign Investment Rs. Crores	Foreign Investment share, %
Grain milling & products	472	6879	1207	17.55
Fruits & vegetables products	2144	8491	1089	12.83
Meat & poultry products	148	1778	267	15.02
Fish processing and aquaculture	305	2688	548	20.39
Fermentation industry	883	11374	1094	9.62
Consumer industry	931	13252	4423	33.38
Milk and milk products	1130	14833	1004	6.77
Other including food beverages	71	934	535	57.28
Edible oil and oil seeds	1675	13416	---	---
Total	7759	73645	10167	13.81

Souce: MOFPI, 2001-02

Table 7. Status of Post Harvest Equipments in India

Name of the post harvest equipment	Number (1991)	Number 2004	Projected for 2010
Flour Units	2,66,000	4,00,000	4,25,000
Cleaners and Graders	1,10,000	3,80,000	4,50,000
Dryers	7,000	25,000	50,000
Rice Mills	1,30,000	1,50,000	1,60,000
Maize shellers	65,000	1,18000	1,25,000
Dal Mills	10,000	20,000	30,000
Oil expellers	2,22,000	4,50,000	5,50,000

Source:
1. Data Book on Mechanization and Agro-processing since Independence, 1997, CIAE, Bhopal
2. Website of Ministry of Food Processing Industry, Annual Report 2002-03.
3. Souvenir of International Conference on Emerging Technology in Agricultural and Food Engineering, held at IIT, Kharagpur during Dec 14-17, 2004

The food-processing sector in India has gained importance due to consumers preferences for ready to cook (RTC) and ready to eat (RTE) foods, besides increased demand for snack foods and beverages. According to a CII-Mc Kingsey report the size of India's food industry

Role of Processing Industry for Food Security in India 99

is estimated at Rs.250, 000 Crores which is expected to double by 2005. Of this, value added processed foods are forecasted to rise three times from the present Rs.80, 000 Crores to Rs.225, 000 Crores during the same period. As much as 42% of the food industry is in the organized sector and 33% in the small scale, tiny and cottage sectors. Since liberalization in 1991 more than 7759 licenses have been issued to food processing sector till November 2001 with a total investment of Rs.73, 645 Crores of which foreign share was Rs.1067 Crores (13.81%) (Table-6). About Rs. 3162 crores foreign investment has already been made till December 2002 (Alam and Singh, 2003) in this sector. Consumer products, grain milling, fermentation and fruits and vegetables are the main sector that has recorded higher investment. Table-7 gives status of use of processing machinery that improves the marketing quality of the agro-produce.

SECTOR WISE AGRO-PROCESSING

Rice Milling

The food grains production has increased to more than 211 million tones of which share of paddy is 90.75 million tonnes. There are a total of 1,50,000rice mills including 35,088 modern-cum-modernized rice mills (Table-8). Conventional rice hullers with a population of 91287 are very popular for milling of rice in rural areas. In conventional rice hullers, bran and husk are produced together and cannot be separated. The by-product is generally burnt. The modern rice mills have separate processing mechanism for dehusking and polishing of the paddy. The husk can be utilized for energy and for industrial products like furfural, and the bran for extraction of edible and non-edible grades oil. These mills also have better recovery and lower energy consumption compared to conventional hullers (Table- 9). The modernization of rice mills in India started in 1970s. Andhra Pradesh (12995), Tamil Nadu (3922), Karnataka (3674), Kerala (2533), Punjab (1965), Maharashtra (1759), Madhya Pradesh (1761), Uttar Pradesh (1415), Gujarat (1045), Haryana (990) and West Bengal (926) have the largest number of modern-cum-modernized rice mills. As a result of adoption of modern rice mills the production of bran has increased to 3.4 million tonnes from which 5.25 lakhs tonnes oil has been recovered of which about 3 lakhs tonnes is edible grade oil.

Table 8. Rice and dal milling industries

Type	Number 2004	Projected 2010
Huller	90,000	95,640
Sheller	5,000	6005
Huller-cum-Sheller	8385	8500
Modern rice mill	35088	38400
Dal mill	20,000	30,000

Increased export of Basmati rice has increased the foreign investment in installation of modern rice mills with hydraulic jet polishing, and colour sorters that provide better quality grains. The installation of rice mills in the country was regulated by Rice Milling Industry (Regulation) Act 1958 and Rice Milling Industry (Regulation and Licensing) Rules 1959;

with effect from May, 1997 these have been repealed as a result no licensing is required. Use of small capacity modern rice mills in rural areas may increase the head rice recovery and the by-product could be utilized for oil extraction and as a cattle feed.

Table 9. Performance of rice mills

Rice mills	Yield recovery %	Grain breakage %	Energy consumption KW/tonne
Huller	65	20	40
Sheller	68	12	-
Modernized	71	10	20

WHEAT MILLING

Wheat production in the country has increased to more than 73.53 million tonnes. Burr mills (*chakki*) with a population of 2,66,000 units are very common for milling of wheat in rural and urban areas (Table-10). These mills process about 80 million tonnes of wheat, gram etc. and provide fresh flours for making *chapatti*. Besides additional income, these mills provide large-scale employment to the rural people. There are about 750 modernized roller mills with a capacity to process 19.5 million tonnes. Uttar Pradesh, Maharashtra, Karnataka, Tamil Nadu, Andhra Pradesh, Bihar, West Bengal, Punjab, haryana, Madhya Pradesh, Assam, Gujarat, kerala have the largest number of roller folur mills. At present flours made by the roller mills are sold to institutional buyers like defence, hotels etc, and the household purchase is limited to only 1% due to absence of open policy for sale of wheat flour through public distribution system. The cost of flour processed by roller mills is also high due to use of obsolete machinery that uses high energy. Sale of soy blended and branded wheat flour is likely to increase due to better quality flour, and thus scope of organized wheat milling will increase in future.

Table 10. Wheat processing mills during 2009

Type	Processing capacity	Number 2004	Projected for 2010
Rollers mills	10	750	1000
Chakki	80	2,66,000	3,00,000

Pulse Milling

Pulses are the major sources for protein for the vegetarians in India. In a total world production of 60 million tonnes of pulses, India China, Brazil, Turkey, and Mexico account for 2/3rd of the output. The production in India is about 14 million tonnes. The pulses are made into *dal* by dehusking and dehulling. *Dal* milling used to be a domestic activity in the rural areas but now around 20,000 *dal* mills have come up with a capacity to process 14 million tonnes. These mills are processing more than 10.5 million tonnes of *dal*. The by-product from these mills is converted into cattle feed. Pulse milling was earlier reserved for small-scale sector has now been de-reserved, as a result, no licensing is required for setting up of pulse mills. Traditional designs of dal mills require immediate modernization as they pollute the environment with fine particles and dust besides, have low dal recovery. The Government provides financial incentive for installation of dust control devices and drier.

OIL EXTRACTION

The Technology Mission on oilseeds has helped in increasing the oilseeds production to 24.5 million tonnes. Oil extraction has been a cottage level activity in the country through *Kolhus* and *Oil Ghanis*. The country has about 2.5 lakhs *ghanis and kolus*, 50,000 mechanical oil expellers, 15,500 oil mills, 843 solvent extraction plants, 300 oil refineries and over 175 *Vanaspati* plants (Singh and Alam, 2003). The introduction of high capacity mechanical expellers and solvent extraction technology has brought in modernization. Small capacity oil expellers have been developed which could be installed in rural areas for promoting agri-business and that might provide more employment. Soybean is not only a good source of oil but also rich in protein. The cultivation has increased to more than 6 million hectares. India is now the fifth largest producer of soybean at a global level with more than 5.2 million tonnes; there are more than 155 solvent extraction plants and 64 soy food-manufacturing units. Soymilk analogues, nuggets, and soy-blends are being marketed. With value of export of oilcake increasing to more than Rs.2231 Crores, soy meal is one of the major oilseed cake that can be utilized for animal and poultry feeds.

Processing of Commercial Crops

Sugarcanes, jute, cotton, tea, coffee and tobacco are major commercial crops grown in India. The country has made remarkable progress in production of these commercial crops. The production of sugarcane has increased to more than 299 million tonnes in 2001-02. About 50 per cent sugarcanes is estimated to be processed by sugar mills and the balance by small scale Gur & Khandsari units. Although, the efficiency of Gur & Khandsari sector is low compared to sugar mills, but these units provide more employment opportunities to rural work force and therefore, cannot be ignored and requires special attention. Improved sugarcane crushers and furnaces have been developed for producing hygienic *gur* for domestic and export market.

FRUITS AND VEGETABLE PROCESSING

India is the world's second largest producer of fruits and vegetables. It has potential to grow all types of temperate, sub-tropical and tropical fruits and vegetables because of varied agro-climatic diversity. The total production of fruits and vegetable is 45 million tonnes and 85 million tonnes respectively in 2002-03 (Agricultural Research Data Book, 2005). The losses are estimated to the extent of 20-40% due to lack of proper harvesting, processing and storage facilities, which is valued at Rs.230 billion as per the estimate of Planning

Commission. The crop wise losses in percentage are; apple-14, banana-20-80, lemon and orange-20-95, grapes-27, papaya-40-100, cabbage-37, cauliflower-49, onion-16-35, and potato-5-50. Simple techniques to increase shelf life like wax coating and polythene/ cellophane packaging will reduce the losses considerably. These activities can also be undertaken at the farm level and therefore will reduce the cost of transport, besides providing employment opportunities. The other priorities in the horticultural sector are; grading, regulated ripening, packaging, proper storage, controlled temperature containers and cold chain facility. Modern technology such as colour grading, osmotic dehydration, cold and cryo-grinding, etc is yet to be adopted in large scale by the Indian industries. The total storage facility available in the country for the perishables is 8 lakhs tonnes. There is a plan to create additional cold storage facility of 12 lakhs tonnes and 4.5 lakhs tonnes for onion. The CWC plans to have a controlled atmospheric (Oxygen, Nitrogen and Carbon) storage of 8000 tonnes in Sirhind in Punjab.

People generally prefer fresh fruits and vegetables in India due to abundance of seasonal fruits through out the year available at low price. The production of *pickles & chutneys* has traditionally been rural level cottage industrial activity. However, in the recent years, processed foods in the form of canned fruits such as pineapple, litchi, mango slices and pulps, grapes, apple, peaches etc have increased considerably. The uses of fruits in the form of concentrated juice, dry powder, jam and jelly have also increased. The percentage of processes fruits and vegetable is given in Table-11.

Table 11. Percentage of production of processed fruits and vegetables

Processed products	Percentage (%)
Fruits juice and fruit pulp	27
Jams and jellies	10
Pickles	12
Ready to serve beverages	13
Synthetic syrups	08
Squashes	04
Tomato products	04
Canned vegetables	04
Others	18

The estimated installed capacity of fruit and vegetable-processing industries has increased from 20.8 lakh tonnes in 1998 to 21.0 lakh tonnes in 1999-2000 (Fig.1). The number of license issued under Fruit Product Order (FPO), 1955 has increased to 5198 till 2004 (Table-12). But the production of processed product increased from 0.24 million tonnes to 1.03 million tonnes only, recording a growth rate of 14% per annum. This is only 0.8% of the total fruits and vegetables produced (130 million tonnes) in the country as against 83% in Malaysia, 80% in South Africa, 78% in Philippine, 70% in Brazil and 65% in USA. The main fruits that enter the export market are mangoes, grapes, apples, citrus but other fruits identified for export are bananas, sapota, litchis etc. The total exports of fruits and vegetables are about 933.654 thousand tonnes valued at Rs. 10867.5 million (2002-03). The main destinations for export of fruits being Middle East, U.K, Europe and to some extent Singapore, Malaysia etc. The important vegetables exported are potatoes (28.0%) onions (7.1%) cauliflower and Cabbage (4.0% each), okra (3.0%), peas (3.0%), Other (50.0%). The exports are limited to Middle East, Europe, U.K. and Singapore etc. There has been significant increase in the investment in fruits and vegetables processing sector during the last 10 years. It has increased to Rs.8491 Crores of which 12.83 % is foreign investment. The processing of fruits and vegetables is regulated by Fruits Products Order (FPO) 1955 under essential commodities Acts. All processing units are required to obtain licence. The Government of India in the budget for 2001-02 has exempted the processed products of fruits and vegetables completely from Central Excise Duty of 16 %. This is likely to increase the growth further. At rural level solar assisted dehydrators could be promoted for preparation of ethnic food products like resins, mango papad, onion flakes and powder, chips, vegetables etc.

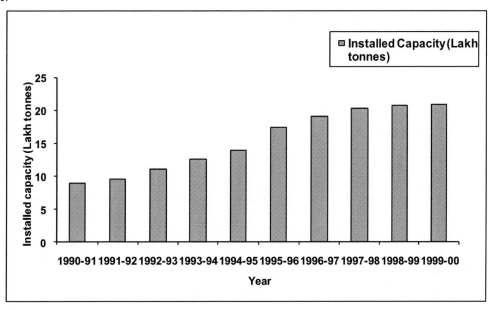

Figure 1. Installed capacity of fruit and vegetable processing units.

(Source: Ministry of Food Processing Industries).

Table 12. Fruit and Vegetable processing licensed industries

Year	Number
1990	3,629
1991	3,846
1992	3,925
1993	4,057
1994	4,132
2004	5198
2010	6200

PACKED AND CONVENIENCE FOOD

Modern packed and convenience foods such as bread, biscuit, confectionary, chocolates, ready to eat like nodules, cereal flakes etc have become popular in recent years especially in urban areas (Table-13), although traditional foods have been used in the country in the form of roasted, puffed, sweat meat and baked products. There are more than 60,000 bakeries and 20,000 traditional foods units. The production of bread and biscuits has increased to 3.70 million tons/year of which about 40% is produced through the organized large-scale sector and the remaining 60% in the small scale and unorganised sector. About 1.15 million tonnes of biscuits are manufactured in the organized and about 0.38 million tonnes in the unorganised sector. The output of flakes has increased to 14,000 tons/year. The extruded foods are largely produced in the unorganised sector. The traditional ethnic ready to eat (RTE) foods prepared in hygienic conditions and marketed with better packaging has plenty of domestic and exports market.

Table 13. Packed and convenience food

Food items	Production, tonnes/year (2001-02)
Macaroni/ noodles	16 500 tonnes
Pearled barley	1240 tonnes
Corn flakes	14000 tonnes
Bread	2.17 million tonnes
Biscuits	1.53 million tonnes

Source: Ministry of Food Process Industries, 2001-02

Fishing and Fish Processing

The fish production of India has increased from 3.8 MT in 1990-91 to 6.2 MT in 2002-03 (Agricultural Research Data book, 2005). With the liberalized policy, fish-processing sector has been attracting more foreign investments. Processed IQF marine products fetch better price than conventional block frozen materials in the foreign markets. Different IQF products suitable for export are: shrimp, lobster, cattle fish, clams and fish fillets. Marine fishes found in India include prawns, shrimps, tuna, cuttlefish, squids, octopus, red snappers, ribbon fish, mackerel, lobsters, cat fish and countless other varieties. Sixty per cent of the production of fish in India is from marine sources. Processing of produce into canned and frozen forms is carried out almost entirely for the export market (Fig.2) (Agricultural Statistics at a Glance, 2004). Processed fish products for export include: conventional block frozen products, individual quick frozen products (IQF), minced fish products like fish sausage, cakes, cutlets, pastes, surimi, texturised products and dry fish etc..

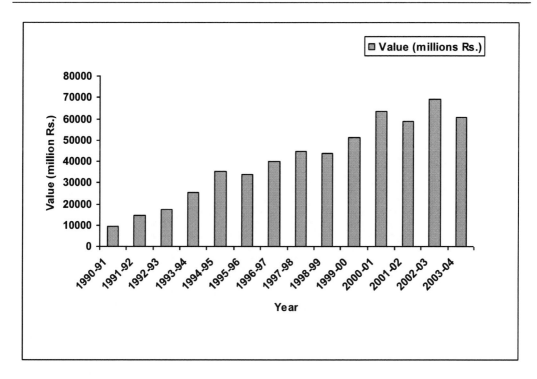

Figure 2. Export of marine product from India.

Meat and Poultry Processing

The production of meat and meat products has shown an impressive growth. The details of production of meat and meat products from 1994 to 1998 is given in Table-14.

Table 14. Production of meat and meat product (in thousand tonnes)

Products	1994	1995	1996	1997	1998
Mutton and Goat Meat	637	647	669	670	675
Pork Meat	366	420	420	420	420
Poultry Meat	422	578	480	580	600
Cattle Meat (Beef)	1290	1292	1202	1292	1295
Buffalo Meat	1200	1204	1204	1205	1210

There is a large potential for setting up of modern slaughter facilities and development of cold chains in meat and poultry processing sector. The market has not been taken tapped tally for ready-to-eat and semi-processed meat products in the domestic market as well as for exports to neighboring countries especially to the Middle East (Fig.-3). Buffalo meat is surplus in the country and has good export potential.

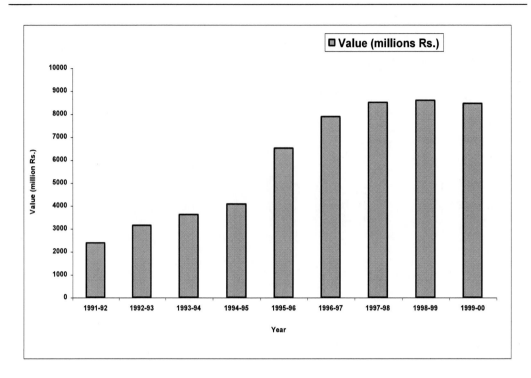

Figure 3. Export of meat and poultry products from India.

Poultry production and egg processing industries have come up in the country in a big way and are exporting egg powder, frozen egg yolk, albumin powder to Europe, Japan and some other countries. Meat products have a growth rate of 10% whereas the growth rate of eggs and broilers are 16% to 20% respectively. Most of the production of meat and meat products continues to be in unorganised sector. However, some branded products have also come up in the domestic market. At present, poultry export from India is mostly to Maldives and Oman. Some other markets can be explored for export of poultry meat products like Japan, Malaysia, Indonesia and Singapore.

Milk and Milk Products

India has one of the largest livestock populations in the world. Fifty percent of the buffaloes and twenty percent of the cattle in the world are found in India, most of which are milch cows and milch buffaloes. India is the second largest milk producing country with the production of 87.3 Million tonnes in 2002-03. The milk surplus states in India are Uttar Pradesh, Punjab, Haryana, Rajasthan, Gujarat, Maharashtra, Andhra Pradesh, Karnataka and Tamil Nadu. The manufacturing of milk products is concentrated in these milk surplus States. The production of milk products i. e milk products including infant milk food, malted food, condensed milk & cheese stood at 3.07 lakh tonnes in 1999-2000. Production of milk powder including infant milk-food has risen to 2.25 lakh tons in 1999-2000, whereas that of malted food is at 65000 tons. Cheese and condensed milk production stands at 5000 and 11000 tonnes respectively. Some plants are coming-up for producing lactose, casein and improved cheese varieties.

FUTURE PERSPECTIVES

Majority of the rural people earn their livelihood through agriculture. Agriculture production and processing provide employment and income. The Indian climate is favourable for growing any temperate and tropical crops. The fresh and processed products have very good domestic and export markets. Increase in self-life and value addition of agro-produce, and rural agri-entrepreneurial development are important. The Indian agriculture and industries will have to play a major role in producing quality agro-produce meeting internationally acceptable quality, hygiene and packaging. Reducing post harvest losses, regulated ripening, precision cleaning, grading, controlled temperature containers, controlled atmospheric and aseptic packaging, cold chain facility, modern material handling equipment and services are some of the issues that need scientific and technical in puts. The NCAER study in 2001 has estimated an investment of Rs.38,531 Crores in food processing sector during 9th plan. With a growth rate of 8% against the achieved growth rate of 8.5% during 1996-97 to 2001-02, the estimated investment during 10th plan may increase to Rs.92, 208 Crores. Agro-processing in rural catchments areas may equally have to be promoted to increase the employment and income of rural people. But that would require infrastructure like assured electricity, roads, communication, and marketing network. The future perspectives in agro-processing therefore may include;

Diversification of Agriculture

To promote agri-business, commercial agriculture need to be given more emphasis for production of quality produce under controlled environment. In the field of medicinal and aromatic plants there is lot of scope for export of perfumery, food flavour, cosmetics and toiletry goods. Better methods of collection, curing, cleaning, grading, packaging etc. for quality products are required as per international standards.

Enhancing Export of Value Added Processed Foods

As a result of GATT agreement and the quality standards stipulated under the agreement of WTO, the Indian agriculture and industry will have to play a major role in producing quality agro-produce meeting internationally acceptable quality, hygiene and packaging. To increase the export of processed oil meals, cereals, fish and marine products, fruits and vegetables, cut flowers, tea, coffee, spices, textiles, etc.

Modernized Handling and Processing of Horticultural Produce

Processing technology is needed which will help in maintaining quality of produce in terms of appearance, colour, texture, flavour and taste. The priorities in the horticultural sector are; grading, regulated ripening, packaging, proper storage, controlled temperature containers and cold chain facility, controlled atmospheric and aseptic packaging, modern material handling equipment and services. Simple techniques to increase shelf life of fruits like wax coating, polythene/cellophane packaging etc. will reduce the losses considerably which should be adopted at farm level.

Agro-Processing and Other Agro-Based Industries in Rural Areas

For promotion of rural employment primary processing facilities need to be created in rural areas for on farm processing of agricultural commodities. This will render enhanced availability of raw materials for processing in the catchment areas at reduced cost, and better utilization of by-products after value addition as animal feed, compost, biogas feed etc

AGRO PROCESSING TECHNOLOGY

Drying and Dehydration Technology

Sun drying to reduce the moisture of the produce to an accepted level is the most common method of drying for long storage but the the quality of the produce especially fruits and vegetables is affected. Numerous designs of solar and mechanical dryers in different capacities are available. These dryer need to be extended for on-farm use. Osmotic dehydration and freeze-drying technology need to studied and promoted for few specific crops.

COLOUR SEPARATOR/OPTICAL GRADER

The colour separator separates the fruits, vegetables or grains due to difference in colour orbrightness. The material is compared with a selected background or colour range, and is separated into fractions according to difference in colour.

CEREAL PROCESSING

Economical small capacity modern rice mills for better head rice recovery and clean bran for animal feed and/or oil recovery is required. Energy efficient flourmills are also required at rural level for promotion of agro processing activities as demand for branded flour has increased. Dry milling of maize, pearling of coarse grains (sorghum, bajra, etc.) has been undertaken at low scale. Pearling of coarse cereals that removes the husk from the outer layer improves the quality. Economical size equipments need to be developed for adoption by small-scale industries. Puffed/roasted breakfast and snack foods using cereals and pulses and their blends may increase the agri-business. These are already popular in the country but prepared in traditional manner in unorganised sector.

Processing of Oilseeds

Oil extraction used to be a cottage level activity in the country through *Kolhus* and Oil *ghanis*. The high capacity mechanical expellers and solvent extraction technology has brought in the modernization of oilseeds, and tiny oil mills located in the rural areas are more or less getting disappeared. The system leaves 10-15 % oil in the cake. Screw press or mechanical oil expellers leaves 6-8 percent oil in the cake. For better quality oil meal, which is major source of export earning, it is widely felt that edible oil extraction industry needs urgent modernization to improve its level of efficiency particularly quality of oilcake.

Bio-Processing

Processing of fruits, vegetables and food grains so far has been confined through chemical and physical processing and to certain extent use of bio-processing was undertaken for animal products, and alcohol. Bio-technology provides immense scope to improve the quality of agro-produce and therefore may have to be addressed.

Modernized Handling and Processing of Horticultural Produce

The priorities in the horticultural sector are; grading, regulated ripening, packaging, proper storage, controlled temperature containers and cold chain facility, controlled atmospheric and aseptic packaging, modern material handling equipment and services. Simple techniques to increase shelf life of fruits like wax coating, polythene/cellophane packaging etc. will reduce the losses considerably which should be adopted at farm level.

Storage Technology

In adequate storage facilities with the Government Sector is causing avoidable huge losses. Scientific on farm storage of food grains in storage bins, or even CAP storage provides an opportunity to the farmers to store the produce and to reduce distress sales. Decentralized storage of food grains in catchments areas may provide some relief for safe storage and reduced handling losses. Not much scientific and economic analysis data is available to support the proposal. The scientists therefore need to have techno-economic study for decentralized storage. Modified Atmosphere Packaging (MAP) and Controlled Atmosphere Storage have prospects for perishables.

EXPORT PROMOTIONAL MEASURES

The Central Government has so far sanctioned 41 Agri Export Zones (AEZ) across the country through an integrated and focused approach. Entailing an investment of around Rs. 1143 crores, these AEZs would lead to a projected export of around Rs. 8595 crores during the next 5 years. For export of onion AEZ has been proposed for the state of Maharastra, which is one of the major onion producing area with around 25% of all production and nearly 70% of all exports from the country. The state of Madhya Pradesh presents an enormous opportunity to export quality spices, Durum and organic wheat from the region owing to its largest area of production. The zone covers the districts of Ujjain, Ratlam, Mandsaur, Neemuch, Indore, Dhar, Shajapur, Dewas, Bhopal, Sehore, Vidisha, Raisen, Hoshangabad, Harda and Narsinghpura. Similarly, Orissa offers a huge base for export quality ginger and turmeric Uttaranchal has about 60,000 hectare of area under Basmati cultivation and produces about 1.8 Lakhs MT of Basmati Rice. The state produces one of the premier varieties, having the distinctive long grains and subtle aroma. Malda is one of the major mangos growing belt in West Bengal, which ranks fourth in terms of area and seventh in terms of production in the country. Tamil Nadu aims to promote the exports of fruits and specially Mango from the districts of Madurai, Theni, Dindigul, Virudhunagar, & Tirunelveli. West Bengal and Jharkhand have substantial marketable surplus for traditional vegetables and vegetables like Okra, Bitter gourd and Green Chili covering the districts of Nadia, Murshidabad and 24 parganas in West Bengal and Ranchi, Hazaribagh and Lohardaga in Jharkhand.

REFERENCES

Agricultural Research Data Book. 2005. *Indian Agricultural Statistics Research Institute.* Indian Council of Agricultural Research, New Delhi.

Agricultural Statistics at a Glance. 2004. *Directorate of Economics and Statistics, Ministry of Agriculture*, Government of India.

Alam, A. and Singh, G. 2003. *Status and future needs of farm mechanization and agro-processing in India.* Central Institute of Agricultural Engineering (ICAR), Bhopal.

Desai, P.B. 1986. *Relevance of population change for balanced development of Industry and Agriculture.* The Eight World Economic Congress, New Delhi. Vol. 17. p4.

Iqbal, B.A. 1951. *Financing agro-Industries in U.P.* Since 1951. Aligarh.p3.

Mangaraj, S. and Singh, R. (2006). Concept and guideline for establishing agro processing canter and its economic perspectives. *Agricultural Engineering Today*. 30 (1&2): 64-70.

Mangaraj, S. and Singh, R. (2008). Status of food processing industry in India and its future outlook (technique review). *Indian Food Industry,* 27 (3): 35-42.

Sahay, K.M. and Singh, K.K. (1994*). Unit operations of agricultural processing.* Vikas Publication House PVT Ltd.

Shukla, B.D. 1993. Problems and Priorities in establishing agro-processing centres. *Seminar on Appropriate Technologies for Agro-processing.* Abohar CIPHET Campus. June 28-29,1993.

Singh, G. 1997. *Data Book on Mechanization and Agro-processing since Independence,* CIAE, Bhopal

Souvenir of International Conference on Emerging Technology in Agricultural and Food Engineering, held at IIT, Kharagpur during Dec 14-17, 2004.

Website of Centre for monitoring Indian Economy, 2002-03.

Website of Ministry of Food Processing Industry, Annual Report 2001-02.
Website of Ministry of Food Processing Industry, Annual Report 2002-03.

In: Progress in Food Science and Technology, Volume 1   ISBN: 978-1-61122-314-9
Editor: Anthony J. Greco   © 2012 Nova Science Publishers, Inc.

Chapter 5

CHARACTERISTICS AND UTILIZATION OF BLACK RICE

Bipin Vaidya and Jong-Bang Eun[*]

Department of Food Science and Technology and Functional Food Research Center,
Chonnam National University, Gwangju, Korea

INTRODUCTION

Rice cultivation feeds more than half of the global population, as it is found on all continents except Antarctica. Among the 24 recognized species of rice, *Oryza sativa* and *Oryza glaberrima* are the most cultivated. Asian cultivated rice, *Oryza sativa*, has migrated across the globe with the human population, whereas *O. glaberrima* is cultivated only in certain areas of western Africa [1]. A few varieties of *Oryza sativa* have unique characteristics in terms of their aroma, kernel color, and chemical composition [2]. Among them, pigmented rice is a unique variety and has been consumed for a long time in Asia, especially in China, Japan, Korea Laos, Thailand, and Burma. Pigmented rice with black and red colored pericarp is also planted in places such as the United States, Italy, Greece, and Africa. Among these varieties, the most famous is black rice, which is generally used as an ingredient in different snacks and desserts. Rice with colored pericarp is considered to be a health food. For example, whole grain pigmented rice is categorized as a potent functional food since it contains high amounts of phenolic compounds, especially anthocyanins in pericarp [3, 4].

China, which is responsible for 62% of global production of black rice [2],, has developed more than 54 modern black varieties with high yield characteristics and multiple resistances. Nutrients such as protein, minerals (Ca, P, Fe, and Zn) and dietary fiber content are also higher in black rice [5]. Demand for "black rice" is growing fast in the USA and EU due to its value as a health food and organic food color.

---

[*] Corresponding Author: Jong-Bang Eun, Department of Food Science and Technology and Functional Food Research Center, College of Agriculture & Life Science, Chonnam National University, 300 Yongbong-dong, Buk-gu, Gwangju, Korea, Tel: +82-62-530-2003 (2145); Fax: +82-62-530-2004 (2149), email: jbeun@chonnam.ac.kr

VARIETIES

The number of rice varieties is great; there are 85,000 varieties in the research stocks of the International Rice Research Institute (IRRI, the Philippines) and estimated over 1,20,000 cultivars available in world, a number so large that the current international nomenclature system is insufficient [6]. All black rice are categorized as either *O. sativa* or *O. glaberrima* and can be found in the *indica* and *japonica* sub-species of the *sativa* species.

O. sativa; a common rice of Asian origin is the predominant type of black rice. However, *O. glaberrima* is still grown by Saramaccan Maroons as food and for ritual use. Saramaccan informants claim that their forefathers first collected the "black rice" from a mysterious wild rice swamp and cultivated the seeds afterwards [7].

Common varieties of popular Korean black rice are the C3Hi line, Dragon Eyeball, Heugjinmi, Heughyang, Heugjinjubyeo, Heugnambyeo, Heuggwang, Ilpumbyeo, Jeonbuk, Josaengheugchal, Jeonbuk, Kilimheugmi, Sanghaehyanghyeolla, Suwon 405, Suwon 420, Suwon 425, Suwon 415, and Suwon 512.

PHYSICAL PARAMETERS

With respect to the physical properties of kernel, weight, length, volume of waxy black rice are higher than those of waxy rice but width and density are similar [**Table 1**] [8].

Table 1. Physical parameters of waxy and black rice[a]

Varieties	Weight (mg)	Length (mm)	Width (mm)	L/W	Volume (mm^3)	Hardness (kgf)	Density (mg/mm^3)
Black rice	24.8±0.4	5.8±0.3	2.1±0.1	2.8±0.2	103.5±11.4	13.8±3.0	0.243
Waxy rice	21.7±0.3	5.1±0.2	2.0±0.1	2.5±0.1	87.5±6.9	12.9±2.3	0.248

[a]Adopted from ref (*8*)

PROXIMATE COMPOSITION

The proximate composition of black rice is presented in **Table 2**. Protein, lipid, ash, and carbohydrate contents in different varieties of rice are in the range of 10.30-12.06, 2.92-3.40, 1.65-1.87, and 83.21-84.62, respectively [9].

Black rice, Heugjinjubyeo showed the highest concentration of protein (14.1±0.1%), whereas Jeonbuk 2 displayed the lowest (10.7±0.0%). However, protein contents are not significantly different between the Korean black rice varieties [**Table 3**]. Korean black rice varieties contain higher protein content, i.e. 12.0% on average, compared to other rice varieties having approximately 7.0% protein content. Oil contents range from 2.1 to 2.9 in black rice. Among the black rice varieties, Jeonbuk 2 shows the highest lipid content while Heugnambyeo exhibits the lowest [10].

Table 2. Proximate composition (%w/w, dry basis) of black rice[a]

Varieties	Crude protein	Crude lipid	Crude ash	Carbohydrate[b]
Suwon 415	11.52	3.40	1.87	83.21
Iksan 427	10.30	3.30	1.78	84.62
Sanghaehyanghyella	12.06	2.92	1.65	83.37

[a]Adopted from ref (9), [b]Calculated by subtracting protein, lipids and ash content

Table 3. Protein and lipid content of Korean black rice varieties[a]

Varieties	Protein	Lipid
Jeonbuk	10.7±0.0	2.9±0.0
Heugjinjubyeo	14.1±0.1	2.3±0.0
Heugnambyeo	11.9±0.1	2.1±0.0
Sanghaehyanghyeolla	12.0±0.6	2.5±0.2
Josaengheugchal	11.8±0.1	2.6±0.0
Dragon Eyeball	10.7±0.3	2.2±0.0
Heughyang	11.6±0.1	2.3±0.0
Jeonbuk	12.2±0.2	2.5±0.0
Heuggwang	12.5±0.1	2.5±0.0
Suwon 512	12.6±0.1	2.6±0.1

[a] Adopted from ref (10)

Carbohydrates and Starch

Amylose content varies significantly among the black rice varieties. Suwon 415 contains 19.6% amylase content, whereas Iksan 427 and Sanghaehyeolla contain very low amylase content, i.e. 0.4 and 0.6%, respectively (9).

The major particle size of black rice starch is 6.27 microns. In an iodine reaction, the maximum absorbance wavelength (λ max) and absorbance at λ max were 521 and 0.184, respectively. Intrinsic viscosity and water holding capacity at 25oC were observed to be 183 mL/g and 103.3%, respectively. The swelling power of black rice starch increases from 4.7 to 53.0 at 55 and 90oC, respectively, and solubility from 0.9 to 12.2 at 55 and 90oC, respectively [11]. The swelling power, solubility, and degree of acid hydrolysis of black rice starch are all lower than those of white rice [12].

The hot water soluble content of black rice starch is 16.6% when heated at 98°C for 8 min. The debranched limit (by β-amylase) of raw starch is 63.4 [13, 14]. The water binding capacity of waxy black rice flour is higher than those of other glutinous rice flour while the solubility of black rice flower is lower than those of waxy glutinous rice flour [15]. The water uptake rate constant of black rice increases during storage in the temperature range of 20~50°C up to 40 mins of soaking time. The water uptake rate constant of black rice (0.033 $min^{-1/2}$) is lower than those of other waxy rice (0.058 $min^{-1/2}$) during storage at 4°C. The hardness reduced rate constant of black rice (0.022 min^{-1}) is lower than that of waxy rice (0.048 min^{-1}) during storage at 4 °C [8]. Additionally, the water uptake rate constant and hardness reduced rate constant of black rice were lower than those of glutinous rice. Using a Rapid Visco Analyzer, the peak viscosity of black rice was found to be lower than glutinous rice at a level of 4 g [16]. The crystalline regions of black rice starch disappear when the temperature is increased to 65°C, and the starch changes into an amorphous form [17].

The total dietary fiber contents of the Korean black rice varieties Kilimheugmi, Sanghaehyanghyella, Heuginjubyeo, and C3Hi are 5.5, 5.8 4.4, and 5.8, respectively [12]. However, Kilimheugmi contains a high amount (7.5%) of dietary fiber, 5.80% hemicellulose, and 1.8% uronic acid on average, which are more than double compared to white glutinous rice [18].

Protein and Amino Acids

Crude protein content of black rice (Sanghaehyanghyeolla variety) flour is reported to be 9.01, which is higher than that of glutinous rice flour (7.54%)[15]. The major amino acids are glutamic acid, aspartic acid, and arginine in the Sanghaehyanghyeolla variety [**Table 4**] [15].

Table 4. Amino acid content (mg %) of black rice[a]

Amino Acids	Suwon 415	Iksan 427	Sanghaehyanghyella
Glutamic acid	1949	1533	1651
Aspartic acid	614	410	596
Arginine	1088	876	860
Leucine	1129	1052	886
Alanine	745	684	579
Valine	837	784	668
Serine	342	169	224
Phenylalanine	683	609	535
Glycine	684	578	524
Proline	724	675	617
Lysine	495	443	377
Methionine	285	163	208
Isoleucine	577	537	478
Tyrosine	647	462	531
Histidine	254	189	193
Threonine	298	209	226
Total Amino acid	11378	9373	9153

[a] Adopted from ref (9)

Lipid and Fatty Acids

The major fatty acids are oleic and linoleic acid for the three varieties of black rice [**Table 5**] [12]. Similarly, Oh, G.-S [16] reported that the presence of oleic (38.57%) and linoleic (38.60%) acid. However, there is no significant difference in fatty acid content.

Table 5. Composition of fatty acids in black rice varieties [a]

Varieties	Fatty acids (%)				
	Palmitic	Stearic	Oleic	Linoleic	Linolenic
Kilimheugmi	16.9	1.9	38.1	42.0	1.5
Sanghaehyanghyeolla	18.3	1.8	37.1	41.8	1.7
Heuginjubyco	17.4	1.9	36.8	42.7	1.5
C3Hi line	18.2	5.9	36.5	41.7	1.7

[a] Adopted from ref [12].

Minor Constituents

Anthocyanins

Black rice contains anthocyanin pigments with anti-oxidative and anti-inflammatory properties that hold potential as nutraceutical or functional food formulations [19]. Anthocyanin acts as an anti-oxidant against reactive oxygen species (ROS), inhibiting peroxidation of linoleic acid and exhibiting superoxide anion and hydrogen peroxide scavenging activities. Extracted anthocyanin exhibits 88.8% inhibition of the peroxidation of linoleic acid, 55.2% DPPH free radical scavenging activity, 55.0% superoxide anion radical scavenging activity, and 72.7% hydrogen peroxide scavenging activity. It also shows ferrous ion reducing capability. Thus, anthocyanin extracted from black rice could be utilized as a possible anti-oxidant agent against ROS [20].

Anthocyanin compounds from black rice include cyanidin 3-O-glucoside, peonidin 3-O-glucoside, malvidin 3-O-glucoside, pelagonidin 3-O-glucoside, and delphinidin 3-O-glucoside [20]. Several studies reported the presence of two major anthocyanins, cyanidin 3-glucoside and peonidin 3-glucoside in black rice and their contents have been reported in the range between 88-95% and 4.7-7.1%, respectively [3,4]. Hiemori M et. al. reported that the predominant anthocyanins are cyanidin-3-glucoside (572.47 µg/g; 91.13% of total) and peonidin-3-glucoside (29.78 µg/g; 4.74% of total) in *Oryza sativa L. japonica var. SBR*. Three cyanidin-dihexoside isomers and one cyanidin hexoside are present as minor constituents [21]. The total anthocyanin content of black rice is 3,276 µg/g, and cyanidin-3-glucoside (2,013 µg/g; 88%) and peonindin-3-glucoside (162 µg/g; 7.1%) are the two most abundant anthocyanins [22].

The two predominant anthocyanins in 10 varieties of black rice found in Korea are cyanidin-3-glucoside (0.0-470 mg/g) and peonindin-3-glucoside (0.0-40 mg/g) [**Table 6**] [4]. Black rice Heuginjubyeo and Dragon Eyeball 100 exhibited the highest amounts of anthocyanin, with a great predominance of cyanidin-3-O-glucoside (160.10±8.5 and 105.79±45.1 mg/mg) relative to peonidin-3-O-glucoside (82.6±1.2 and 71.2±9.2 µg/ mg), respectively [10]. Cyanindin-3-glucoside (85%) and peonindin-3-glucoside (15%) have also been identified in Oryza sativa L. japonica [23].

Anthocyanins as functional ingredients of black rice have a potential role in preventing chronic and degenerative disease due to their superoxide and hydroxyl radical scavenging activities [24, 25], anti-oxidative [24], hypolipidemic [26], anti-diabetic [27], hepatopotective [28], atherosclerotic plaque stabilization [29], anti-inflammatory [19] and cardioprotective effects [30].. Peonidin 3-glucoside and cyanidin 3-glucoside, major anthocyanins extracted from black rice (Oryza sativa L. indica), show inhibitory effects on the invasion of various cancer cells [31]. Cyanidin 3-glucoside (C3G) isolated from black rice has a stronger scavenging effect (91%) on superoxide anion radicals than any other natural or synthetic anti-oxidant, suggesting that C3G from black rice possesses hepatoprotective effects *in vitro* due its superoxide anion scavenging activity [28].

Table 6. Anthocyanin content of Korean black rice varieties

Varieties	Anthocyanin (mg/100g grain weight)				
	C3G	P3G	Total	C3G/Total (%)	ref
Suwon #415	470	23	493	95.3	(4)
Kilimheugmi	240	26	266	90.2	(4)
Suwon #425	206	40	246	83.7	(4)
Heugjinmi	200	32	232	86.2	(4)
Sanghaehyanghyella	50	5	55	90.9	(4)
Hongmi	30	6	36	83.3	(4)
Suwon # 405	16	4	20	80.0	(4)
Suwon # 420	10	ND+	10	-	(4)
Jagwangdo	10	ND	10	-	(4)
Ilpumbyeo	ND	ND	ND	ND	(4)
Jeonbuk 2*	43.04	4.03	47.07	91.4	(10)
Heugjinjubyeo*	160.10	8.26	168.36	95.1	(10)
Heugnambyeo*	12.66	Trace	12.66		(10)
Josaengheugchal*	10.36	Trace	10.36		(10)
Dragon Eyeball 100*	105.79	7.12	112.91	93.7	(10)
Heughyang*	5.21	ND	5.21		(10)
Jeonbuk 1*	37.35	4.91	42.26	88.4	(10)
Heuggwang*	7.78	ND	7.78		(10)
Suwon 512*	32.16	3.70	35.86	89.7	(10)

+ND: Not detectable

Black rice extract can prevent hyperlipidemia and insulin resistance induced by a high-fructose diet, as indicated by the lower concentrations of plasma thiobarbituric acid reactive substances and blood oxidized glutathione. The results indicate that black rice extract is effective for preventing diabetes, and the underlying mechanism may be related to the inhibition of oxidative stress and improvement of the plasma lipid profile [32].. Extract of black rice pigmented fraction containing cyanidin 3-glucoside and peonidin 3-glucoside exhibits marked anti-oxidant and free radical scavenging activities *in vitro*. The addition of black rice extract to the diet significantly decreases the level of total cholesterol, LDLs, and triacylglycerols in the plasma of rats fed control diets. However, there was not any observed difference in the level of HDLs. Liver crude lipids and total cholesterol levels were also significantly decreased in test groups compared to the control group [33].

Radical-Scavenging Components in Black Rice

The DPPH scavenging activity of black rice extracts ranged from 3.1 to 10.8 μmol-Trolox equiv. mL^{-1} of extract. Free radical scavenging activity increases upon increased anthocyanin content with a high correlation coefficient (0.987). Two major anthocyanins, peonidin- 3-glucoside (Pn-3-Glc) and Cy-3-Glc, are observed in eight black rice varieties, but with a slight compositional variation in anthocyanin content. However, the sum of Cy-3-Glc and Pn-3-Glc comprises more than 85% of total anthocyanin. Thus, the dominant free radical scavenging activity of black rice is due to Cy-3-Glc and Pn-3-Glc [34]. Additionally, DPPH free radical scavenging assay found that 2-Arylbenzofuran having strong antioxidant activity is present in the bran of black rice [35].

The bran extracts treated at 260°C for 5 min show inhibition against the autoxidation of linoleic acid [36]. Bran oil of black rice can improve cholesterol metabolism and anti-oxidant activity. The contents of free and bound phenolic compounds extracted from black rice are 845.4 and 401.6 mg per 100 g of sample weight, respectively. Free phenolic compounds have higher anti-oxidative ability compared to bound phenolic compounds. Solvent fractionation of free phenolic compounds shows that the butanol fraction has the highest content of phenolic compounds as well as higher anti-oxidative activity compared to other solvent fractions. Lipid peroxidation inhibition ability of the butanol fraction is lower than that of α-tocopherol or BHT, whereas the free radical scavenging ability is higher than that of α-tocopherol or BH [37].

Contents of Polyphenol, Flavonoids, Γ-Oryzanol, Phytic Acid

Black rice bran contains most of the antioxidant compounds, including phytic acid, γ-oryzanol, anthocyanins, and vitamin E isomers. There are also higher levels of free polyphenols and flavonoids in the bran than in the endosperm fractions [**Table 7**]. Two cultivars of whole black rice (*Oryza sativa* cv. *Heugjinjubyeo* and *O. sativa* L. *Heugkwangbyeo*) contain higher levels of free polyphenols (98.5 and 81.0 mg, respectively) and flavonoids (19.8 and 15.1 mg, respectively) than the endosperm fraction [38].

Table 7. Antioxidant compounds in whole grain; and endosperm, and bran of two Korean black rice varieties [a]

Milling fraction	Polyphenols	Flavonoids	γ-Oryzanol	Phytic acid
Heugjinjubyeo				
Whole grain	18.2±0.59	3.39±0.05	0.75±0.02	47.5±0.51
Endosperm	2.73±0.02	0.32±0.01	0.07±0.00	7.38±1.19
Rice bran	108±3.65	21.8±0.23	3.45±0.09	227±1.35
Heugkwangbyeo				
Whole grain	13.3±0.31	2.13±0.02	0.60±0.00	33.4±0.56
Endosperm	1.73±0.03	0.15±0.00	0.03±0.00	3.96±0.74
Rice bran	89.2±2.97	16.4±0.58	3.94±0.27	202±0.32

[a]Adopted from ref (*38*)

Total polyphenol content of black rice is higher than that of polished white rice. Black rice has 86 times more total phenolic compounds than does red rice. Black rice extract has significantly stronger ability to inhibit LDL oxidation than white rice extract. It also shows significant increases in superoxide dismutase (SOD) and catalase activities (161.6% and 73.4%, respectively) compared to white rice extract. The major components responsible might be cyanidin-3-O-glucoside chloride and peonidin- 3-O-glucuside chloride, which are responsible for free radical scavenging and anti-oxidative activities [39].

The constituents of black rice extract have cumulative health implications. The extract prevents DNA strand cleavage induced by ROS such as peroxyl and hydroxyl radicals and suppresses oxidative modification of human LDL. Black rice extract reduces the formation of nitric oxide by suppressing inducible nitric oxide synthase expression in murine macrophage cells without introducing cell toxicity [19]. Upon supplementation of natural rice bran oil from black rice, the concentrations of total cholesterol and LDL-cholesterol are lowered in the plasma levels of hypercholesterolemic rat. Bran oil also decreases plasma and hepatic oxidative stress upon the increase in hepatic thiobarbituric acid reactive substances associated with the elevation of hepatic superoxide dismutase, catalase, and glutathione peroxidase activities and plasma tocopherol. Black rice bran oil may also improve cholesterol metabolism and protect against oxidative liver damage caused by lipid peroxidation [40]. Ethanol-water extract from black rice bran is more effective than extract from non-pigmented rice cultivar than inhibiting allergic reactions by suppressing the release of histamine and β-hexosaminidase from basophilic cells [41].

Among the flavonoids, quercetin and isorhamnetin are present in the bran of Thai black rice varieties between the range of 1.08–2.85 and 0.05–0.83 µg/g of rice bran, respectively [42]. Smyristic, caproic and pelargonic acids, and pentadecanal and nonanal are the main volatile compounds in black rice bran. Guaiacol, present at 0.81 mg/100 g in black rice bran, is responsible, along with 2-acetyl-1-pyrroline (2-AP), for the characteristic aroma of black rice [43, 44].

Phytic acid is distributed in the bran fractions or aleurone layers. The highest amount of phytic acid is found in the bran portions of Heuginjubyeo (227 mg/g sample), followed by Heugkwangbyeo (202 mg/g sample).

Inhibition of tyrosinase activity has also been observed using black rice bran. The ethyl acetate extract has the most potent inhibitory activity against tyrosinase (80.5% at a concentration of 0.4 mg/mL), and it is identified as a protocatechuic acid methyl ester [45].

Mineral Content

Total mineral content of black rice is 783.74 mg/100g on average [**Table 8**]. Chaudhary R. C et al. also reported selenium content along with other minerals in two varieties of (Yangxianheimi and Heizhenmi) black rice.

Table 8. Mineral contents of Korean black rice varieties[a]

Varieties	Mineral element (mg/100gm)									
	Na	Ca	Fe	K	Mg	P	Mn	Zn	Total	Mean
Kilimheugmi	2.03	18.20	1.12	340.6	103.9	361.3	4.02	1.45	832.62	783.74
Sanghaehyanghyeolla	3.40	18.15	0.98	336.4	107.3	287.4	4.51	1.42	759.56	
Heugjinjubyeo	3.00	15.92	1.18	314.5	119.6	303.4	3.36	1.81	762.77	
C3GHi line	3.13	18.35	1.24	336.9	113.2	301.5	3.38	1.74	780.04	

aAdopted from ref (12)

The highest iron content is found in black rice, followed by red rice, rice, and glutinous rice; millet has the lowest iron content. Black rice also contains the highest relative amounts of other elements (zinc, calcium, copper, manganese) followed by rice and red rice, compared to the low content of glutinous and millet. Therefore, iron nutrition may be improved by increasing the supply of black rice [46]. Similarly, iron content in milled black brown rice is higher than in milled red or white brown rice. However, Zn content of milled black brown rice is lower than that of milled red brown rice [47].

Vitamin content

Three Korean black rice varieties (Suwon, Iksan 427, and Sanghaei) contain vitamin B_1, B_2, and B_3 in the range of 0.23-0.36, 0.07-0.09, and 3.49-5.14 mg/100mg, respectively [**Table 9**] [9]. Trans-β-carotene is present in the bran layer of Thai black rice varieties within the range of 33.60–41.5µg/g of rice bran. However, the bran of most white rice varieties lack vitamin A and β-carotene. Therefore, vitamin A deficiency is a major problem in geographical areas where rice is a staple food [42].

Table 9. Vitamin B contents of Korean black rice varieties (mg/100mg, dry basis) [a]

Varieties	B_1	B_2	B_3
Suwon 415	0.23	0.09	4.66
Iksan 427	0.34	0.07	5.14
Sanghaehyanghyeolla	0.36	0.08	3.49

[a]Adopted from ref (9)

Content of Tocopherol and Homologues

α-tocopherol is major isomer in bran fractions, present at levels of 60.6 and 78.3 µg/g in Heugkwangbyeo and Heugjinjubyeo, respectively, whereas β-tocopherol is only present in small amounts (1.91 and 2.72 µg/g, respectively). Among the tocotrienol isomers, the highest levels of γ - and α -tocotrienol (92.5 and 44.2 µg/g, respectively) are present in the bran fraction of Heugkwangbyeo. In the endosperm fractions, low levels of α-tocopherol, α-tocotrienol, and γ-tocotrienol have been detected. These results show that tocopherols and tocotrienols are mainly located in rice bran, including the rice germ. δ-tocopherol and β-tocotrienol have not been detected in any milling fractions or whole grains from the two cultivars [**Table 10**] [38].

Table 10. Tocopherol and tocotrienol content (µg/g, dry basis) of milling fractions and whole black rice from two Korean varieties[a]

Fraction	α-T	β-T	γ-T	δ-T	α-T3	β-T3	γ-T3	δ-T3	Total
Heugjinjubyeo									
Whole grain	14.1	–	1.28	–	8.97	–	19.0	4.78	48.1
Endosperm	1.87	–	–	–	2.62	–	7.15	–	11.6
Rice bran	78.3	2.72	9.10	–	43.7	–	85.6	31.6	251
Heugkwangbyeo									
Whole grain	11.90	–	1.51	–	9.46	–	21.7	4.20	48.7
Endosperm	2.22	–	–	–	2.06	–	8.29	0.73	13.3
Rice bran	60.6	1.91	10.01	–	44.2	–	92.5	31.1	241

[a]Adopted from ref (38)

PROCESSING

Milling

Hunter 'L' value and water soluble index (WSI) of black rice extracts increase with decreasing particle size, whereas Hunter 'a' value decreases. Total polyphenol contents (91.4-115.8 mg/100 g) and total amino acids (75.0-96.4 mg/100 g) of three types of black rice extracts are higher compared to those of polished rice flour (6.3 mg/100 g and 30.5 mg/100 g, respectively). Physicochemical properties such as moisture content, color parameters, and WSI have are closely related with the particle size of black rice extract prepared under different milling conditions [48].

Cooking

Cooked milled rice added to 9% waxy black rice shows the highest values for cohesiveness, taste, and acceptability, whereas 11% waxy black rice shows the highest values for flavor, color, glossiness, crispiness, hardness, and adhesiveness. The correlation between the sensory parameters of cooked milled rice added to black rice and glutinous rice has revealed a significant relationship between taste and acceptability [16].

Effect of Cooking on Volatile Compound

Table 11. Volatiles Identified in Cooked Black rice[a]

Aromatics	N-containing compounds	Aliphatic aldehydes
Toluene	2-methylpyridine	Hexanal
p-xylene	2-acetyl-1-pyrroline	(E)-2-hexenal
Benzaldehyde	Benzothiazole	Heptanal
2-pentylfuran	Indole	Octanal
Phenylacetaldehyde	Aliphatic Alcohols	(E)-2-octenal
Guaiacol	3-methyl-1-butanol	Nonanal
1,2-dimethoxybenzene	(S)-2-methyl-1-butanol	(E)-2-nonenal
Naphthalene	1-pentanol	Decanal
2-methylnaphthalene	1-hexanol	(E)-2-decenal
4-vinylguaiacol	1-heptanol	(E,E)-2,4-decadienal
Terpenoids	1-octen-3-ol	
d-limonene	Aliphatic/Alicyclic Ketones	
(Z)-linaloloxide	3-octen-2-one	
	2-nonanone	
	(E)-geranylacetone	

[a] Adopted from ref [49]

Black rice is occasionally mixed with white rice prior to cooking in order to enhance flavor, color, and nutritional value. Thirty-five different volatile compounds have been observed in cooked black rice: 10 aromatic, 4 nitrogen-containing, 6 alcohol, 10 aldehyde, 3 ketone, and 2 terpenoid compounds. Among them, 25 are considered odor active; 21 are commonly found in white rice, and 4 are unique to black rice 2-acetyl-1-pyrroline (2-AP), guaiacol, toluene, and 4-vinyl guaiacol. 2-AP and guaiacol are major contributors to the characteristic flavor of black rice due to their intensity and unique characteristics. The eight detected aromatic compounds included 3 benzenes, 2 phenols, 1 furan, and 2 naphthalenes [**Table 11**] [49].

The phenolic compounds 1-nonanol, 2-methylpyridine, guaiacol, and indole have been detected in black rice. 1-nonanol, guaiacol, and indole have higher odor intensities while 2-AP, 2-nonenal, and nonanal contribute as odoring compounds in black rice [44]. Along with 2-AP and guaiacol, two other compounds, toluene and 4-vinylguaiacol have also been reported as critical odorants in black rice [49].

Effects of Cooking on Anthocyanins in Black Rice

Quantitative comparison of anthocyanin content between presoaked and non-presoaked rice indicate that soaking has no significant impact on anthocyanin stability or retention. The content of cyanidin-3-glucoside is decreased significantly by all cooking methods (commercial rice cooker, pressure cooker, or gas range). Pressure cooking results in the greatest decreases (79.8%), followed by the rice cooker (74.2%) and gas range (65.4%). Conversely, levels of protocatechuic acid are increased 2.7 to 3.4 times in response to all cooking methods. These findings indicate that cooking black rice results in the thermal degradation of cyanidin-3-glucoside and the concomitant production of protocatechuic acid. Peonidin-3-glucoside also decreases in the similar fashion [21].

Anthocyanidin dihexoside shows the highest thermal stability among all anthocyanidins, based on the fact that anthocyanin stability increases with an increased number of methoxyl groups in the β-ring and decreases as the number of free hydroxyl groups in the β-ring is increased. The content of total anthocyanins in raw rice was reported to be 628.17 $\mu g/g$ dry weight, whereas the total content in cooked rice ranged from 130.67 to 221.50 $\mu g/g$ dry weight (21-35% of raw rice anthocyanins), indicating that there was significant loss of anthocyanins during thermal processing. Boiling rice on a gas range at low temperature resulted in the highest retention of total anthocyanins compared to cooking with either a rice cooker or a higher pressure cooker. The level of cyanidin-3-glucoside decreases as that of protocatechuic acid is concomitantly increased by across all cooking methods. These results were also observed for the individual anthocyanins cyanidin-3-glucoside and peonidin- 3-glucoside [**Table 12**]. These findings suggest that the loss of anthocyanins from black rice may be due to the degradation or decomposition of anthocyanins caused by thermal processing.

Cyanidin-glycosides undergo deglycosylation during heating and degrade into phloroglucin-aldehyde and protocatechuic (3,4-dihydroxybenzoic) acid [50]. Correspondingly, the content of total anthocyanins in cooked rice is almost one-third lower than that in raw rice, whereas the levels of protocatechuic acid are about three times higher after cooking [Table 13]. Cyanidin-3-glucoside in black rice degrades into protocatechuic acid during cooking, but phloroglucinaldehyde is not detected in cooked black rice. Protocatechuic acid is an *in vitro* anti-coagulatory, anti-oxidative, and anti-inflammatory compound [51].

Table 12. Anthocyanin and protocatechuic acid content (μg/g dry weight) of cooked black rice[a]

Anthocyanin	Raw	Rice cooker	Pressure cooker	Gas range
Protocatechuic acid	120.44 ± 2.81	395.64± 22.17	405.40± 26.90	321.87±11.24
Cyanidin-dihexoside	4.06 ± 0.12	3.03 ± 0.46	2.43 ± 0.25	2.98 ± 0.18
Cyanidin-dihexoside	9.02 ± 0.41	0.91 ± 0.14	0.93 ± 0.14	1.84 ± 0.84
Cyanidin-dihexoside	8.82 ± 0.46	4.56 ± 0.13	3.07 ± 0.31	4.50 ± 0.24
Cyanidin-3-glucoside	572.47± 20.70	147.62± 10.01	115.68± 14.09	198.24± 12.66
Cyanidin-hexoside	4.02 ± 0.11	0.50± 0.07	0.25 ± 0.34	0.76 ±0.27
Peonidin-3-glucoside	29.78± 2.21	10.93 ± 0.52	8.31 ± 0.89	13.18 ± 1.06

Content of each anthocyanin and protocatechuic acid are reported as the mean ± SD.
[a]Adopted from ref [21]

Germination

Germination increases the content of γ- aminobutyric acid to 55.36 mg/100g, 3.9 fold higher than the initial content, after immersion in $CaCl_2$ solutions for 24 hrs at 28oC and germination for 30 hr at 29oC [52]. Similarly, germination increases the level of γ-aminobutyric acid of other Korean rice varieties such as Ilpum, Goami 2, Keunnun, and Heugkwang by 2.4, 2.5, 6.1, and 3.4 fold, respectively and that of γ -oryzanol by 0.8, 1.1, 1.5, and 1.2 fold, respectively. In addition, α-tocopherol, α-tocotrienols, and γ-tocotrienols also increase significantly [53]. Sprouted black rice oligopeptides may improve skin elasticity and thus can be effective anti-aging ingredient for cosmetic purposes [54].

UTILIZATION OF BLACK RICE

In previous studies, the utilization of black rice has been reported in wide range of processed foods.

Biscuits and Cookies

Incorporating black rice into cookies enhances their functional quality. Due to the addition of black rice, total dietary fiber and total polyphenol content increases in cookies while the color changes to a darker one. As the textural parameter, the hardness decreases and crispness increases upon increased black rice flour content. In a sensory test, the appearance, aroma, and texture of the cookies containing added black rice flour are significantly lower than those of the wheat flour cookies. However, the taste and overall acceptance of the cookies containing black rice are not altered. According to the results of the sensory evaluation, the aroma, flavor, crispness, graininess, and color (gray-violet) of cookies are significantly increased upon increased black rice flour content. The optimal ratio for the addition of black rice flour to cookies is recommended to be 10% with respect to the sensory evaluation and functional counterpart from black rice [55]. In addition, Kim YS et. al.

reported that spread factor and sensory flavor of cookies increases with the addition of black rice flour [56].

Dough Properties

Another previous study reported that the addition of black rice flour to wheat flour dough decreases water absorption, development time, elasticity, and extensibility of the dough as revealed thorough the farinograph shown in **Table 13**. The increase in concentration of black rice flour results in a decrease in protein and gluten content, whereas increases in ash content; decreases in water absorption, stability, development time, elasticity, and valorimeter value of the dough, whereas increases in weakness of dough, as revealed thorough farinogram; decreases in extensibility, resistance to extension, and energy, whereas increase in the ratio of resistance and extensibility (R/E ratio), as revealed through extensogram [57].

Table 13. Ferinograph properties of wheat flour and black rice-wheat flour mixtures[a]

Sample[1]	Water absorption (%)	Stability (min)	Development time (min)	Elasticity (BU)	Weakness (BU)	V/V^2 (units)
Control	72.0	31.0	6.0	130	20	69
BR-10	69.5	16.0	6.5	125	50	65
BR-20	66.8	15.2	5.8	125	55	63
BR-30	64.2	14.7	5.3	125	55	61

[a]Adopted from ref [57].

[1] Control: wheat flour 100%+black rice flour 0%, BR-10: wheat flour 90%+black rice flour 10%, BR-20: wheat flour 80%+black rice flour 20%, BR-30: wheat flour 70%+black rice flour 30%,

[2] Valorimeter/Value

Bread Quality

Bread volume decreases from 2250cc to 1700cc as black rice flour increases from 0% to 30% [58]. Similarly, the bread volume decreases from 2,096.7 cc to 1,703.4 cc as the whole black rice flour content increases from 0% to 40% [59]. For color values, lightness (L value) decreases and redness (a value) increases with increased black rice flour content. For textural characteristics, hardness, gumminess, and chewiness increase with addition of the flour, whereas springiness and cohesiveness are decreased [58]. However, the hardness of bread decreases with 10% and 20% black rice flour and increases with 40% black rice flour [59]. Sensory parameters such as color, taste, and overall eating quality are at their highest levels with 20% black rice flour, although the appearance of grain and texture quality is decreased[58]. Similarly, in another study of sensory evaluation, sensory scores decrease for appearance, external color, crumb color and grain with increasing black rice flour content. Overall acceptability of bread containing 10% and 20% black rice flour does not vary significantly with respect to the score of the control group [59].

Utilization of Black Rice in Korean Food

Rice cakes prepared with whole grain black rice are not satisfactory with respect to integrity [60, 61] Preliminary studies in the laboratory indicate that fractured black rice, such as black rice flour or particulates, can produce a denser cake with better integrity than whole grain black rice. The specific volume of black rice cakes increases at higher tempering moisture, heating temperature, and heating time. The hardness of puffed black rice cake decreases as tempering moisture and heating time are increased, and the effect on hardness is more pronounced using black rice content than other processing parameters, such as heating temperature [61].

Julpyun (Korean rice cake) prepared with 20% black rice flour shows high scores for color, flavor, and overall acceptability, excluding mouthfeel. Decreased hardness, gumminess, and chewiness indicates the low rate of retrogradation of black rice [62]. Addition of whole grain black rice containing a high amount of phytochemicals increases the storage stability of rice cake.

Color values such as lightness and yellowness of black rice cake decrese steadily with increasing black rice content, whereas these two color values are not significantly affected by changes in tempering moisture (16-20%), heating time (4-6 sec), and temperate (250-270°C). However, the integrity of rice cake is enhanced with increasing black rice content [63], as is the water holding capacity. However, alkaline water retention capacity, Pelshenke value, sedimentation value, and viscosity are all decreased. Increased black rice flour content also reduces the loaf volume. However, replacing up to 15% of wheat flour with black rice flour has no significant affect. The texture of crumbs is also influenced by addition of black rice flour [64].

The addition of black rice to soybean-based "D*asik*" alters the chewiness, gumminess, cohesiveness, and hardness but not springiness and adhesiveness. Chewiness and cohesiveness are reduced significantly upon the addition of black rice to a level of over 40%, whereas hardness is decreased significantly [65].

The nutritious black-color biscuits, *Yukwa* (Korean puffed rice snack), and rice wine are also made by incorporating black rice.

Packaging

Related to the storage of black rice using flexible packaging, water activity, acidity, and color values are not changed significantly by three different packaging materials: polyethylene (PE) film, polypropylene (PP) film, and laminated film with PE and PP (PE/PP). However, the rate of increase of haxanal content in the PE/PP combination is smaller than either packaging material individually [66].

Ingredients for Nutraceuticals

Black rice has remarkable nutritional and therapeutic benefits and can be used in varieties of food formulations for its hyperlipidemia and arteriosclerosis effects. Most color components experience beneficial effects from black rice in rice bran. Therefore, whole grain black rice should be used more often.

Constraints for Utilization and Future Implication

Although black rice is good as a food ingredient when it is used directly as cooked rice, additional research should be carried out to overcome its uncomfortable mouth-feel.

The functional compounds of black rice have health improvement effects, but the exact compounds and their specific beneficial effects remains unclear. Previous research findings favor the application of black rice for the possibility of developing cereal-based functional foods due to the presence of phytochemicals. However, this resource has still been underutilized.

REFERENCES

[1] Khush, G. S., Origin, dispersal, cultivation and variation of rice. *Plant Molecular Biology* 1997, 35, (1-2), 25-34.

[2] Chaudhary, R., Speciality rices of the world: effect of WTO and IPR on its production trend and marketing. *J Food Agr Environ* 2003, 1 34-41.

[3] Choi, S.; Kang, W.; Osawa, T., Isolation and identification of anthocyanin pigments in black rice. *Foods and Biotechnology (Korea Republic)* 1994, 3, (3), 131.

[4] Ryu, S.; Park, S.; Ho, C., High performance liquid chromatographic determination of anthocyanin pigments in some varieties of black rice. *Journal of Food and Drug Analysis* 1998, 6, (4), 729-736.

[5] Meiyu, G., A Study on special nutrient of purple black glutinious rice. *Scientia Agricultura Sinica* 1992, 5, 36-41.

[6] Roche, J., *The international rice trade.* Woodhead Pub Ltd: 1992.

[7] Van Andel, T., African Rice (*Oryza glaberrima* Steud.): lost crop of the enslaved Africans discovered in Suriname. *Economic Botany* 2010, 64, (1), 1-10.

[8] Oh, G.; Kim, K.; Park, J.; Kim, S.; Na, H., Physical properties on waxy black rice and waxy rice. *Korean journal of food science and technology* 2002, 34, (2), 339-342.

[9] Chung, H.; Lim, S., Pasting and nutritional characteristics of black rices harvested in Korea. *Journal of Food Science and Nutrition* 1999, 4, (4), 231-235.

[10] Lee, J., Identification and quantification of anthocyanins from the grains of black rice (Oryza sativa L.) varieties. *Food Science and Biotechnology* 2010, 19, (2), 391-397.

[11] Choi, G.; Na, H.; GS, O.; SK, K.; K, K., Physicochemical properties on Shinsun (Waxy) and black rice starch. *Journal of the Korean Society of Food Science and Nutrition* 2003, 32, (7), 953-959.

[12] Lee, H. H.; Kim, H. Y.; Koh, H. J.; Ryu, S. N., Varietal difference of chemical composition in pigmented rice varieties *Korean Journal of Crop Science* 2006, 51, (spe.), 113-118.

[13] Choi, G.; Na, H.; Oh, G.; Kim, S.; Kim, K., Hot-water soluble and insoluble materials of waxy black rice starch. *Journal of the Korean Society of Food Science and Nutrition* 2005, 34, (2).

[14] Choi, G.-C., Physiological and structural properties of black glutinous rice starch. *MS dissertation, Dept of food science and technology, Chonnam national University, Korea,* 2002.

[15] Oh, G.-S.; Kim, K.; Na, H.-S.; G-C, C., Comparision of physicochemical properties on waxy black rice and glutinous rice. *J. Korean Sco. Food Sci. Nutr.* 2002, 31 (1), 12-16.

[16] Oh, G.-S., Properties of waxy black and glutinous rice and textures of cooked milled rice added waxy black and glutinous rice. *Ph. D. Thesis* 2001, Dept. of food science and technology, Chonnam National university.

[17] Choi, G.; Na, H.; Oh, G.; Kim, S.; Kim, K., Gelatinization properties of waxy black rice starch. *J Korean Soc. Food Sci Nutr* 2005, 34, (1), 87-92.

[18] Chang, J.; HJ., R.; Kimiko O; Lee OK, Non-starch polysaccharides of cell walls in glutinous rice, rice and black rice. *Journal of Korean home economics association* 2001 39, 91-102

[19] Hu, C.; Zawistowski, J.; Ling, W.; Kitts, D., Black rice (*Oryza sativa* L. *indica*) pigmented fraction suppresses both reactive oxygen species and nitric oxide in chemical and biological model systems. *J. Agric. Food Chem* 2003, 51, (18), 5271-5277.

[20] Park, Y.; Kim, S.; Chang, H., Isolation of anthocyanin from black rice (Heugjinjubyeo) and screening of its antioxidant activities. *Korean Journal of Microbiology* 2008, 36, (1).

[21] Hiemori, M.; Koh, E.; Mitchell, A. E., Influence of cooking on anthocyanins in black rice (*Oryza sativa* L. *japonica* var. SBR). *Journal of agricultural and food chemistry* 2009, 57, (5), 1908-1914.

[22] Abdel-Aal, S.; Young, J.; Rabalski, I., Anthocyanin composition in black, blue, pink, purple, and red cereal grains. *Journal of agricultural and food chemistry* 2006, 54, (13), 4696.

[23] Robert, Y.; Shinji, T.; Naofumi, M., Identification of phenolic components isolated from pigmented rices and their aldose reductase inhibitory activities. *Food Chem.* 2007, 101, 1616.

[24] Ichikawa, H.; Ichiyanagi, T.; Xu, B.; Yoshii, Y.; Nakajima, M.; Konishi, T., Antioxidant activity of anthocyanin extract from purple black rice. *Journal of Medicinal Food* 2001, 4, (4), 211-218.

[25] Nam, S. H.; Choi, S. P.; Kang, M. Y.; Koh, H. J.; Kozukue, N.; Friedman, M., Antioxidative activities of bran extracts from twenty one pigmented rice cultivars. *Food Chemistry* 2006, 94, (4), 613-620.

[26] Mingwei, Z.; Ruifen, Z.; Baojiang, G., Hypolipidemic and antioxidative effects of black rice pericarp extract accompanied by its components analysis. *Scientia Agricultura Sinica* 2006.

[27] Sasaki, R.; Nishimura, N.; Hoshino, H.; Isa, Y.; Kadowaki, M.; Ichi, T.; Tanaka, A.; Nishiumi, S.; Fukuda, I.; Ashida, H.; Horio, F.; Tsuda, T., Cyanidin 3-glucoside ameliorates hyperglycemia and insulin sensitivity due to downregulation of retinol binding protein 4 expression in diabetic mice. *Biochem. Pharmacol.* 2007, 74, 1619.

[28] Shim, S.; Chung, J.; Lee, J.; Hwang, K.; Sone, J.; Hong, B.; Cho, H.; Jun, W., Hepatopotective effects of black rice on superoxide anion radicals in HepG2 cells. *Food Science and Biotechnology* 2006.

[29] Xia, X.; Ling, W.; Ma, J.; Xia, M.; Hou, M.; Wang, Q.; Zhu, H.; Tang, Z., An anthocyanin-rich extract from black rice enhances atherosclerotic plaque stabilization in apolipoprotein E–deficient mice. *J. Nutr.* 2006, 136, 2220.

[30] Wang, Q.; Han, P.; Zhang, M.; Xia, M.; Zhu, H.; Ma, J.; Hou, M.; Tang, Z.; Ling, W., Supplementation of black rice pigment fraction improves antioxidant and anti-inflammatory status in patients with coronary heart disease. *Asia Pacific Journal of Clinical Nutrition* 2007, 16, (1), 295-301.

[31] Chen, P.-N.; Kuo, W.-H.; Chiang, C.-L.; Chiou, H.-L.; Hsieh, Y.-S.; Chu, S.-C., Black rice anthocyanins inhibit cancer cells invasion via repressions of MMPs and u-PA expression. *Chemico-Biological Interactions* 2006, 163, (3), 218-229.

[32] Guo, H.; Ling, W.; Wang, Q.; Liu, C.; Hu, Y.; Xia, M.; Feng, X.; Xia, X., Effect of anthocyanin-rich extract from black rice (Oryza sativa L. indica) on hyperlipidemia and insulin resistance in fructose-fed rats. *Plant Foods for Human Nutrition (Formerly Qualitas Plantarum)* 2007, 62, (1), 1-6.

[33] Zawistowski, J.; Kopec, A.; Kitts, D. D., Effects of a black rice extract (*Oryza sativa* L. indica) on cholesterol levels and plasma lipid parameters in Wistar Kyoto rats. *Journal of Functional Foods* 2009, 1, (1), 50-56.

[34] Oki, T.; Masuda, M.; Nagai, S.; Take'ichi, M.; Kobayashi, M.; Nishiba, Y.; Sugawara, T.; Suda, I.; Sato, T., Radical-scavenging activity of red and black rice. *Rice is life: scientific perspectives for the 21st century* 2005, 256.

[35] Han, S. J.; Ryu, S. N.; Kang, S. S., A new 2-arylbenzofuran with antioxidant activity from the black colored rice (*Oryza sativa L.*) bran. *ChemInform* 2005, 36, (18).

[36] Wiboonsirikul, J.; Hata, S.; Tsuno, T.; Kimura, Y.; Adachi, S., Production of functional substances from black rice bran by its treatment in subcritical water. *LWT-Food Science and Technology* 2007, 40, (10), 1732-1740.

[37] Chung, Y.; Lee, J., Antioxidative properties of phenolic compounds extracted from black rice. *Journal of the korean Society for Horticultural Science* 2003.

[38] Kong, S.; Lee, J., Antioxidants in milling fractions of black rice cultivars. *Food Chemistry* 2010, 120, (1), 278-281.

[39] Chiang, A.; Wu, H.; Yeh, H.; Chu, C.; Lin, H.; Lee, W., Antioxidant effects of black rice extract through the induction of superoxide dismutase and catalase activities. *Lipids* 2006, 41, (8), 797-803.

[40] Yean Ju Nam; Seok Hyun Nam, S.; Mi Young Kang, M., Cholesterol-lowering efficacy of unrefined bran oil from the pigmented black rice (*Oryza sativa* L cv. Suwon 415) in hypercholesterolemic rats. *Food Science and Biotechnology* 2008, 17, (3), 457-463.

[41] Choi, S.; Kang, M.; Koh, H.; Nam, S.; Friedman, M.; USDA, A., Antiallergic activities of pigmented rice bran extracts in cell assays. 2007.

[42] Nakornriab, M.; Sriseadka, T.; Wongpornchai, S., Quantification of carotenoid and flavonoid components in brans of some Thai balck rice cultivars using supercritical fluid ectraction and high-performance liqid chromatography-mass spectrometry. *Journal of Food Lipids* 2008, 15, (4), 488-503.

[43] Sukhonthara, S.; Theerakulkait, C.; Miyazawa, M., Characterization of volatile aroma compounds from red and black rice bran. *Journal of Oleo Science* 2009, 58, (3), 155-161.

[44] Yang, D. S.; Shewfelt, R. L.; Lee, K.-S.; Kays, S. J., Comparison of odor-active compounds from six distinctly different rice flavor types. *Journal of agricultural and food chemistry* 2008, 56, (8), 2780-2787.

[45] Miyazawa, M.; Oshima, T.; Koshio, K.; Itsuzaki, Y.; Anzai, J., Tyrosinase inhibitor from black rice bran. *J. Agric.Food Chem.* 2003, 51, (24), 6953-6956.

[46] Meng, F.; Wei, Y.; Yang, X., Iron content and bioavailability in rice. *Journal of Trace Elements in Medicine and Biology* 2005, 18, (4), 333-338.

[47] Jiang, S.; Wu, J.; Thang, N.; Feng, Y.; Yang, X.; Shi, C., Genotypic variation of mineral elements contents in rice (Oryza sativa L.). *European Food Research and Technology* 2008, 228, (1), 115-122.

[48] Choi, B.; Kum, J.; Lee, H.; Park, J., Physicochemical properties of black rice flours (BRFs) affected by milling conditions. *Korean journal of food science and technology* 2006, 38, (6), 751.

[49] Yang, D. S.; Lee, K.-S.; Jeong, O. Y.; Kim, K.-J.; Kays, S. J., Characterization of volatile aroma compounds in cooked black rice. *Journal of agricultural and food chemistry* 2007, 56, (1), 235-240.

[50] Tanchev, S.; Ioncheva, N., Products of thermal degradation of the anthocyanins cyanidin-3-glucoside, cyanidin-3-rutinoside and cyanidin-3-sophoroside. *Food / Nahrung* 1976, 20, (10), 889-893.

[51] Lin, C.-Y.; Huang, C.-S.; Huang, C.-Y.; Yin, M.-C., Anticoagulatory, antiinflammatory, and antioxidative effects of protocatechuic acid in diabetic mice. *Journal of agricultural and food chemistry* 2009, 57, (15), 6661-6667.

[52] Mingwei, Z.; Encheng, C.; Yan, Z.; Ruifen, Z.; Jianwei, C.; Zhencheng, W.; Xiaojun, T., Effect of technological conditions on -aminobutyric acid accumulation in germinated indica black rice [J]. *Transactions of the Chinese Society of Agricultural Engineering* 2007, 3.

[53] Lee, Y.-R., LYR; Kim, J.-Y., KJY; Woo, K.-S., WKS; Hwang, I.-G., HIG; Kim, K.-H., KKH; Kim, K.-J., KKJ; Kim, J.-H., KJH; Jeong, H.-S., JHS, Changes in the chemical and functional components of Korean rough rice before and after germination. *Food Science and Biotechnology* 2007, 16, (6), 1006-1010.

[54] Sim, G.-S., Lee, Dong-Hwan, Kim, Jin-Hwa, Lee, Bum-Chun, Ahn, Sung-Kwan, Choe, Tae-Boo, Pyo, Hyeong-Bae Sprouted black rice oligopeptide induces expression of hyaluronan synthase in HaCaT keratinocytes and improves skin elasticity *Journal of the Society of Cosmetic Scientists of Korea* 2006, 32, (1), .7-15

[55] Lee, J.; Oh, M., Quality characteristics of cookies with black rice flour. *Korean J Food Cookery Sci* 2006, 22, 193-203.

[56] Kim, Y.; Kim, G.; Lee, J., Quality characteristics of black rice cookies as influenced by content of black rice flour and baking time. *J Korean Soc. Food Sci. Nutr.* 2006, 35, (4), 499-506.

[57] Jung, D.; Eun, J., Rheological properties of dough added with black rice flour. *Korean J. food sci. technol.* 2003, 35, (1), 38-43.

[58] Jung, D.; Lee, F.; Eun, J., Quality properties of bread made of wheat flour and black rice flour. *Korean journal of food science and technology* 2002, 34, (2), 232-237.

[59] Kim, W.; Kim, T.; Lee, Y., A Study on the rheological and sensory properties of bread added waxy black rice flour. *Korean Journal of Food and Cookery Science* 2008.

[60] Kim, J.-D., Physicichemical characteristics of black rice varieties and puffing of rice cake using black rice and medium-grain brown rice *MS thesis* 1998, Gwangju, Chonnam National University.

[61] Kim, J.; Lee, J.; Hsieh, F.; Eun, J., Rice cake production using black rice and medium-grain brown rice. *Food Science and Biotechnology* 2001, 10, (3), 315-322.

[62] Yoon, G., Effect of partial replacement of rice flour with black or brown rice flour on textural properties and retrogradation of Julpyun. *J Korean Home Economics Association* 2001, 39, 103-111.

[63] Lee, J.; Kim, J.; Hsieh, F.; Eun, J., Production of black rice cake using ground black rice and medium-grain brown rice. *International journal of food science & technology* 2007, 43, (6), 1078-1082.

[64] Park, Y.; Chang, H., Quality characteristics of sponge cakes containing various levels of black rice flour. *Korean journal of food science and technology* 2007.

[65] Cho, M., The characteristics of soybean Dasik in addition of black pigmented rice. *Korean J Food & Nutr* 2006, 19, 58-61.

[66] Kim, J.; Kim, K.; Eun, J., Storage of black rice using flexible packaging materials. *KoreanJ. Food Sci. Technol.* 1999, 31, (1), 158-163.

In: Progress in Food Science and Technology, Volume 1   ISBN: 978-1-61122-314-9
Editor: Anthony J. Greco   © 2012 Nova Science Publishers, Inc.

Chapter 6

IN-SEASON WHEAT YIELD PREDICTION IN THE SEMIARID PAMPA OF ARGENTINA USING ARTIFICIAL NEURAL NETWORKS

*A. Bono[1], J. De Paepe[2] and R. Álvarez[*2]*
[1]EEA Anguil INTA, Anguil La Pampa, Argentina
[2]Facultad de Agronomía, Universidad de Buenos Aires-CONICET,
Buenos Aires - Argentina

ABSTRACT

More than 40 % of wheat is produced in the Argentinean Pampa under semiarid conditions, in the so called Semiarid Pampean Region. In this region, seeded area rounded 2.2 Mha with and average crop yield of 2000 kg ha^{-1} during the last years. Yield varied depending on soil fertility, texture, water content, rainfall and management. The objectives of this study were 1) to generate an explicative wheat yield model that accounted for the effect of environmental and cultural variables affecting yield in the Semiarid Pampa and 2) to generate a predictive model for in-season crop yield estimation. An analysis was performed from results produced in 85 field experiments from 1996 to 2004, in which the combination of soils x climate x management conditions resulted in 912 yield data. Surface regression and artificial neural networks (ANN) methodologies were tested and compared for data analysis. A surface regression response model was developed that included soil moisture, organic nitrogen, nitrate nitrogen and fertilizer nitrogen, soil phosphorus, soil depth, tillage system, previous crop and rainfall as independent variables that accounted for 78 % of yield variance. An ANN was also fitted to the data accounting for 90 % of the yield variance which used as inputs the same variables included in the regression model. Comparing predicted vs. observed yields resulted in a lower (P = 0.05) RMSE using the ANN than using the regression method (381 vs. 552 kg ha^{-1}). A predictive ANN model that accounted for 88 % of the yield variability could also be developed excluding rainfall as input. In-season prediction of wheat yield can be performed with this later ANN model by farmers and decision-makers.

---

* Facultad de Agronomía, Universidad de Buenos Aires-CONICET. Av. San Martín 4453 (1417) Buenos Aires - Argentina. E-mail: ralvarez@agro.uba.ar

Keywords: wheat yield, artificial neural networks, Semiarid Pampean Region.

1. CHARACTERISTICS OF THE ARGENTINE SEMIARID PAMPA AND ITS WHEAT PRODUCTION

The Argentine Pampa is located between 28-40°S and 68-57°W and it covers approximately 50 Mha (Alvarez and Lavado, 1998). The region is a vast fertile plain, considered as one of the most suitable areas for grain production worldwide (Satorre and Slafer, 1999), with a relief that is flat or slightly rolling and in which grasslands represent its natural vegetation. Mean annual rainfall varies from 200 mm to 1200 mm from west to east and mean annual temperature ranges from 14 °C in the south to 23 °C in the north (Hall et al., 1992; Servicio Meteorológico Nacional, 2010). On well-drained soils, mainly Mollisols formed on loess like materials, from the humid and semiarid portions of the region, a regular feature of land use is agriculture and areas with hydromorphic soils are devoted to pastures (Hall et al., 1992). A texture gradient exists, with coarser soils in the southwestern area and finer textures in the northeast area, resulting from the past aeolian sediment deposition. Following the pattern of rainfall decrease, organic matter content of the plough layer shows marked geographical changes, diminishing westwards from more than 4% to 1.5 % (Hall et al., 1992; Moscatelli et al., 1991). At present, wheat (*Triticum aestivum* L.) is one of the main crops widespread over the region with an annual seeded surface of nearly 5 Mha (MinAgri 2010; Hall et al., 1992).

Wheat is the single most important crop on a global scale in terms of total production used for human and animal nutrition (Evans, 1998). During the 20th Century, global wheat yield increased considerably, from less than 1000 to ca. 2500 kg ha^{-1} (Slafer et al., 1996). Two factors were largely responsible for the production levels achieved by farmers worldwide: better management practices and new improved genotypes, and their importance vary spatially and temporally (Slafer and Kernich, 1996, Cassman 1999). The components of the technological improvement are diverse and go from optimal sowing dates and seed densities to the amount of fertilizers, pesticides and irrigation applied (Calderini and Slafer, 1999). The contribution of the genetic improvement on wheat yield is mainly due to changes in harvest index (Austin et al., 1980). Modern cultivars are more responsive to environmental conditions than older ones, and a negative relationship between yield and its stability has already been reported (Calderini and Slafer, 1999, Verón et al., 2004).

In the Pampa, during the last century, important changes took place in the wheat production technology (MinAgri, 2010; Viglizzo et al., 2001). At the beginning of the 20th Century, wheat crops occupied an area of ca. 4.5 Mha and yielded in average 650 kg ha^{-1} per year (Barsky and Gelman, 2001). By the end of the century wheat production was 4.5 times greater than in 1990, mainly because of a four-fold increase in yield and a rather small growth of the sown area devoted to this crop (MinAgri, 2010). The major changes in the sown area occurred in the humid portion of the Pampa (Paruelo et al., 2006). Wheat crops were mostly grown as part of a 4-year agriculture/6-year pasture rotation with an increasing area of continuous agriculture that can include different crop rotations (Satorre and Slafer, 1999, Viglizzo et al., 2001). In the Pampean Region wheat crop is a winter crop. Depending on the genetic material and location, sowing starts late May to early August and flowering takes place between late September and early November (Hall et al., 19992, Satorre and Slafer, 1999, MinAgri, 2010).

The Semiarid Pampa Region (between 36° S and 63° W) occupies around 40 % of the whole Pampean Region (Figure 1). The area lacks a fluvial network, and the flat landscape is only broken by ridges of fixed sand dunes (Soriano, 1991). The combination of high maximum temperatures (ca. 40 °C) with fairly brief droughts during the summer months, can lead to loss of summer crops through poor pollination (Hall et al., 1992). The region is characterized by the variability of its annual rainfall, varying between 500 and 700 mm depending on the latitude, and by water deficit throughout the year (Hall et al., 1992, Soriano 1991). Soils are deep with predominant sandy and loamy textures and its representative soils are Entic Haplustolls and Typic Ustipsaments (Hall et al., 1992). In some soils a petrocalcic layer is present that can limit crop yield (Sadras and Calviño, 2001). In the Semiarid Pampa, wheat is the most important winter crop with an average seeded surface of 2.2 Mha and a mean production of 2000 kg ha^{-1}, which vary depending on soil moisture content, rainfall during the crop growing period, management practices and soil fertility (Lorda et al., 2004).

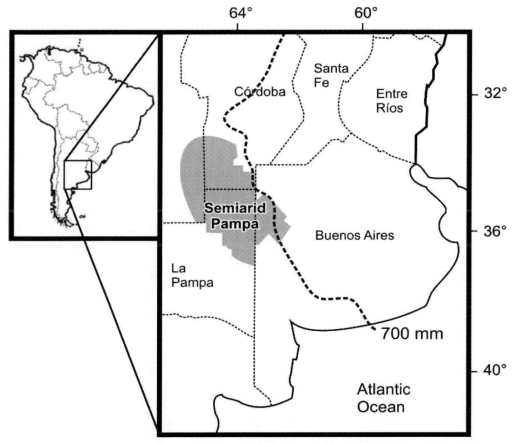

Figure 1. Map of the provinces of Buenos Aires, La Pampa, Córdoba, Santa Fe and Entre Ríos that integrate the Argentine Pampa. The 700 mm isoline is shown which separates the semiarid region from the more humid region. The gray area denotes the experimental network area.

2. FACTORS AFFECTING WHEAT YIELD IN THE PAMPA

At a regional scale, under optimal soil moisture conditions and without nutritional limitations, 50 % of the variability of wheat yield is explained by the photothermal quotient (ratio temperature/radiation) (Magrin et al., 1993). It has not been possible to demonstrate that this quotient is able to explain such percentage of yield variability under water constrained scenarios and/or nutritional limitations. In fact, under field conditions across the region, Verón et al. (2004) found that only 13% of the interannual wheat yield variability is explained by the photothermal quotient. Rainfall is another climatic source of variation and environmental control factor of wheat yield in pampean agroecosystems (Satorre and Slafer, 1999). As a substantial part of rainfall occurs after wheat harvest, it is not useful for the crop (Quiroga et al., 2005).

When analysing soil data from the whole Pampa Region, across a very wide range of environmental variability, it has been demonstrated that 39 % of interannual wheat yield variation is explained by the water storage capacity of the soil and 26 % of the variation is accounted for by its organic carbon content, both considered up to a depth of one meter (Alvarez, 2009). Crop yield is influenced by soil organic matter mainly because of its function as a nitrogen source (Alvarez et al., 2002). In field experiments in the Semiarid Pampa, comparing plots with a similar agricultural history, previous crop (always sunflower), crops managed with tillage and without fertilizer application, organic matter content in the upper 20 cm of the soil profile has been pointed out as the main factor related to wheat yield, accounting for 48 % of its variability (Díaz-Zorita et al., 1999). In the mentioned study, wheat yield followed a lineal-plateau trend with a critical level at 72 t ha^{-1} organic matter. A loss of 1 Mg of organic matter per hectare represented a loss of 40 kg grain per hectare. Nevertheless, for the same region, it has been demonstrated that when the soil water holding capacity increases, as a consequence of finer soil textures or deeper soil profiles, the organic matter content increases too (Alvarez et al., 2006, Quiroga ad Funaro, 2004). This result can be the consequence of two factors: a greater crop residue input and the protective effect of clay from mineralization (Quiroga ad Funaro, 2004). This implies that in the research of Díaz-Zorita et al. (1999), a possible confounding effect could exist between soil organic matter and soil water holding capacity effects on yield. Loveland et al. (2003) supports this questioning by posting that until present, no organic matter threshold has been determined successfully as a yield limiting factor in different worldwide studies. In the humid portion of the Pampa Alvarez et al. (2008) could not demonstrate a relationship between soil organic matter content in the 0-20 cm soil layer and crop yield.

Management practices also have a strong impact on wheat yield. In the Semiarid Pampa, it has been demonstrated that longer fallow periods improve the soil water content and nitrate concentration compared to the contents resulting from shorter fallow periods, affecting crop productivity (Quiroga et al, 2005). Limited tillage systems, and especially no-till practices, have been widespread in recent years, occupying around 70 % of the surface devoted to annual crops (Alvarez, 2009, Díaz-Zorita et al., 2002). No-till practices are characterized by chemical weed control and by leaving 15 % or more of the previous crop aboveground biomass residues on soil surface, decreasing soil erosion (Unger, 1994, Díaz-Zorita et al., 2002). Moreover, under no-till, soil water content is higher because of the higher water infiltration rate in the region (Panigatti et al., 1998). Usually, soils have higher nitrate levels under intensive tillage systems than under conservational systems, mainly attributed to the fact that in tilled systems, crop residues are incorporated into the soil and the decomposition process is faster, releasing nitrogen during fallow (Abascal et al., 2003). In a review of the effect of the tillage system on wheat yield at pampean scale, it was demonstrated that in the humid portion of the region yields were 13 % lower under no-till systems than under tilled managements without nitrogen fertilization (Alvarez and Steinbach, 2009). Only a few researches exist on the effect of the tillage system on crop yield in the Semiarid Pampa. The mentioned review found a tendency in the results of four experiments to obtain higher yields (6 %) under no-till systems with nitrogen fertilization than under conventional tillage managements. The adoption of conservation tillage systems in the Argentine Pampa creates a generalized need to increment the use of nitrogen fertilizer to maintain optimal wheat yield (Alvarez and Steinbach, 2009).

Fertilization has not been common until recent years in the Pampa Region and the mean application rate to wheat during the 1980s rounded 30 kg ha^{-1} (Obschatko and del Bello, 1986). The application of fertilizers in the Pampa was low because of the high natural fertility of its soils, but in the last 15 years their use incremented almost eight-times (Satorre, 2005, Alvarez et al., 2003). Some information is available on the response of wheat yield to fertilization but it has not been integrated into a recommendation model until now in the semiarid portion of the region (Bono and Quiroga, 2003, Bono et al., 2004).

3. DEVELOPMENT OF WHEAT YIELD MODELS FOR THE SEMIARID PAMPA

3.1. Introduction

In the a humid portion of the Pampa explicative wheat yield models have been developed linking climate, soil and management factors (Alvarez and Grigera, 2005, Calviño et al., 2002), but there exist no such models for the Semiarid Pampa. In this region it is important to identify yield predictive variables in order to determine soil productivity and perform trade decisions before crop harvest. In this study, we first assess the generation of wheat yield models using field experiment data and then we attempt to produce a model that is useful for in-season yield prediction.

3.2. Data Generation

Results of an analysis from 85 field experiments conducted in the Semiarid Pampa during the period 1996-2004 were performed. The experiments were installed across the region (Figure 1) on different soil types, with contrasting site characteristics and management conditions (Table 1). Two tillage systems were used, that are typical for the region, no till and conventional tillage using disk plow.

Table 1. Number of experiments performed as a function of soil type, texture, depth, tillage system and previous crop. Sandy soils are soils with more than 70 % sand and loamy soils have less than this percentage. Deep soils have a depth of more than 60 cm and shallow soils are less than 60 cm deep. Conventional tillage used in the Semiarid Region is usually disk plow. The previous crops were grouped as pastures (lucerne) and soybean or other crops (wheat, sunflower or sorghum)

Soil type	Soil texture		Soil depth		Tillage system		Previous crop	
	Sandy	Loamy	Deep	Shallow	No-till	Conventional tillage	Pasture-Soybean	Other crops
Calciustolls	0	2	0	2	0	2	1	1
Ustipsaments	4	0	4	0	1	3	3	1
Hapludolls	5	7	12	0	8	4	17	50
Haplustolls	16	51	24	43	11	56	7	5

The fertilized treatments varied as different fertilizer rates and application times were used (Table 2). Nitrogen fertilizer was applied broadcasted at sowing or during the tillering period, and phosphorus was incorporated banded at sowing. The experimental network generated 912 wheat yield data through the combination of sites across the region, fertilization treatments and years. Different wheat cultivars and sowing dates were used and the management practices were determined by local farmers.

Table 2. Fertilization treatments contrasted in the experimental network

Fertilization treatments contrasted	Nutrient sources	N rate ($kg\ ha^{-1}$)	P rate ($kg\ ha^{-1}$)	Time of N application	Number of experiments
Different times of N applied with variable rates of N and P	Urea and superphosphate	0, 50 and 100	0, 15 and 30	Spread: sowing and tillering	35
N rates combined with different P sources	Urea, superphosphate or diammonium phosphate	variable	0 and 30	Sowing	10
N sources and rates	Urea, ammonium sulfate or urea ammonium nitrate	0, 50, 75 and 100	0	Sowing or tillering	4
N rates	Urea	0, 40 and 80	0	Sowing	4
N rates combined with P application	Urea and superphosphate	0, 50 and 100	0 and 30	Sowing	9
Time of N application and rates	Urea	0, 50 and 100	0	Sowing or tillering	23

Plot size was 100 m^2 (10 x 10 m) and the experimental design was variable. Most of the experiments were blocked with three replicates at random. Before sowing, soil was sampled (0-20 cm) and organic matter determined using the Walkey and Black method (Walkley and Black, 1934), organic nitrogen by the Kjeldahl procedure (Bremmer, 1965), and extractable phosphorus using the Bray and Kurtz N° 1 technique (Bray and Kurtz, 1945). Bulk density was determined with 20 cm steel cylinders with a volume of 250 cm^3 in layers of 20 cm up to a depth of 140 cm or up to the presence of a petrocalcic layer. Soil water was determined by the gravimetric method on bulk density samples. Soil nitrates were determined by the colorimetric method of cromotropic acid (**Bremmer, 1996**) in layers of 20 cm up to a depth of 60 cm or up to the presence of a petrocalcic layer. At harvest, grain yield was evaluated (14 % water content). Rainfall was registered during the growing season (Table 3).

Table 3. Means and ranges of some soil properties, fertilizer rates, soil moisture content, rainfall, and wheat yield along the experimental network

Parameter	Organic matter 0 - 20 cm (%)	Extractable P 0 - 20 cm (ppm)	Fertilizer P (kg ha^{-1})	Nitrate N 0 - 60 cm (kg ha^{-1})	Fertilizer N (kg ha^{-1})	Moisture content at sowing (mm)	Rainfall during growing season (mm)	Yield (kg ha^{-1})
Mean	1.8	18	12	55	69	240	398	2100
Maximum	3.2	52	30	220	166	450	713	6900
Minimum	0.60	2.4	0	16	0	66	48	180

3.3. Regression Techniques

The quadratic polynomial regression method was used. Linear and quadratic terms incorporated in this multiple regression technique, assess for linear and curvilinear effects of independent variables, and the interaction terms between independent variables were also tested (Neter et al., 1990). In agronomic experiment evaluation, this technique has been of common use, with expected positive linear effects and negative quadratic effects (Colwell, 1994). The effect of each independent variable on yield can be assessed separately along with the interaction between them (Laird and Cady, 1969; Nelson et al., 1985). The polynomial response surface model has this form:

$$\text{Yield} = a_0 + a_1 v_1 - a_2 v_1^2 + a_3 v_2 - a_4 v_2^2 + a_5 v_1 v_2 + \ldots + a_{n-2} v_x - a_{n-1} v_x^2 + a_n v_x v_{x-1}$$

Where:

a_0 to a_n: regression coefficients
v_1 to v_x: independent variables

In order to obtain the simplest model and the one with the highest R^2, a combination of forward, backward and stepwise regression adjustments were used. The final regression model was selected if it was significant at $P = 0.01$ determined by the F test and it included only statistically significant terms (at $P = 0.05$). The VIF (Variance Inflation Factor) value was used to check the autocolinearity of independent variables. Only values lower than 5 were accepted (Neter et al., 1999).

For assessing the generalization ability of the fitted models, a cross validation technique was used and the dataset was randomly partitioned into two datasets, a training set (70 % of data) for the model construction and an independent validation set (remaining 30 % of data) to test models. A hierarchical approach was tested to combine independent variables for calculation of other variables with the purpose of including the effects of the variables in the first level and allowing the simplification of the selected models (Schaap et al., 1998).

Categorical variables like previous crop, tillage system, soil depth, texture and fertilization time were tested as dummy variables in the models. The determination coefficients of training and validation sets were contrasted (Kleinbaum and Kupper, 1979). The ordinates and the slopes of the regression of observed vs. predicted data were tested by the t test using IRENE (Fila et al., 2003).

The best regression model fitted could explain 78 % of wheat yield variability (Figure 2). Yield was positively impacted by soil moisture content at sowing, rainfall, organic nitrogen, soil phosphorus and nitrates and nitrogen fertilizer rate. Wheat yield was also higher in deeper soils of finer textures, when the crop was managed under no till and with soybean or pastures as previous crops.

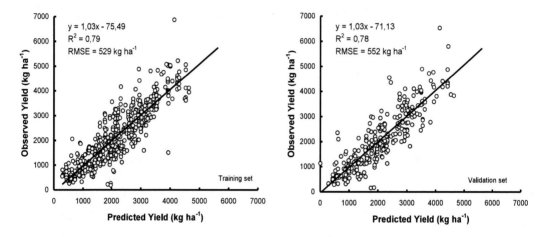

Figure 2. Regression of observed vs. predicted wheat yield estimated by a regression model.

The model has a good generalization capacity because no significant differences were detected between the R^2 of the training and validation sets. Ordinates were not different from 0 and slopes from 1. Soil moisture content at sowing and variables related to water storage capacity of soil had a strong influence on yield as depicted by the model. Soil depths of more than 60 cm along with textures with less than 70 % of sand resulted in higher wheat yields. These results have been previously observed in the region (Bono et al., 1997, Bono and Quiroga 2003).

3.4. Artificial Neural Networks

Due to the fact that they are simpler than process based models and because of their predictive quality, artificial neural networks (ANN) have become a popular technique in biological and agronomic sciences (Joergensen and Bendoricchio, 2001, Özesmi et al. 2006). ANN are adaptive analytical methodologies, based on neural structures and processing of the brain, capable of learning relationships in patterns of information (Joergensen and Bendoricchio, 2001). ANN have the advantage of not assuming an a priory structure for the data, as is the case for many other empirical modeling techniques. They are well suited for fitting complex interactions and non-linear relationships, and can expose hidden relationships among input variables (Batchelor et al., 1997).

The typical ANN structure is of three neural layers: input, hidden and output layer. The information flow starts at the input layer, through the hidden layer, and ends in the output layer. The input layer is the layer in which the number of neurons corresponds to the number of input variables, the hidden layer has a complexity determined empirically during the ANN development, and the output layer has a neuron for each output variable (Figure 3). Usually, it is the backpropagation algorithm that fits the weights during the learning process, starting at the output layer and all through the input layer (Kaul et al., 2005). The learning process consists in adjusting weights associated to the transfer functions between neurons of the different layers and comparing the ANN outputs with observed data by an iterative procedure (Joergensen and Bendoricchio, 2001). Commonly, a lineal function passes information from the input layer to the hidden layer, and a sigmoid transfer function is used between the hidden layer and the output layer (Kaul et al., 2005). The results of the ANN cannot be extrapolated outside the range of input data, and this is a common feature of empirical models.

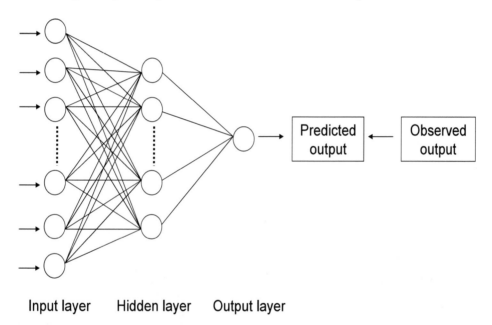

Figure 3. Representation of a feed-forward artificial neuronal network showing layers and connections.

Soil organic carbon stock prediction (Somaratne et al., 2005), fertilization recommendations (Broner and Comstock 1997), soil hydraulic property estimations (Nemes et al., 2003), crop development prediction (Elizondo et al., 1994), epidemic severity evaluation (Batchelor et al., 1997), and yield prediction (Alvarez, 2009. Kaul et al., 2005) are some examples of the agronomic ANN uses (Park and Vlek 2002).

We tested the ability of ANN to predict wheat yield using network results. To avoid overlearning (overfitting) when modelling wheat yield using ANN, the dataset was partitioned into training and validation sets as described for the regression model development. The iteration procedure was stopped when the R^2 of the validation set became smaller than the one from the training set. The back propagation algorithm was used to run the artificial neural network through a supervised learning procedure (Rogers and Dowla, 1994) and the best chosen model had the highest R^2. Input variables tested were: organic nitrogen, nitrogen from nitrates, nitrogen from fertilizers, extractable phosphorus, fertilizer phosphorus, soil moisture at sowing, soil depth and texture, tillage system, previous crop and rainfall during the crop growing cycle. With these inputs, an ANN model was constructed. The wheat yield prediction capacity of the selected network was assessed by contrasting the R^2 of training and validation sets as stated for regression models. The RMSE of the models fitted by regression techniques and ANN were compared by an F test (Snedecor and Cochran, 1967).

The best ANN obtained was able to explain 90 % of the wheat yield variability (Figure 4). The network has a RMSE of 381 kg ha^{-1}, and this value is equivalent to the 20 % of average wheat yield. The regression observed vs. predicted yield had an ordinate non-differing from 0 and a slope similar to 1. There were no significant differences between the R^2 of the training and validations sets. The inputs selected by the ANN were the same included in the regression model and had 12 neurons in the hidden layer. When comparing the yield model generated by regression with the ANN, RMSE of the ANN resulted smaller, both for the training and validation sets (P = 0.05). All continue variables have curvilinear effects and a general positive trend. Yield was also higher, as predicted by the ANN model, in deeper soils of finer textural composition, under no-till or when previous crop were pastures or soybean. The model allows a good estimation of yield but needs rainfall data to perform its predictions, meaning that it has no use for yield predictions before harvest.

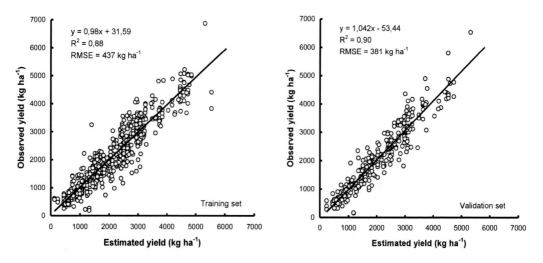

Figure 4. Regression of observed vs. predicted wheat yield estimated by an artificial neural network.

Soil moisture at sowing had a positive effect on wheat yield as it was a main determinant of crop water availability in our semiarid soils and was the variable that explained the greatest part of wheat yield variability. The ANN estimated a positive effect of rainfall on yield, and this has been demonstrated before in this region (Díaz-Zorita et al., 1994). Soil depth and fine texture also had positive effects on wheat yield because of their impact on the soil water storage capacity. The ANN also estimated a positive effect on wheat yield of organic nitrogen, soil nitrates and fertilizer nitrogen, being the latter the most important. Investigations in Semiarid Pampa have demonstrated that nitrogen fertilizers influence wheat yield positively when water is a non-limiting factor (Galantini et al., 2000). Wheat yield was also positively influenced by the extractable phosphorus content; although in the Semiarid

Pampa, medium to high levels of soil phosphorus (> 20 ppm) were observed in many experiments. Rainfall accounted for a small effect on yield (3 %). When this input was excluded from the ANN, the R^2 only dropped to 0.88. Eliminating rainfall as an input allowed the development of a predictive yield model for in-season prediction. Some previous studies performed in the Semiarid Pampa showed significant effects of rainfall during the wheat growing season on yield (Cantamutto and Möcket, 1990, Gallo Candolo, 2001, Galantini et al., 2000). If rainfall deficit conditions occur, crop water requirements should be provided by the soil moisture availability (Scian and Bouza, 2005), and specifically in this region the initial soil water content at sowing time (Travasso et al., 1994) and the soil water storage capacity (Díaz-Zorita et al., 1991, Sadras and Calviño, 2001) have been pointed out as of particular importance. Our results showed that wheat yield is mainly controlled by soil water content and its water storage capacity with a minor contribution of rainfall.

The model uses variables that can be available at sowing and with an RME equivalent to the 22 % of the average experimental network wheat yield. This model can be used by technicians and producers for the early estimation of wheat yield. The methodology can also be useful for other crops in other regions to improve production estimations that at present are performed based on field surveys nearly before harvest. Different methodologies have been tested in other cropping areas worldwide for in-season yield prediction but these techniques are not available at present in the Pampean Region; for instance the use of the NDVI index (Freeman et al. 2003), or the application of agro-climatic models (Potgieter et al., 2005).

CONCLUDING REMARKS

In this chapter, it was demonstrated that ANN can generate estimating wheat yield models with a better performance than multiple regression techniques. In fact, the fitted regression model could explain up to 78 % of the wheat yield variability while the ANN explained 90 % of the same variability and its RMSE was significantly smaller. The most important independent variable affecting wheat yield was soil moisture content at sowing. A simplified ANN model, which excludes rainfall as input, allowed a good in-season prediction of yield. It is important to note that this methodology can be applied in other cropping areas worldwide and for other crops.

REFERENCES

Abascal, S. A., D. E. Buschiazzo and H. R. Mirasson. 2003. Almacenaje de agua y nitratos por barbechos para girasol en un suelo Haplustol Éntico de la Región Semiárida Pampeana Central. Cultivos de cosecha gruesa. *Boletín de Divulgación Técnica 77 INTA* Anguil: 85-89.

Alvarez, C., N. Peinemann and A. R. Quiroga. 2006. *Sistemas de labranza, propiedades edáficas y rendimiento de maíz en molisoles diferenciados por el régimen hídrico.* Publicación Técnica EEA Anguil INTA N° 67: 1 - 10.

Alvarez, R. 2008. Predicción del rendimiento y la producción de trigo en la Región Pampeana usando una red neuronal artificial. *Congreso Nacional de Trigo. INTA.* Santa Rosa, La Pampa, Argentina: pp. 5.

Alvarez, R. 2009. Predicting average regional yield and production of wheat in de Argentine Pampas by an artificial neural netword approach. *Europ. J. Agronomy* 30: 70-77.

Alvarez, R. 2009. Predicting average regional yield and production of wheat in the Argentine Pampas by an artificial neural network approach. *Eur. J. Agron.* 30, 70-77

Alvarez, R. and H. S. Steinbach. 2009. A review of the effects of tillage systems on some soil physical properties, water content, nitrate availability and crops yield in the Argentine Pampas. *Soil and Tillage Research* 104 (1): 1-15.

Alvarez, R. and H. Steinbach. 2006. *Efecto de la agricultura sobre el nivel de materia orgánica, Editorial Facultad de Agronomía* (UBA), editor R. Alvarez, capítulo 4.

Alvarez, R. and R. S. Lavado. 1998. Climate, organic matter and clay content relationships in the Pampa and Chaco soils, Argentina. *Geoderma* 83: 127-141.

Alvarez, R. and S. Grigera. 2005. Analysis of soil fertility and management effects on yields of wheat and corn in the Rolling Pampa of Argentina. *J. Agronomy & Crop Science* 191: 321-329.

Alvarez, R., C. R. Alvarez and H. Steinbach. 2002. Association between soil organic matter and wheat yield in humid pampa of Argentina. *COMMUN. SOIL SCI. PLANT ANAL.* 33 (5&6): 749-757.

Alvarez, R., H. Steinbach, C. R. Alvarez and S. Grigera. 2003. Recomendaciones para la fertilización nitrogenada de trigo y maíz en la Pampa Ondulada. *Informaciones Agronómicas del Cono Sur* 18 (Junio): 1-12.

Andriulo, A. and G. Cordone. 1998. Impacto de labranzas y rotaciones sobre la materia orgánica de suelos de la Región Pampeana Húmeda. Siembra Directa. Eds JL Panigatti, H. Marelli, D. Buschiazzo, R. Gil. *Buenos Aires, Argentina, Hemisferio* Sur: 65-96.

Austin, R.B., Bingham, J., Blackwel, R., Evans, L., Ford, M., Morgan, C., Taylor, M., 1980. Genetic improvement in winter wheat since 1900 and associated physiological changes. *J. Agric. Sci. Camb.* 94, 675–689.

Bakker, M. M., G. Govers, F. Ewert, M. Rounsevell and R. Jones. 2005. Variability in regional wheat yields as a function of climate, soil and economic variables: Assessing the risk of confounding. *Agriculture, Ecosystems & Environment* 110 (3-4): 195-209.

Barberis, L., A. Nervi, H. Del Campo, M. Conti, S. Urricariet, J. Sierra and P. Daniel. 1983. Análisis de la respuesta del trigo a la fertilización nitrogenada en la Pampa Ondulada y su predicción. *Ciencia del Suelo* 1: 57-64.

Barsky, O. and Gelman, J., 2001. *Historia del Agro Argentino.* Grijalbo Mondador, Buenos Aires.

Batchelor, W. D., X. B. Yang and A. T. Tschanz. 1997. Development of a neural network for soybean rust epidemics. *Trans. ASAE* 40: 247-252.

Bending, G. D., M. K. Turner, F. Rayns, M.-C. Marx and M. Wood. 2004. Microbial and biochemical soil quality indicators and their potential for differentiating areas under contrasting agricultural management regimes. *Soil Biology and Biochemistry* 36 (11): 1785-1792.

Berardo, A. 1994. Aspectos generales de la fertilización y manejo del trigo en el área de influencia de la estación experimental INTA-Balcarce. Boletín Técnico N°128. , *Estación Experimental Balcarce INTA:* 34.

Bolinder, M., D. Angers, E. Gregorich and M. Carter. 1999. The response of soil quality indicators to conservation management. *Can. J. Soil Sci.* 79: 37-45.

Bono A and A Quiroga. 2003. Avances en el ajuste de la fertilidad en el cultivo de trigo en la región semiárida y sughúmeda pampeana. In: Trigo: Actualización 2003. *Boletín de Divulgación Técnica N° 76.* EEA Anguil INTA. pp 17-32.

Bono A and A Quiroga; R Jouli; A Corro Molas. 2004. Estrategias para el manejo de la fertilidad en suelos de la Región Semiárida Pampeana. In: Trigo: Actualización 2004. *Boletín de Divulgación Técnica N° 83.* EEA Anguil INTA. pp 29-33.

Bono A; JC Montoya; P Lescano and FJ Babinec. 1997. *Fertilización de trigo con nitrógeno y fósforo el la región semiárida pampera.* Campaña 1997. Publicación Técnica N° 47. EEA Anguil INTA. 22 pag.

Bono, A. and Alvarez, R., 2006. Rendimiento de trigo en la Región Semiárida y Subhúmeda Pampeana: un modelo predictivo de la respuesta a la fertilización nitrogenada. In: *XX Congreso Argentino de la Ciencia del Suelo*, Proceedings on CD, p. 5.

Bono, A. and Alvarez, R., 2008. Estimación del rendimiento de trigo en la Región Semiárida y Subhúmeda Pampeana usando redes neuronales artificiales. In: *VII Congreso Nacional de Trigo*, Proceeding on CD, p. 16.

Bray, R.H., and L.T. Kurtz, 1945. Determination of total, organic and available forms of phosphorus in soils. *Soil Sci.* 59:39-45.

Bremmer, J.M. 1965. Organic forms of soil nitrogen. P. 1238-1255. IN C.A. Black et al. (ed.) Methods of soil analysis. Part 2. *Agron. Monogr.* O ASA and SSSA, Madicson, WI.

Bremmer, J.M. 1996. Nitrogen-total. P. 1058-1121. In D.L. Sparks et al. (ed) Methods of soil analysis. Part 3. *Chemical methods.* SSSA Book Ser. 5. SSSQ and ASA. Madison, WI.

Broner, I. and Comstock, C.R., 1997. Combining expert systems and neural networkds for learning site-specific conditions. *Comp. Elec. Agric.* 19, 37-53.

Buschiazzo, D. E., J. L. Panigatti and P. W. Unger. 1998. Tillage effects on soil properties and crop production in the subhumid and semiarid Argentinean Pampas. *Soil and Tillage Research* 49 (1-2): 105-116.

Buschiazzo, D., E., A. Quiroga, R. and K. Stahr. 1991. Patterns of Organic Matter *Accumulation in Soils of the Semiarid Argentinian Pampas.* Zeitschrift für Pflanzenernährung und Bodenkunde 154 (6): 437-441.

Calderini, D. F. and G. A. Slafer. 1999. Changes in yield and yield stability in wheat during the 20th century. *Field Crops Research* 57: 335-347.

Calviño, P. and V. Sadras. 2002. On-farm assessment of constraints to wheat yield in the south-eastern Pampas. *Field Crops Research* 74: 1-11.

Cantamutto, M.A., and F.E. Möckel. 1990. *Factors conditioning wheat response to nitrogen and phosphorus fertilizers in the southern semiarid Pampas.* p. 439-450.

Cassman, K. G. 1999. Ecological intensification of cereal production systems: yield potencial, soil quality, and precision agriculture. *Proc. Natl. Acad. Sci.* USA 96: 5952-5959.

Colwell, J. 1994. *Estimating fertilizer requirements. A quantitative approach.* UK, CAB International.

Davidson, E. A. and I. L. Ackerman. 1993. Changes in soil carbon inventories following cultivation of previously untilled soils. *Biogeochemistry* 20: 161-193.

Debaeke, P., T. Aussenac, J. L. fabre, A. Hilaire, B. Pujol and L. Thuries. 1996. Grain nitrogen content of winter bread wheat (*Triticum aestivum L.*) as erlated to crop manaegement and to the previous crop. *Eur. J. Agron 5*: 273-286.

Díaz-Zorita, M., D. E. Buschiazzo and N. Pienemann. 1999. Soil organic matter and wheat productivity in the Semiarid argentine Pampas. *Agron. J.* 91: 276-279.

Díaz-Zorita, M., D.E. Buschiazzo and N. Peinemann. 1994. Soil properties and wheat productivity in the semiarid Argentinian Pampas. vol. 5b: Com. IV. P. 462-463. *Trans. Int. Soil Sci. Soc.* Congr. 15th. Acapulco. Mexico. 10-17 July 1994. ISSS, Vienna.

Díaz-Zorita, M., G. A. Duarte and J. H. Grove. 2002. A review of no-till systems and soil management for sustainable crop production in the subhumid and semiarid Pampas of Argentina. *Soil and Tillage Research* 65 (1): 1-18.

Elizondo, D.A., McClendon, R.W. and Hoogenboom, G., 1994. Neural network models for predicting flowering and physiological maturity of soybean. Trans. ASAE 37, 981-988.

Evans, L. T. 1998. Feeding the Ten Billion. *Plants and Population Growth.* Cambridge University Press, Cambridge, UK, 247 pp.

Fila, G., G. Bellocchi and M. Donatelli. 2003. IRENE: a software to evaluate model performance. *Europ. J. Agronomy* 18: 369-372.

Follet, R., J. Kimble and R. Lal. 2001. *The potential of US grazing lands to sequester carbon and mitigate the greenhouse effect.* Washington DC, Lewis Publishers.

Freeman, K.W., Raun, W.R., Jonson, G.V., Mullen, R.W., Stone, M.L. and Soilie, J.B., 2003. Late-season prediction of wheat yield and grain proteína. *Commun. Soil. Sci. Plant Anal.* 34, 1837-1852.

Frye, W., S. Ebelhar, L. Murdock, and R. Blevis. 1982. Soil erosion effects on properties and roductivity of two Kentucky soils. *Soil Sci. Soc. Am. J.* 46: 1051-1055.

Galantini, J. and R. Rosell. 2006. Long-term fertilization effects on soil organic matter quality and dynamics under different production systems in semiarid Pampean soils. *Soil and Tillage Research* 87 (1): 72-79.

Gallo Candolo E (2001) Zona semiarida. In: Cuaderno de actualiazación 63. *Consorcio Regional de Expermientación Agrícola (CREA),* Buenos Aires, pp 139-142.

Hall, A. J., C. Rebella, C. Guersa and J. Culot. 1992. Field-crop system of the Pampas. In: Pearson CJ, editor. *Field crop ecosystems.* Amsterdam, Elsevier.

Islam, K. and R. Weil. 2000. Soil quality indicator properties in mid-Atlantic soils as influenced by conservation management. *Journal of Soil and Water Conservation* 55: 69-78.

J.R. Sims and G. D. Jackson, 1971. Rapid analysis of soil nitrate with chromotropic acid. *Soil Sci. Soc. Am. J.* 35: 603-606.

Joergensen, S. E. and G. Bendoricchio. 2001. *Fundamentals of Ecological modelling*, Third edition. Oxford, UK, pp. 530.

Kaul, M., R. Hill and C. Walthall. 2005. Artificial neural networks for corn and soybean yield prediction. *Agric. Sys.* 85: 1-18.

Kleinbaum, D.G, and L.L Kupper. 1979. *Applied regression analysis and other multivariable methods.* Duxbury Press, Massachusetts, USA, pág. 555.

Kobayashi, K. and M. U. Salam. 2000. Comparing simulated and measured values using mean square deviation and its components. *Agron. J.* 92: 345-352.

Laird, R. and F. Cady. 1969. Combined analysis of yield data from fertilizer experiments. *Agron. J.* 61: 829-834.

Lal, R. 2004. Soil carbon sequestration impacts on global climate change and food security. *Science* 304: 1623-1626.

Lal, R., 1980. Crop residue management in relation to tillage techniques for soil and water conservation. Organic Recycling in Africa. *FAO Soils Bull.* 43. FAO, rome, pp. 72-78.

Lorda H., Lucchetti P., Bellini Saibene Y., Farell M., Zinda R., Sipowicz A. and Coma C. 2004. Caracterización Productiva y Tecnológica del cultivo de trigo. In: *Trigo Actualización 2004 Boletín de Divulgación Técnica N° 83. EEA INTA Anguil.* Pag 1-24.

Loveland, P. and J. Webb. 2003. Is there a critical level of organic matter in the agricultural soils of temperate regions: a review. *Soil and Tillage Research* 70 (1): 1-18.

Magrin, G., A. J. Hall, G. Baldi and M. Grondona. 1993. Spatial and interannual variations in the phototermal quotient: implications for the potential kernel number of wheat crops in Argentina. *Agricultural and Forest Meteorology* 67: 29-41.

Magrin, G., M. Travasso and G. R. Rodriguez. 2005. Changes in climate and crop production during the 20th century in Argentina. *Climatic Change* 72: 229-249.

Malhi, S.S., Gran, C.A., Johnston, A.M., Gill, K.S. and 2001. Nitrogen fertilization management for no-till cereal production in the Canadian Great Plain: a review. *Soil Till. Res.* 60, 101-122.

Martens, D.A., 2000. Nitrogen cycling under different soil management systems. *Adv. Agron.* 70, 143-191.

Ministerio de Agricultura, G. y. P. 2010. *Estimaciones Agrícolas.* Buenos Aires.

Montoya, J. C. and A. Gili. 2005. Barbechos químicos: eficicacia en el control de malezas y acumulación de agua y nitratos en el perfil edáficos. *Campaña* 2004-2005. Capítulo 13. Cultivos de cosecha gruesa. Actualización 2005. Publicación Técnica N° 61. 63-69 pp.

Moscatelli, G. 1991. Los suelos de la Región Pampeana. En: El desarrollo agropecuario pampeano. *INDEC-INTA-IICA.* Buenos Aires: Pp. 1-76

Nelson, L., R. Voss and J. Pesek. 1985. Agronomic and statistical evaluation of fertilizer response. En: Fertilizer technology and use. O. Englestad. Madison, WI, USA, *Soil Science Society of America*: 53-90.

Nemes, A., M. G. Schaap and J. H. M. Wosten. 2003. Functional Evaluation of Pedotransfer Functions Derived from Different Scales of Data Collection. *Soil Sci Soc Am J* 67 (4): 1093-1102.

Neter, J., W. Wasserman and M. Kutnet. 1990. *Applied linear statistical models.* Illinois, USA, Ed. Irwin.

Obschatko, E.S. and del Bello, J.C.k, 1986. Tendencias productivas y estrategia tecnológica para la agricultura Pampeana. *CISEA*, Buenos Aires.

Özesmi, S. L., C. O. Tan and U. Özesmi. 2006. Methodological issues in building, training, and testing artificial neural networks in ecological applications. *Ecological Modelling* 195: 83-93.

Panigatti, J. L., H. Marelli, D. E. Buschiazzo and R. GIl. 1998. Siembra Directa, Ed. *Hemisferio Sur. Buenos Aires*, 333 pág.

Park, S.J. and Vlek, P.L.G., 2002. Environmental correlation of three-dimensional soil spatial variability: a comparison of three adaptive techniques. *Geoderma* 109. 117-140.

Paruelo, J. M., J. P. Guerschman, G. Piñeiro, E. G. Jobbagy, S. R. Verón, G. Baldi and S. Baeza. 2006. Cambios en el uso de la tierra en la Argentina y Uruguay: Marcos conceptuales para su análisis. *Agrociencia* X: 47-61.

Potgieter, A. B., G. L. Hammer, A. Doherty and P. de Voil. 2005. A simple regional-scale model for forecasting sorghum yield across North-Eastern Australia. *agricultural and Forest Meteorology* 132 (1-2): 143-153.

Quiroga, A. R. and D. O. Funaro. 2004. Materia orgánica, factores que condicionan su utilización como indicador de calidad en Molisoles de las Regiones Semiárida y Subhúmeda Pampeana. *XIX Congreso Argentino de la Ciencia del Suelo*. Actas en CD.

Quiroga, A. R., D. E. Buschiazzo and N. Peinemann. 1996. Soil Organic Matter Particle Size Fractions in Soils of the Semiarid Argentinian Pampas. *Soil Science* 161 (2): 104-108.

Quiroga, A. R., D. O. Funaro, R. Fernández and E. Noellemeyer. 2005. Factores edáficos y de manejo que condicionan la eficiencia del barbecho en la Región Pampeana. *Ciencia del Suelo* 23 (1): 79-86.

Reganold, R., A. Palmer, J. Lockhart and A. Macgregor. 1993. Soil quality and financial performance of biodynamic and conventional farm in New Zealand. *Science* 160: 344-349.

RIAP. *Trigo. Manual de campo*. 2005. Ediciones INTA.

Rogers, L. L. and F. U. Dowla. 1994. Optimization of groundwater remediation using artificial neural networks with parallel solute transport modeling. *Water Res. Res.* 30: 457-481.

Romano N. and Z. Roberto. 2007. Contenido de fósforo extractable, pH y materia orgánica en los suelos del este de la provincia de La Pampa. *Informaciones Agronómicas del Cono Sur*. IPNI. 33: 1-7.

Ron M.M. and T. Loewy, 2000. Effect of phosphorus placement of wheat yield and quality in Southwestern Buenos Aires (Argentina). *Communications in Soil Science and Plant Analysis,* 1532-2416, Volume 31, Issue 17

Sadras, V. O. and P. A. Calviño. 2001. Quantification of grain yield response to soil depth in soybean, maize, sunflower, and wheat. *Agronomy Journal* 93: 577-583.

Sanchez, S. R., G. A. Studdert and H. E. Echeverría. 1998. Dinámica de la mineralización del nitrógeno de residuos de cosecha en descomposición en un Argiudol Típico. *Ciencia del Suelo* 16: 1-6.

Satorre, E. and G. Slafer. 1999. Wheat production systems of the Pampas. En: *Wheat. Ecology and physiology of yield determination*. E. T. H. Press. New York: 333-348.

Satorre, E. H. 2005. Cambios tecnológicos en la agricultura argentina actual. *Ciencia Hoy 15* (87 (Junio-Julio)): 24-31.

Schaap, M. G. and W. Bouten. 1996. Modeling water retention curves of sandy soils using neural networks. *Water Res. Res.* 32: 3033-3040.

Scian B.V. and Bouza M.E. 2005. Environmental variables related to wheat yields in the semiarid pampa region of Argentina. Journal or Arid Environments 61: 669-679.

Servicio Meteorológico Nacional 2010. *www.smn.gov.ar*

Silgram, M. and Sherpherd, M.A., 1999. The effects of cultivation on soil nitrogen mineralization. *Adv. Agron.* 65, 267-311.

Sims, J.R., and G.D. Jackson. 1971. Rapid analysis of soil ntrate with chromotropic acid. *Soil Sci. Soc. Am. Proc.* 35: 603-606.

Slafer, G. A. and G. Kernich. 1996. Have changes in yield (1990-1992) been accompanied by a decreased yield stability in Australian cereal production? *Aust. J. Agric. Res.* 478: 323-334.

Snedecor, G.W., and G. Cochran, 1967. *Statistical methods. Sixth Edition*. The Iowa State University Press, Ames, Iowa, USA, 575 pág.

Somaratgne, S., G. Seneviratne and U. Coomaraswamy. 2005. Prediction of soil organic carbon across different land-use patterns: a neural network approach. *Soil Sci. Soc. Am. J.* 69: 1580-1589.

Soriano, A. 1991. *Río de la Plata grasslands. En: Ecosystems of the World,* vol. 8A. Elsevier. Amsterdam, Coupland. R.T. (Ed.): pp. 367-407.

Starrett, S.D., Starret, S.K. and Adams, G.L., 1997. Using artificial neural network and regression to predict percentage of applied nitrogen leached under turfgrass. *Commun. Soil Sci. Plant Anal.* 28, 497-507.

Studdert, G. A. and H. E. Echeverría. 2000. Crop rotation and nitrogen fertilization to manage soil organic carbon dynamics. *Soil. Sci. Soc. Am J.* 64: 1496-1503.

Travasso, M., G. Magrin and M. Grondona. 1999. *Relations between climatic variability related to ENSO and maize production in Argentina.* Proceedings 10th Symposium on Global Change Studies. T. Dallas, AMS. Boston, MA: 67-68.

Unger P.W., 1994. Residue management strategies-Great Plains. In: Hatfield, J. L. (Ed.), Crops Residue Management, *Advances in Soil Science.* CRC Press, Inc., Boca Ratón, pp. 37-61.

Unger, P.W., and T. M. McCalla. 1980. Conservation tillage systems. *Adv. Agron.* 33: 1-58.

Vázquez, M., L. Berasategui, E. Chamorro, L. Taquini and L. Barberis. 1990. Evolución de la estabilidad estructural y diferentes propiedades químicas según el uso de los suelos en tres áreas de la Pradera Pampeana. *Ciencia del Suelo* 8: 203-210.

Vega, C. R. C., V. O. Sadras and F. H. Andrade. 2000. Reproductive allometry in soybean, maize and sunflower. *Ann. Bot.* 85: 461-468.

Verón, S. R., J. M. Paruelo and G. A. Slafer. 2004. Interannual variability of wheat yield in the Argentine Pampas during the 20th century. *Agriculture, Ecosystems and Environment* 103: 177-190.

Viglizzo, E. F., F. Lértora, A. J. Pordomingo, J. N. Bernardos, Z. E. Roberto and H. Del Valle. 2001. Ecological lessons and applications from one century of low external-input farming in the pampas of Argentina. *Agriculture, Ecosystems and Environment* 83: 65-81.

Viglizzo, E. F., Z. E. Roberto, M. C. Filippin and A. J. Pordomingo. 1995. Climate variability and agroecological change in the Central Pampas of Argentina. *Agriculture, Ecosystems and Environment* 55: 7-16.

Walkley, A. and Black, I. A. 1934. An Examination of the Degtjareff method for determining soil organic matter, and a proposed modification of the chromic acid titration method. *Soil Science* 37 (1): 29-38.

In: Progress in Food Science and Technology, Volume 1   ISBN: 978-1-61122-314-9
Editor: Anthony J. Greco   © 2012 Nova Science Publishers, Inc.

Chapter 7

LEGUMES AS COVER CROPS OR COMPONENTS OF INTERCROPPING SYSTEMS AND THEIR EFFECTS ON WEED POPULATIONS AND CROP PRODUCTIVITY

I. S. Travlos[*]

Laboratory of Agronomy, Department of Crop Production,
Agricultural University of Athens, Athens, Greece

ABSTRACT

Legumes are the second family after the Poaceae in importance to humans. The more than 18,000 species of legumes include important species. In recent years, there has been remarkably increased interest in alternative agricultural production systems in order to achieve high productivity and promote sustainability over time. Among them, intercropping and cover crop mulches seem quite promising since they affect several agronomic traits. Cover crops have long been used to reduce soil erosion and water runoff and improve several soil parameters, while in many cases are also used to manage weeds in several crops. Several annual legume species have already been studied for their weed control benefits, while in the same time they biologically fix nitrogen that subsequently becomes available during residue decomposition. This further improves crop yield, making the use of cover crops economically profitable by reduced inputs and increased yield. Additionally, intercropping legumes for forage or grain production with cereals or other crops is widely used in many parts of the world. The presence of legumes as a major component of these systems contributes into soil conservation, efficient resource utilization, higher forage protein level and better weed management. It is also noticeable that some legumes, such as sweetclover, suppress weeds both during growth and after death. In addition to competition-based or physical weed suppression; certain legumes used as cover crops or in intercropping (hairy vetch, red clover etc) are also known to suppress weeds through their allelopathic activity. Future research should focus on legumes which can suppress weeds without reducing main or next crop yield, in order to maintain and enhance the profitability of these sustainable agriculture systems.

---

[*] Laboratory of Agronomy, Department of Crop Production, Agricultural University of Athens, 75, Iera Odos st., 11855 Athens, Greece. Corresponding author's E-mail: htravlos@yahoo.gr

Legumes and Their Importance

Among all plant families, Fabaceae is one of the most important, as long as legumes have been essential in the development of several civilizations, forming a main part of the basic diet (as sources of proteins, carbohydrate etc). In terms of economic importance, the Leguminosae is the most important family in the Dicotyledonae (Harborne, 1994). Legumes are second only to the grasses (cereals) in providing food crops for world agriculture. In comparison to cereal grains the seeds of legumes are rich in high quality protein, providing man with a highly nutritional food resource. The major staple foods such as beans (*Phaseolus* ssp.), soybean (*Glycine max*), lentils (*Lens culinaris*), pea (*Pisum sativum*) and chickpeas (*Cicer arietinum*) are all legumes. Grain and forage legumes account for 27% of the worl'd primary crop production, with grain legumes contributing 33% of the dietary protein nitrogen (N) needs of humans (Vance et al., 2000). In rank order, bean, pea, chickpea, broad bean (*Vicia faba*), pigeon pea (*Cajanus cajan*), cowpea (*Vigna unguiculata*), and lentil constitute the primary dietary legumes (National Academy of Sciences, 1994). Moreover, soybean and peanut (*Arachis hypogeae*) provide more than 35% of the world's processed vegetable oil, while they are also rich sources of protein.

Additionally, legumes are commonly used as soil-improving biofertilizers, because of their biological nitrogen fixation. Some legumes are better at fixing nitrogen than others. Common beans are poor fixers and fix less than their nitrogen needs, while other grain legumes, such as peanuts, cowpeas, soybeans and faba beans are good nitrogen fixers. Perennial and forage legumes, such as alfalfa, sweetclover, clovers and vetches, may also fix significant amounts of nitrogen. A perennial or forage legume crop only adds significant nitrogen for the following crop if the entire biomass (stems, leaves, roots) is incorporated into the soil. If forage is cut and removed from the field, most of the nitrogen fixed by the forage is removed. Therefore, legumes used as cover crops or manures may play a dual role in agroecosystems by protecting the soil from erosion and by enriching it with organic matter and N through *Rhizobium* symbiosis (Caamal-Maldonado et al., 2001).

It has also to be noted that the inclusion of legumes is even more critical for sustainable production of the infertile savannah soils and agroecosystems of the tropics and subtropics but still being lagged due to lack of information, seed costs etc (Graham and Vance, 2003). Species from several genera (*Aeschynomene, Arachis, Centrosema, Desmodium, Macroptilum* and *Stylosanthes*) offer promise for improved tropical pasture systems (Thomas and Sumberg, 1995). Moreover, presently underutilized and neglected crop and pasture crops could still emerge. Plants such as marama bean (*Tylosema esculentum*), sword beans (*Canavalia gladiata*) and annual medics could offer solutions (Dakora et al., 1999; Howieson et al. 2002; Travlos et al., 2007). For example, marama bean, a wild perennial legume indigenous in southern Africa, produces protein- and oil-rich seed (comparable to soybean and groundnut respectively) and tubers of relatively high protein and carbohydrate content. Moreover, it is a typical drought-avoiding species since it uses its tubers as water reservoirs and its stomata and leaflet closure to save water (Travlos et al., 2008). Therefore, it has been characterized of great potential for use as human food and animal fodder (Dakora et al. 1999; Travlos et al. 2007). According to Graham and Vance (2003) "legumes play a critical role in natural ecosystems, agriculture, and agroforestry, where their ability to fix N in symbiosis makes them excellent colonizers of low-N environments."

It seems that it is now time to use legumes properly and extensively, since they have many things to give us and their inclusion in agricultural systems seems crucial for their sustainability. Within that point of view, this review article was written in order to highlight several cases of legumes used at intercropping systems or as cover crops and their effects on crop growth and productivity and weed communities.

LEGUMES AS COVER CROPS OR COMPONENTS OF INTERCROPPING SYSTEMS

Intercropping can be broadly defined as a system where two or more crop species are grown in the same field at the same time during a growing season (Ofori and Stern, 1987). Intercropping can stabilize grain yield and reduce pest problems (Anil et al., 1998) and globally, many organic and conventional farmers are already familiar with this practice (Entz et al., 2001). Among grain intercrops, mixtures of cereals and pulses have perhaps received the most attention. Bulson et al. (1997) observed that wheat and bean (*Phaseolus vulgaris* L.) planted at 75% of the recommended plant population density under organic management used the land area more efficiently than monocultures of the crops and that this practice was more profitable. Another form of intercropping involves cover crops, where understory crops are used not for an economic yield but for other benefits such as weed suppression (Liebman, 1986) and N supply to following crops (Thiessen Martens et al., 2005).

Theophrastus (372–287 BC), the Greek student of Plato and the successor of Aristotle in the Peripatetic school gave us his botanical treatises, *Enquiry into Plants*, and *On the Causes of Plants*, which constitute the most important contribution to botanical science during antiquity and the Middle Ages, the first systemization of the botanical world; on the strength of these works we call him the "father of Botany". The noticeable is that Theophrastus was also one of the first that pointed out the importance of legumes as mulches.

Because quality of cereal forage is usually lower than legumes, cereal forages (barley and oat) are often mixed with field pea and other legumes in many countries to increase protein content with no negative effect on total yield (Anil et al., 1998; Chapko et al., 1991; Hall and Kephart, 1991). Other legumes occasionally intercropped with cereal forages include vetches (*Vicia* spp.) (Anil et al., 1998). The cereal provides support for the legume to aid in harvesting and drying in the windrow (Jedel and Helm, 1993). Other benefits of these mixtures include greater use of light, greater uptake of water and nutrients, enhanced weed suppression, and increased soil conservation (Anil et al., 1998).

Additionally, using a winter cereal grain as a companion crop during legume establishment can provide a cash grain and straw (Exner and Cruse, 2001) and reduce soil erosion (Kaspar et al., 2001), nitrate losses (Strock et al., 2004), and weed competition (Van Heemst, 1985). Hesterman et al., 1992; Singer and Cox, 1998). Red clover is one of the best choices for winter cereal grain intercrops because it tolerates shading (Blaser et al., 2006) and has similar feed value to alfalfa (Broderick et al., 2001). Moreover, some potential benefits to the farming system of intercropping a legume in sunflower are nitrogen fixation, soil erosion control, and improvement of the soil structure and organic matter content (Biederbeck and Bouman, 1994). Intercropping may also improve snow trapping and green manure production during the year after legume establishment (Lilleboe, 1991).

Furthermore, cover crops have long been used to reduce soil erosion and water runoff, reduce herbicide inputs, and improve water infiltration, soil moisture retention, soil tilth, organic carbon and nitrogen (Teasdale 1996; Yenish et al. 1996). Among the commonly used and studied cover crops there are many annual legumes such as crimson clover, hairy vetch and subterranean clover (Teasdale and Daughtry, 1993; Yenish et al. 1996). On the contrary, many farmers avoid using legumes (i.e. *Pisum sativum*, *Vicia* spp.) that are known hosts of diseases like *Sclerotinia minor* (Koike et al., 1996).

Likewise, many tropical bean plants are used worldwide for similar purposes. These bean plants are annual vines or shrubs that grow vigorously for a short time after being sowed, and cover the soil with a thick layer of growth, thus reducing light, preventing extreme temperature changes, and impairing weed growth (Lanini, 1987; Radosevich and Holt, 1984). Some of these legumes also have allelopathic potential that affects the growth of other weeds (Hart, 1986; Vandermeer, 1989; Lathwell, 1990; Trenbath et al., 1990; Fujii, 1999). In tropical American countries, these bean plants are intercropped with corn, sorghum (*Sorghum bicolor*), and millet (*Pennisetum americanum*). Velvetbean (*Mucuna deeringiana, M. cochinchinensis*, and *M. pruriens*) are among the important widely used bean plants because of their fast growth and abundant biomass (Buckles and Barreto, 1994). Some are used also as fodder (Chacón and Gliessman, 1982; Anaya et al., 1992).

THE EFFECTS OF LEGUMES ON OTHER CROPS AND WEEDS

Competition is one of the factors that can have a significant impact on yield of mixture compare with pure stands. Higher yields have been reported when competition between the two species of the mixture was lower than competition within the same species (Vandermeer, 1990). Interplant competition usually includes competition for soil water, available nutrients, and solar radiation (Buxton and Fales, 1993). Potential benefits of intercropping include increased yields, increased protein and forage quality, N contributions from legumes, greater yield stability, and reduced incidence of pests, weeds, and diseases (Anil et al., 1998; Vasilakoglou and Dhima, 2008). However, farmers are unlikely to adopt the increased cost and complexity of managing intercrops without demonstrated evidence of potential advantages over monocropping (Ross et al., 2004).

Competition between crops can vary both temporally and spatially, but it is often difficult to distinguish between spatial and temporal effects in crop mixtures (Willey, 1979). Temporal complementarity occurs when component crops have life cycles that differ in timing (e.g., early vs. late maturing crops), whereas crops that differ in their architecture could show spatial complementarity. For example, an intercrop of tall and short crops could intercept more light than the mean of the sole crops (Trenbath, 1986; Ramakrishna and Ong, 1994). In addition, component crops could use different pools of the same resource [e.g., N in the form of nitrate (NO_3^-) by a nonlegume and dinitrogen (N_2) by a legume], thus reducing competition for soil N (Vandermeer, 1989; Szumigalski and Van Acker, 2006).

According to Dhima et al. (2007), common vetch intercropped with wheat or oat had a yield advantage for exploiting the resources of the environment and therefore these intercropping systems were found to be more profitable. Moreover, intercropping hairy vetch (*Vicia villosa*) at specific growth stage (V4) of sunflower appears superior because it did not reduce sunflower yield, provided soil cover adding between 540 and 2400 kg ha^{-1} above ground dry matter to the system, and increased NO_3^-–N levels at the beginning of the subsequent wheat season in several environments (Kandel et al., 2000). Intercropping berseem clover with cereals has increased the yield and quality of cereal forage crops in India (Singh et al., 1989), increased total biomass production without reducing cereal grain yields in Mexico (Reynolds et al., 1994) and USA (Ghaffarzadeh, 1997; Holland and Brummer, 1999), and improved forage quality, reduced fertilizer needs, and increased subsequent crop yields in British Columbia (Stout et al., 1997) and Iowa (Ghaffarzadeh, 1997). Intercropped with Italian ryegrass (*Lolium multiflorum* Lam.) in Spain, berseem clover provided a N equivalence of 80 kg N ha^{-1} per year (Caballero et al., 1994). Berseem clover improved the forage quality and yields of barley–ryegrass–legume intercrops more than did annual *Medicago* and *Lespedeza* species (Stout et al., 1997). Berseem clover was better adapted than crimson clover (*Trifolium incarnatum* L.) or white clover (*Trifolium repens* L.) for intercropping with wheat (*Triticum aestivum* L.) or barley to improve the ground cover, N use efficiency, and productivity of low-input systems (Reynolds et al., 1994). Intercropping of berseem clover (*Trifolium alexandrinum*) with spring cereals grown for silage may improve forage quality and yield.

Intercropping generally reduces yields of component crops from what they would be in monoculture, but intercrops have the potential to more effectively use the land area (i.e., higher land equivalent ratio) (Anil et al., 1998; Pridham and Entz. 2008). In a study conducted under herbicide-free production conditions in Manitoba, a wheat-pea intercrop treatment produced total yields comparable to sole crop treatments of wheat and pea, though in some cases total yield was reduced compared with monoculture (Szumigalski and Van Acker, 2005). The advantage of pulses in an intercrop are often greater under low soil N conditions (Lunnan, 1989). Carr et al. (2004) observed higher total aboveground plant biomass production for barley-pea intercrops in low N soils; no benefit of pea in the barley crop was observed under high soil N conditions. Much previous intercropping research has focused on forage production (Carr et al., 2004). The few intercropping studies in industrialized countries aimed at grain production often focused on N economy (Waterer et al., 1994). Additionally, and in organic agriculture, plant diseases are controlled through resistant cultivars and crop management, including intercropping (Theunissen, 1997). In a recent Danish study conducted under organic management, Jensen et al. (2005) observed that when barley was intercropped with pea, faba bean (*Vicia faba* L.), or narrow-leaved lupin (*Lupinus angustifolius* L.), disease severity was reduced by at least 20% in all of the intercrop systems compared to the pure stands.

Utilizing a legume for the remainder of the growing season after cereal grain harvest can provide forage for livestock, weed suppression (Blaser et al., 2006; Mutch et al., 2003), and N for subsequent crops (Bruulsema and Christie, 1987). The cereal component in barley-pea and oat-pea mixtures should be sown at a sole-crop or greater seeding rate for maximum forage production. Forage concentration can be increased as the relative proportion of pea seed to cereal kernels sown in a mixture is increased, but forage N yield may not be affected, since the cereal component contributes more to yield than the pea component (Carr et al., 1998). Studies on some other intercrops, such as corn-cowpea and corn-soybean, resulted in greater productivity per unit of land than monocultures of the intercrop components (Allen and Obura, 1983). Despite the competition between the species at intercropping which resulted to lower yield than monocropping, the economics of vetch-wheat and vetch-oat mixtures indicated a significant advantage from intercropping which was attributed to better land use efficiency. Similar mixtures seem promising in the development of sustainable crop production with a limited use of external inputs (Dhima et al., 2007).

Despite the positive effects often produced by winter annual cover crops in corn production, there is also a potential for reduction in corn yield. Spring regrowth of legumes can lower available water in the subsoil creating conditions of moisture stress for corn in years of low precipitation (Badaruddin and Meyer, 1989; Hesterman et al., 1992; Tiffin and Hesterman, 1998). Anil et al. (1998) reviewed literature from the Mediterranean and coastal regions of western Europe that demonstrated a yield reduction when common vetch was intercropped with oat, while in another study, oat-hairy vetch intercrops yielded 12 Mg ha^{-1} when oat comprised a high proportion of the seeding rate (Anil et al., 1998). Similarly, in order to obtain the best results from the use of cover crops, it is necessary to avoid (or minimize) interference (competition and allelopathy) of the cover crops and the companion crops. This could be achieved by using the right density of legumes in the crop field, and sowing them at least two weeks after the corn has been sown.

Intercrops suppress weeds by reducing their available space (Liebman, 1986; Shetty and Rao, 1979; Unamma et al., 1986). Enhanced competition from intercropping can be exhibited through soil shading and resource competition that impedes weed germination and growth (Anil et al., 1998). A former review of the relative literature revealed that in 50% of the studies, weed biomass was lower than the intercrop biomass, intermediate in 42%, and higher in only 8% of the studies (Liebman and Dyck, 1993). Mutch et al. (2003) reported significantly less ragweed (*Ambrosia artemisiifolia* L.) when wheat was undersown with red clover cover crop.

The example of hairy vetch is also representative of the effects of legumes on weeds. The hairy vetch cover crop could reduce weed density in organic no-till corn by at least 50%, with annual weeds being affected more than perennials. Total weed biomass was reduced 31 to 94% (depending on the site and the year) compared with no-cover plots (Mischler et al., 2010). Although hairy vetch provides early-season weed suppression, herbicides generally are required to achieve full-season weed control and optimum corn yields (Johnson et al, 1993; Teasdale 1993), since weed suppression effects of cover crop residue decrease with time after cover crop residue decomposition. Thus, the use of hairy vetch, does not totally eliminate the need for herbicides, but it suppresses certain weed species (such as *Ipomoea* ssp., *Sida spinosa*, *Cyperus esculentus*), reduces total weed dry biomass, and allow management of late-season weeds with postemergence herbicides on an as-needed basis (Reddy and Koger, 2004).

In some cases, the effects of legumes on several noxious weeds are also noticeable. Thus, intercropping sorghum with cowpea (*Vigna unguiculata*), greengram (*Vigna radiata*), and crotalaria (*Crotalaria ochroleuca*), and maize with crotalaria significantly reduced witchweed (*Striga hermonthica*) populations (Khan et al., 2007). Indeed, it has been shown in various studies that intercropping cereals with legumes can reduce the number of *Striga* plants that mature in an infested field (Carsky et al., 1994; Tenebe and Kamara, 2002). However, apart from *Desmodium*, a fodder legume whose species have been shown to uniformly control *Striga* (Khan et al., 2006), there is a wide variation in the ability of different other legumes and their cultivars in suppression of *Striga* (Emechebe and Ahonsi, 2003).

Rotations in organic production systems often include winter annual crops such as rye, hairy vetch, whose maximum growth occurs before the period of low Canada thistle (*Cirsium arvense*) carbohydrates reserves (HDRA 2006). Combining cowpea (*Vigna unguiculata*) with sudangrass (*Sorgum sudanense*) produces a large amount of diverse residue which suppresses weeds (Creamer and Baldwin 2000; Bicksler and Masiunas, 2009).

Fisk et al (2001) conducted a 2-yr study in order to investigate the influence of several annual legume cover crops (annual medics, berseem clover and medium red clover) on weed populations in a winter wheat-corn rotation system. Density of winter annual weeds were between 41 and 78% lower following most cover crops when compared with no cover control, while dry weight was between 26 and 80% lower in all sites. There are a number of mechanisms responsible for the effect of cover crops on weeds. The living cover crop can reduce light (Teasdale and Mohler, 1993) and as cover crops in moisture available to fall germinating seeds. Weeds attempting to establish along with a cover crop would be in competition for resources and may not develop sufficiently to survive the winter. Cover crop residue can modify the conditions under which weeds germinate or regrow in the spring. Such effects could be due to changes in soil temperature, increase in soil moisture, release of allelopathic chemicals, and physical impediments to weed seedlings (Facelli and Pickett, 1991; Teasdale, 1996; Teasdale and Mohler, 1993). Concerning allelopathy, it has been already shown that extracts from broadleaf plants were more inhibitory to weeds than extracts from cereal plants. Among the other legumes, yard-long bean (*Vigna sesquipedalis*) extracts, completely inhibited the germination of downy brome (*Bromus tectorum*) seed, and seems to have the potential for use in the control of downy brome in wheat-based cropping systems. Most legumes (bean, pea, vetch) produce quinolizidine alkaloids that may be responsible for allelopathy (Wink, 2004) while in other legumes such as velvet bean (*Mucuna deeringiana*), L-3,4-dihydroxyphenylalanine (L-DOPA), an intermediate of many alkaloids, was determined to be the allelochemical (Fujii, 1999). Hence, several legumes have the potential to be used for biologically based weed control methods in organic cropping systems (Machado 2007), since the most practical and immediate way to use allelopathy in weed control is to use allelopathic cover crops in rotations, or apply residues of allelopathic weeds or crops as mulches (Rice, 1974; Caamal-Maldonado et al., 2001; Dhima et al., 2006).

It has to be noted that in some cases there are significant differences, not only among species but also among cultivars of the same species used as cover crops, regarding their ability to suppress weed growth. Field trials were conducted in Canada to measure the growth of brown mustard (*Brassica juncea*), in mowed and nonmowed production, as influenced by alsike (*Trifolium hybridum*), balansa (*T. michelianum*), berseem (*T. alexandrinum*), crimson (*T. incarnatum*), Persian (*T. resupinatum*), red (*T. pratense*), and white Dutch (*T. repens*) clover and rye (*Secale cereale*). Although the effects varied with location, timing, and species, the characteristics of berseem clover (upright growth, long stems, high biomass, and late flowering) would support its use as a cover crop or forage in similar regions (Ross et al. 2001). On the contrary, there are different results from other studies. Thus, annual medics and red clover planted after wheat harvest reduced density and dry weight of winter annual weeds before planting no-till corn. Similarly, annual medics and red clover reduced summer annual weed dry weight; however, weed density was only occasionally reduced. The suppressive effect of annual medic residue on summer annual weed density and dry weight was high and consistent across all site years. Berseem clover had no effect, whereas the effect of red clover was not consistent. Residue of all legumes reduced both density and dry weight of perennial weeds (Fisk et al., 2001). Under dryland conditions, large-seeded legumes and sweetclover (*Melilotus officinalis*) grown as green fallow provided better weed suppression than *Trifolium* spp. (Schlegel and Havlin, 1997). Research suggests that competitive abilities vary among clover species and cultivars within a species and with intent of use. Three clover species failed to establish under conditions of severe weed interference and when interseeded as a cover crop with corn (*Zea mays* L.), but crimson clover established well (Abdin et al., 1998). In a study of seven berseem clover cultivars intercropped with oat (*Avena sativa*) cultivars, Holland and Brummer (1999) observed considerable variability in important agronomic traits (forage stand, plant health, maturity, yield, height, and weeds) due to berseem cultivar effects. Nelson et al. (1991) observed less weed biomass with an annual clover (crimson) than with two perennial clovers (red and white) used as spring cover crops. Many studies of cover crops have used the native weed population as part of the experimental design (Nelson et al., 1991; Schlegel and Havlin, 1997; Abdin et al., 1998). Given that the differences between clover species may be small, the use of a fixed weed population may provide greater precision in assessing weed suppression.

Previous research on legumes, investigating the possibility of biological weed control through plant interference has focused on the use of winter annual or perennial species sown in the fall and suppressed or killed with a herbicide application or mechanical disturbance in the spring (Curran et al., 1994; Eberlein et al., 1992; Enache and Ilnicki, 1990; Grubinger and Minotti, 1990; Hoffman et al., 1993; Mohler, 1991; Yenish et al., 1996). Species evaluated include alfalfa *(Medicago sativa)*, pea, hairy vetch and several clovers (crimson, ladino, white and subterranean). Hairy vetch, ladino clover, subterranean clover, and white clover show promise, but none of these species is consistently winterhardy, and spring growth of these species can deplete soil water resources. In addition, all of the species investigated, except subterranean clover, require chemical or mechanical suppression to limit their competitive effects on corn. Therefore, specialized cover or mulch species selected to suppress weeds without reducing main crop yield are widely adopted and known as smoother plants (De Haan et al., 1994). Smother plants include several legumes (such as annual medic) and have the potential to suppress weed growth (De Haan et al., 1994), increase soil water infiltration, decrease soil erosion (Cripps and Bates, 1993), contribute N to the main crop (Decker et al., 1994; Maskina et al., 1993), and reduce economic risk (Hanson et al., 1993). Results from several experiments indicate that the response of corn and weeds to annual medic smother crops varies considerably from one location to another. Thus, it may be difficult to develop a smother plant that performs well under a wide range of soil types, fertility levels, and weed species. Competition between corn and annual medic smother plants could be reduced by selecting annual medic genotypes characterized by small individual plant size, prostrate growth habit, and a short life cycle length. Furthermore, research investigating the influence of soil type, precipitation, weed species, and soil nutrient status on smother plant performance will be necessary to identify the environments in which annual medic and other legumes as smother plants can be successfully integrated into cropping systems (De Haan et al., 1997).

TRENDS AND PROSPECTIVE

Our review highlighted some of the noteworthy attempts of using and managing legumes as living mulches or parts of intercrops in order to provide desirable ecosystem functions in multifunctional cropping systems. As expressed by Reddy (2001), although long-term agronomic, environmental and sustainable monetary benefits of cover crops and intercrops are difficult or impossible to calculate, but should be factored into crop management decisions. The additional cost associated with such systems must be compensated for by increased yield and/or reduced herbicide cost. In addition, if the cover crop is a legume, the atmospheric nitrogen fixed can be a renewable source of soil nitrogen (Reddy, 2001).

Under that point of view, several legumes show a high potential but further research is required in order to suggest the optimal legume cover crops for each region in terms of their productivity and ability to suppress weeds. The use of velvetbean and other legumes, particularly jackbean, as living cover crops or dead mulches could contribute to the reduction of the weed seed bank in soils and in the improvement of corn production, delaying weed appearance. This can be obtained if interference with the legumes used as living cover crops and corn can be eliminated or avoided (Caamal-Maldonado et al., 2001). Many studies show significant interactions, while further experiments with more legumes and in various seeding ratios should be conducted in order to find intercrops and cover crops that are more productive and could be used in the future in sustainable production systems. New resources management initiatives are necessary to restore the soil; increase organic matter and nutrients; control pests; improve crop production; and find adequate techniques to reach a sustainable production (Warren, 1983; Weston, 1996). Although, there is certain need for further research to quantify if chemical control can be reduced or eliminated by the use of annual legume cover crops and intercropping (Fisk et al., 2001).

REFERENCES

Abdin, O., B.E. Coulman, D. Cloutier, M.A. Faris, X. Zhou, and D.L. Smith. 1998. Yield and yield components of corn interseeded with cover crops. *Agron. J.* 90: 63–68.

Allen, J.R., and R.K. Obura. 1983. Yield of corn, cowpea, and soybean under different intercropping systems. *Agron. J.* 75: 1005–1009.

Anaya, A.L., R. Cruz Ortega, and V. Nava Rodríguez. 1992. Impact of allelopathy in the traditional management of agroecosystems in Mexico. p. 271–301. *In* S.J.H. Rizvi and V. Rizvi (ed.) *Allelopathy: Basic and applied aspects.* Chapman and Hall, New York.

Anil, L., J. Park, R.H. Phipps, and F.A. Miller. 1998. Temperate intercropping of cereals for forage: A review of the potential for growth and utilization with particular reference to the UK. *Grass Forage Sci.* 53: 301–317.

Badaruddin, M., and D.W. Meyer. 1989. Water use by legumes and its effect on soil water status. *Crop Sci.* 29: 1212–1216.

Bicksler, A.J., and J.B. Masiunas. 2009. Canada thistle (*Cirsium arvense*) suppression with buckwheat or sudangrass cover crops and mowing. *Weed Technol.* 23: 556–563.

Biederbeck V.O., Bouman O.T. Water use by annual green manure legumes in dryland cropping systems. *Agron. J.* 86: 543–549.

Blaser, B.C., L.R. Gibson, J.W. Singer, and J.L. Jannink. 2006. Optimizing seeding rates for winter cereal grains and frost-seeded red clover intercrops. *Agron. J.* 98: 1041–1049.

Broderick, G.A., R.P. Walgenbach, and S. Maignan. 2001. Production of lactating dairy cows fed alfalfa or red clover silage at equal dry matter or crude protein contents in the diet. *J. Dairy Sci.* 84: 1728–1737.

Bruulsema, T.W., and B.R. Christie. 1987. Nitrogen contribution to succeeding corn from alfalfa and red clover. *Agron. J.* 79: 96–100.

Buckles, D., and H. Barreto. 1994. Aumentando la sustentabilidad de los sistemas de agricultura migratoria con leguminosas de cobertura: Consideraciones técnicas y socioeconómicas. p. 123–138. *In Taller sobre las políticas para una agricultura sustentable en la sierra de Los Tuxtas y Santa Marta, Veracruz. Marzo,* 1994. México.

Bulson, H.A.J., R.W. Snaydon, and C.E. Stopes. 1997. Effects of plant density on intercropped wheat and filed beans in an organic farming system. *J. Agric. Sci.* 128: 59–71.

Buxton, C.L., and S.L. Fales. 1993. *Plant environment and quality.* In: Fahey, Jr., G.C. (Ed.), Forage Quality, Evaluation and Utilization. ASA, CSSA and SSSA, Madison, WI.

Caamal-Maldonado, J.A., J.J. Jiménez-Osornio, A. Torres-Barragán and A.L. Anaya. 2001. The use of allelopathic legume cover and mulch species for weed control in cropping systems. *Agron. J.* 93: 27–36.

Caballero, R., M. Arauzo, and P.J. Hernaiz. 1994. Response to N-fertilizer of Italian ryegrass grown alone and in mixture with berseem clover under continental irrigated Mediterranean conditions. *Fert. Res.* 39: 105–112.

Carr, P.M., R.D. Horsley, and W.W. Poland. 2004. Barley, oat and cereal-pea mixtures as dryland forages in the Northern Great Plains. *Agron. J.* 96: 677–684.

Carsky, R.J., L. Singh, and R. Ndikawa. 1994. Suppression of *Striga hermonthica* on sorghum using a cowpea intercrop. *Exp. Agric.* 30: 349–358.

Chacón, J.C., and S.R. Gliessman. 1982. The use of the non weed concept in traditional agroecosystems of Southeastern Mexico. *Agro-Ecosystems* 8: 1–11.

Chapko, L.B., M.A. Brinkman, and K.A. Albrecht. 1991. Oat, oat–pea, barley, and barley–pea for forage yield, forage quality, and alfalfa establishment. *J. Prod. Agric.* 4: 486–491.

Creamer, N.G., and K.R. Baldwin. 2000. An evaluation of summer cover crops for use in vegetable production systems in North Carolina. *HortScience* 35: 600–603.

Crips, R.W., and H.K. Bates. 1993. Effect of cover crops on soil erosion in nursery aisles. *Weed Technol.* 5: 664–673.

Curran, W.S., L.D. Hoffman, and E.L. Werner. 1994. The influence of a hairy vetch (*Vicia villosa*) cover crop on weed control and corn (*Zea mays*) growth and yield. *Weed Technol.* 8: 777–784.

Dakora, F.D., D.W. Lawlor, and K.P. Sibuga. 1999. Assessment of symbiotic nitrogen nutrition in marama bean (*Tylosema esculentum* L.) a tuber-producing underutilized African grain legume. *Symbiosis* 27: 269–277.

Decker, A.M., A.J. Clark, J.J. Meisinger, F.R. Mulford, and M.S. McIntosh. 1994. Legume cover crop contributions to no-tillage corn production. *Agron. J.* 86: 126–135.

De Haan, R.L., D.L. Wyse, N.J. Ehlke, B.D. Maxwell, and D.H. Putnam. 1994. Simulation of spring-seeded smother plants for weed control in corn (*Zea mays*). *Weed Sci.* 42: 35–43.

De Haan, R.L., C.C. Sheaffer, and D.K. Barnes. 1997. Effect of annual medic smother plants on weed control and yield in corn. *Agron. J.* 89: 813–821.

Dhima, K.V., I.B. Vasilakoglou, I.G. Eleftherohorinos, and A.S. Lithourgidis. 2006. Allelopathic potential of winter cereals and their cover rop mulch effect on grass weed suppression and corn development. *Crop Sci.* 46: 345–352.

Dhima, K.V., A.S. Lithourgidis, I.B. Vasilakoglou, and C.A. Dordas. 2007. Competition indices of common vetch and cereal intercrops in two seeding ratio. *Field Crops Res.* 100: 249–256.

Eberlein, C.V., C.C. Sheaffer, and V.F. Oliveira. 1992. Corn growth and yield in an alfalfa living mulch system. *J. Prod. Agric.* 5: 332–339.

Emechebe, A.M., and M.O. Ahonsi. 2003. Ability of excised root and stem pieces of maize, cowpea and soybean to cause germination of *Striga hermonthica* seeds. *Crop Prot.* 22: 347–353.

Enache, A.J., and R.D. Ilnicki. 1990. Weed control by subterranean clover (*Trifolium subterraneum*) used as a living mulch. *Weed Technol.* 4: 534–538.

Entz, M.H., R. Guilford, and R. Gulden. 2001. Productivity of organic crop production in the eastern region of the Northern Great Plains: A survey of 14 farms. *Can. J. Plant Sci.* 81: 351–354.

Exner, D.N., and R.M. Cruse. 2001. Profitability of crop rotations in Iowa in a stress environment. *J. Iowa Acad. Sci.* 108: 84–89.

Facelli, J.M., and S.T.A. Pickett. 1991. Plant litter: Its dynamics and role of legumes in effects on plant community structure. *Bot. Rev.* 57: 1–32.

Fisk, J.W., O.B. Hesterman, A. Shrestha, J.J. Kells, R.R. Harwood, J.M. Squire, and C.C. Sheaffer. 2001. Weed suppression by annual legume cover crops in no-tillage corn. *Agron. J.* 93: 319–325.

Fujii, Y. 1999. Allelopathy of hairy vetch and *Mucuna*; their application for sustainable agriculture. p. 289–300. *In* C.H. Chou, et al. (ed.) *Biodiversity and allelopathy from organisms to ecosystems in the pacific.* Academia Sinica, Taipei.

Ghaffarzadeh, M. 1997. Economic and biological benefits of intercropping berseem clover with oat in corn–soybean–oat rotations. *J. Prod. Agric.* 10: 314–319.

Graham, P.H., and C.P. Vance. 2003. Legumes: importance and constraints to greater use. *Plant Physiol.* 131: 872–877.

Grubinger,V.P., and P.L. Minotti. 1990. Managing white clover living mulch for sweet corn with partial rototilling. *Am. J. Altern. Agric.* 5: 4–12.

Hall, M.H., and K.D. Kephart. 1991. Management of spring-planted pea and triticale mixtures for forage production. *J. Prod. Agric.* 4: 213–218.

Hanson, J.C., E. Lichtenberg, A.M. Decker, and A.J. Clark. 1993. Profitability of no-tillage corn following a hairy vetch cover crop. *J. Prod. Agric.* 6: 432–437.

Harborne, J.B. 1994. Phytochemistry of the Leguminosae. p. 40–56. In *Phytochemical Dictionary of the Leguminosae,* eds Bisby,F.A. et al. London: Chapman & Hall.

Hart, R. 1986. *Ecological framework for multiple cropping research. In* Ch. Francis (ed.) Multiple cropping systems. Macmillan Publ. Co., New York.

[HDRA] Henry Doubleday Research Association. 2006. Creeping Thistle Management Strategies in Organic Systems. *www.organicweeds*.org.uk.

Hesterman, O.B., T.S. Griffin, P.T. Williams, G.H. Harris, and D.R. Christenson. 1992. Forage legume-small grain intercrops: Nitrogen production and responses of subsequent corn. *J. Prod. Agric.* 5: 340–348.

Hoffman, M.L., E.E. Regnier, and J. Cardina. 1993. Weed and corn (*Zea mays*) responses to a hairy vetch (*Vicia villosa*) cover crop. *Weed Technol.* 7: 594–599.

Holland, J.B., and E.C. Brummer. 1999. Cultivar effects on oat–berseem clover intercrops. *Agron. J.* 91: 321–329.

Howieson, J.G., G.W. O'Hara, and S.J. Carr. 2000. Changing roles for legumes in Mediterranean agriculture: developments from an Australian perspective. *Field Crops Res.* 65: 107–122.

Jedel, P.E., and J.H. Helm. 1993. Forage potential of pulse–cereal mixtures in central Alberta. *Can. J. Plant Sci.* 73: 437–444.

Kandel, H.J., B.L. Johnson, and A.A. Schneiter . 2000. Hard red spring wheat response following the intercropping of legumes into sunflower. *Crop Sci.* 40: 731–736.

Kaspar, T.C., J.K. Radke, and J.M. Laflen. 2001. Small grain cover crops and wheel traffic effects on infiltration, runoff, and erosion. *J. Soil Water Conserv.* 56: 160–164.

Khan, Z.R., C.A.O. Midega, A. Hassanali, J.A. Pickett, and L.J. Wadhams. 2007. Assessment of different legumes for the control of *Striga hermonthica* in maize and sorghum. *Crop Sci.* 47: 730–734.

Khan, Z.R., J.A. Pickett, L.J. Wadhams, A. Hassanali, and C.A.O. Midega. 2006. Combined control of *Striga* and stemborers by maize–*Desmodium* spp. intercrops. *Crop Prot.* 25: 989–995.

Koike, S.T., R.F. Smith, L.E. Jackson, L.J. Wyland, J.I. Inman, and W.E. Chaney. 1996. Phacelia, Lana woolypd vetch, and Austrian winter pea: three new cover crop hosts of Sclerotinia minor in California. *Plant Disease* 80: 1409–1412.

Lanini, T. 1987. *Organic methods of weed control.* p. 90. *In* Organic Farming Training Conf. Proc., Riverside, CA. October 1986. Univ. of California, Riverside, CA.

Lathwell, D. 1990. Legume green manures: Principles for management based on recent research. p. 1–30. *In TropSoils Bulletin.* 90-01. Soil Management Collaborative Research Support Program, North Carolina State Univ., Raleigh, NC.

Liebman, M.Z. 1986. *Ecological suppression of weeds in intercropping systems: Experiments with barley, pea, and mustard.* Ph.D. diss., Berkeley, CA.

Liebman, M., and E. Dyck. 1993. Crop-rotation and intercropping strategies for weed management. *Ecol. Appl.* 3: 92–122.

Lilleboe D. 1991. North Dakotans investigate benefits of sweetclover interseeded in sunflower. *Sunflower* 17: 22–23.

Lunnan, T. 1989. Barley-pea mixtures for whole crop forage. Effect of different cultural practices on yield and quality. *Norw. J. Agric. Sci.* 3: 57–71.

Mischler, R., S.W. Duiker, W.S. Curran, and D. Wilson. Hairy vetch management for no-till organic corn production. *Agron. J.* 102: 355–362.

Mohler, C.L. 1991. Effects of tillage and mulch on weed biomass and sweet corn yield. *Weed Technol.* 5: 545–552.

Mutch, D.L., T.E. Martin, and K.R. Kosola. 2003. Red clover (*Trifolium pratense*) suppression of common ragweed (*Ambrosia artemisiifolia*) in winter wheat (*Triticum aestivum*). *Weed Technol.* 17: 181–185.

Nelson, W.A., B.A. Kahn, and B.W. Roberts. 1991. Screening cover crops for use in conservation tillage systems for vegetables following spring plowing. *HortScience* 26: 860–862.

Pridham, J.C., and M.H. Entz. 2008. Intercropping spring wheat with cereal grains, legumes, and oilseeds fails to improve productivity under organic management. *Agron. J.* 100: 1436–1442.

Radosevich, S., and J. Holt. 1984. Weed ecology. Implications for vegetation management. John Wiley & Sons, New York.

Ramakrishna, A., and C.K. Ong. 1994. Productivity and light interception in upland rice-legume intercropping systems. *Trop. Agric. (Trinidad)* 71: 5–11.

Reddy, K.N., and C.H. Koger. 2004. Live and killed kairy vetch cover crop effects on weeds and yield in glyphosate-resistant corn. *Weed Technol.* 18: 835–840.

Reynolds, M.P., K.D. Sayre, and H.E. Vivar. 1994. Intercropping wheat and barley with N-fixing legume species: A method for improving ground cover, N-use efficiency and productivity in low input systems. *J. Agric. Sci. (Cambridge)* 123: 175–183.

Rice, E.L. 1974. *Allelopathy.* Academic Press, New York.

Ross, S.M., J.R. King, R.C. Izaurralde and J.T. O'Donovan. 2001. Weed suppression by seven clover species. *Agron. J.* 93: 820–827.

Ross, S.M., J.R. King, J.T. O'Donovan, and D. Spaner. 2004. Forage potential of intercropping berseem clover with barley, oat, or triticale. *Agron. J.* 96: 1013–1020.

Schlegel, A.J., and J.L. Havlin. 1997. Green fallow for the central Great Plains. *Agron. J.* 89: 762–767.

Shetty, S.V.R., and A.N. Rao. 1979. *Weed-management studies in sorghum/pigeonpea and pearl/millet/groundnut intercrop systems– Some observations.* p. 238–248. *In* R.W. Willey (ed.) Proc. of the Int. Workshop on Intercropping. ICRISAT, Hyderabad, India.

Singer, J.W., and W.J. Cox. 1998. Agronomics of corn production under different crop rotations in New York. *J. Prod. Agric.* 11: 462–468.

Singh, V., Y.P. Joshi, and S.S. Verma. 1989. Studies on the production of Egyptian clover and oats under intercropping. *Exp. Agric.* 25: 541–544.

Stout, D.G., B. Brooke, J.W. Hall, and D.J. Thompson. 1997. Forage yield and quality from intercropped barley, annual ryegrass and different annual legumes. *Grass Forage Sci.* 52: 298–308.

Strock, J.S., P.M. Porter, and M.P. Russelle. 2004. Cover cropping to reduce nitrate loss through subsurface drainage in the northern U.S. corn belt. *J. Environ. Qual.* 33: 1010–1016.

Szumigalski, A., and R. Van Acker. 2005. Weed suppression and crop production in annual intercrops. *Weed Sci.* 53: 813–825.

Szumigalski, A.R., and R.C. Van Acker. 2006. Nitrogen yield and land use efficiency in annual sole crops and intercrops. *Agron. J.* 98: 1030–1040.

Teasdale, J.R. 1996. Contribution of cover crops to weed management in sustainable agricultural systems. *J. Prod. Agric.* 9: 475–479.

Teasdale, J.R., and C.S.T. Daughtry. 1993. Weed suppression by live and desiccated hairy vetch (*Vicia villosa*). *Weed Sci.* 41: 207–212.

Teasdale, J.R., and C.L. Mohler. 1993. Light transmittance, soil temperature, and soil moisture under residue of hairy vetch and rye. *Agron. J.* 85: 673–680.

Tenebe, V.A., and H.M. Kamara. 2002. Effect of *Striga hermonthica* on the growth characteristics of sorghum intercropped with groundnut varieties. *J. Agron. Crop Sci.* 188: 376–381.

Theunissen, J. 1997. Intercropping in field vegetables as an approach to sustainable horticulture. *Outlook Agric.* 26: 95–99.

Thiessen Martens, J.R., M.H. Entz, and W.J. Hoeppner. 2005. Legume cover crops with winter cereals in southern Manitoba: Fertilizer replacement values for oat. *Can. J. Plant Sci.* 85: 645–648.

Tiffin, P.L., and O.B. Hesterman. 1998. Response of corn grain yield to early and late killed red clover green manure and subirrigation. *J. Prod. Agric.* 11: 112–121.

Travlos, I.S., G. Economou and A.I. Karamanos, 2007. Germination and emergence of the hard seed coated *Tylosema esculentum* (Burch) A. Schreib in response to different pre-sowing seed treatments. *J. Arid Environ.* 68: 501–507.

Travlos, I.S., G. Economou, and A.I. Karamanos. 2008. Inhibitory effects of marama (*Tylosema esculentum*) on the growth of oat (*Avena sativa*) and barley (*Hordeum vulgare*). *J. Biol. Sci.* 8: 970–973.

Trenbath, B.R. 1986. Resource use by intercrops. p. 57–81 *In* C.A. Francis (ed.) *Multiple cropping systems.* Macmillan Publ. Co., New York.

Trenbath, B.R., G.R. Conway, and I.A. Craig. 1990. Threats to sustainability in intensified agricultural systems: Analysis and implications for management. p. 337–365. *In* S. Gliessman (ed.) *Agroecology.* Researching the ecological basis for sustainable agriculture. Springer-Verlag, New York.

Unamma, R.P.A., L.S.O. Ene, S.O. Odurkwe, and T. Enyinnia. 1986. Integrated weed management for cassava intercropped with maize. *Weed Res.* 26: 9–17.

Van Heemst, H.D.J. 1985. The influence of weed competition on crop yield. *Agric. Sys.* 18: 81–93.

Vance, C.P., P.H. Graham, and D.L. Allan. 2000. *Biological nitrogen fixation. Phosphorus: a critical future need.* pp 506-514. In F.O. Pedrosa, M. Hungria, M.G. Yates, W.E. Newton, eds., Nitrogen Fixation: From Molecules to Crop Productivity. Kluwer Academic Publishers, Dordrecht, The Netherlands.

Vandermeer, J. 1989. The ecology of intercropping. Cambridge Univ. Press, Cambridge.

Vandermeer, J.H. 1990. Intercropping. *Agroecology.* pp. 481–516. In McGraw-Hill (Eds.), New York.

Vasilakoglou, I., and K. Dhima. 2008. Forage yield and competition indices of berseem clover intercropped with barley. *Agron. J.* 100: 1749–1756.

Warren, G.F. 1983. *Technology transfer in no-tillage crop production in third world agriculture.* p. 25–31. *In* I.O. Akobundu and A.E. Deutsch (ed.) No-tillage Crop Production in Tropics, Symp., Monrovia, Liberia. 6–7 Aug. 1981. International Plant Protection Center, Oregon State Univ., Corvallis, OR.

Waterer, J.G., J.K. Vessey, E.H. Stobbe, and R.J. Soper. 1994. Yield and symbiotic nitrogen-fixation in a pea mustard intercrop as influenced by N fertilizer addition. *Soil Biol. Biochem.* 26: 447–453.

Weston, L.A. 1996. Utilization of allelopathy for weed management in agroecosystems. *Agron. J.* 88: 860–866.

Willey, R.W. 1979. Intercropping-its importance and research needs. Part 1. Competition and yield advantages. *Field Crop Abstr.* 32: 1–10.

Yenish, J.P., A.D. Worsham, and A.C. York. 1996. Cover crops for herbicide replacement in no-tillage corn (*Zea mays*). *Weed Technol.* 10: 815–821.

In: Progress in Food Science and Technology, Volume 1   ISBN: 978-1-61122-314-9
Editor: Anthony J. Greco   © 2012 Nova Science Publishers, Inc.

Chapter 8

OPTIMIZING BARLEY GRAIN USE BY DAIRY COWS: A BETTERMENT OF CURRENT PERCEPTIONS

A. Nikkhah[*]

[a]Department of Animal Sciences, University of Zanjan, Zanjan, Iran

ABSTRACT

Dairy cows are irreplaceable major suppliers of amino acids, unique unsaturated fatty acids, energy, and calcium to humans. Optimizing ruminal and small intestinal starch and protein assimilation has been a challenge in modern dairy farms. Dietary barley inclusion rate and processing type manipulation leads our efforts in improving starch utilization. Grinding has traditionally been considered a risk to diet palatability and healthy rumen conditions whilst steam-processing has been rationalized to improve diet palatability and reduce barley fermentation rate, aiming to reduce the risk of subacute rumen acidosis. However, no conclusive research has thoroughly compared processing techniques at different dietary inclusion rates. The main objective of the first experiment was to determine effects of feeding either (1) finely ground, (2) steam-rolled, (3) finely dry-rolled, or (4) coarsely dry-rolled barley grain on rumen fermentation, digestibility and milk production. Eight multiparous midlactation Holstein cows were used in a replicated 4×4 Latin square design study with four periods of 21 d. Diets contained 256 g barley grain/kg DM. Processing type did not affect milk yield and composition, DM intake, rumen pH and volatile fatty acids (VFA), fecal and urine pH, and apparent total tract nutrient digestibility. Results suggested that finely ground barley grain is no different than dry-rolled and steam-rolled barley grains in stimulating feed intake and productivity of midlactation cows, with 256 g/kg barley grain. The objective of the second experiment was to compare effects of grinding versus steam-rolling of barley grain at 30% or 35% of diet dry matter on feed intake, chewing behavior, rumen fermentation, and milk production. Eight multiparous Holstein cows (85 ± 9 d in milk) were used in a replicated 4 × 4 Latin square design experiment with four 21-d periods. Treatments included grinding (GB) or steam-rolling (SB) of barley grains at either 35% or 30% of dietary DM. Diets were prepared as a mixed ration and delivered twice daily at 0730 and 1600 h. Neither processing method nor dietary barley grain use affected dry

---

[*] Correspondence: Department of Animal Sciences, University of Zanjan, Zanjan 313-45195 Iran; Tel: +98-241-5283202; Fax: +98-241-5283202; E-mail: anikkha@yahoo.com

matter intake, daily eating, ruminating and chewing times, rumen pH and major VFA molar percentages, or milk percentages and yields of fat and protein. Energy-corrected milk yield increased for SB compared with GB at 35% but not at 30% barley grain. Feed efficiency was increased by SB, but was unaffected by dietary barley grain level. Results suggest that at 30% dietary barley grain, GB resulted in similar lactation performance as SB and that SB did not affect productivity when dietary barley grain increased from 30 to 35%. Regardless of barley grain level, grinding effectively maintained feed intake and rumen pH at 4 h post-feeding, whereas steam-rolling increased feed efficiency. Increasing barley grain from 30% to 35% of diet dry matter did not improve feed intake and milk production.

Keywords: barley grain, grinding, steam-rolling, lactating cow

INTRODUCTION

Steam-rolling was recently shown to offer no digestive and productive advantages over grinding at 26% dietary use of barley grain (*Hordeum Spp.,* Sadri et al., 2007). Whole barley grain, rich in highly fermentable starch and protein (Herrera-Saldana et al., 1990; Ørskov, 1986), is not optimally digested by lactating cows (Valentine and Wickes, 1980). As a result, grinding and steam-rolling are commonly used to increase barley starch and protein availability to rumen microbes (Mathison, 1996; Yang et al., 2000). Unlike corn and sorghum, barley starch is not extensively integrated with slowly degradable protein matrices and possesses greater effective DM degradability (e.g., 70% vs. 40%, Herrera-Saldana et al., 1990; Nikkhah et al., 2004). Feeding processed barley grains at high dietary inclusion rates may increase the risk of subacute rumen acidosis **(SARA)**(Owens et al., 1997) and asynchronize ATP, carbon skeleton and nitrogen compounds release (Hall and Huntington, 2008). The elevated rumen VFA release by feeding high amount of barley grain can raise blood insulin and depress milk yield (Ørskov, 1986). As conventionally believed, finely ground grains may depress feed intake by increasing ration dustiness and hastening ruminal release of organic acids (Mathison, 1996; Morrison, 1935; Nikkhah and Ghorbani, 2003). Grinding, however, is an economical processing technique easily accessible to almost all dairy producers. We intuit that the dietary inclusion rate of barley grain is more critical than processing method for reducing the risk of SARA, optimum immune function and economical milk production. Feeding high barley starch and low effective NDF diets has lowered DMI and rumen pH, and compromised immunity (Emmanuel et al., 2008; Yang et al., 2000). Based on rumen in situ studies, treating barley kernel with heat and moisture during steam-rolling process may reinforce protein-starch and lipid-starch bonds and reduce the initial rumen degradation rate of barley endosperm (Arieli et al., 1995; Ljøkjel at al., 2003a,b; Mathison, 1996; Nikkhah and Ghorbani, 2003). Also, the coarser particles produced by steam-rolling may decrease degradation rate of barley starch (Fiems et al., 1990; Nikkhah and Ghorbani, 2003; Tothi et al., 2003). However, these considerations have not conclusively been examined by definitive in vivo comparisons of steam-rolling **(SB)** and grinding **(GB)** of barley grains. Our objectives were to determine effects of grinding vs. steam-rolling of barley grain at different barley inclusion rates on DMI, eating, ruminating and chewing times, rumen fermentation, and lactation performance of high-producing cows.

MATERIALS AND METHODS

Cows, Treatments, and Experimental Designs

In the first study, 8 multiparous lactating Holstein cows (85 ± 15 days in milk and 32 ± 4 kg daily milk yield) were used in a replicated 4×4 Latin square design experiment. Cows were housed in individual tie-stalls from February to May 2003. The experiment consisted of 4 periods of 21d, with 14 d of adaptation. Individual stalls (2.17×1.55 m) were equipped with concrete feed bunkers and automatic metal drinkers. Clean wood shavings were used as bedding and refreshed 3 times daily to minimize the risk of mastitis. Cows were allowed 2 h daily exercise prior to the afternoon milking. Cows were offered a total mixed ration (TMR) 3 times daily with forage to concentrate ratio of 43:57 at 0700, 1500, and 2300 h. Dietary ingredients and their chemical composition are given in Table 1. The treatments were diets containing 1) finely ground (GB), 2) finely dry-rolled (DB720), 3) coarsely dry-rolled (DB810), and 4) steam-rolled (SRB) barley grains. The processing extent of dry-rolled and steam-rolled grains was expressed by "processing index" (PI; Yang et al., 2000). The PI was the ratio of processed barley grain density to whole barley grain density. Densities of grains were measured in triplicate by weighing a standard volume (2 litre glass cylinder) of processed and whole barley grains. An average density was then calculated from the three estimates for each treatment to determine PI. For example, if the density of rolled grains was 420 g/l and the density of whole grains was 600 g/l, the PI calculated was 700 g/kg or $[420/600] \times 1000$ g.

In the second study, 8 multiparous Holstein cows (85 ± 9 days in milk; 620 kg BW; 43 ± 2 kg milk yield, mean $\pm$ SE) were used in a duplicated 4×4 Latin square design experiment with four 21-d periods. The Latin square was balanced for carry over effects (Cochran and Cox, 1995). Cows were housed in individual free boxes (4×4 m) and were allowed 1-h daily exercise. Treatments were grinding (GB) or steam-rolling (SB) of barley grain at 30% or 35% of dietary DM (Table 1). In increasing barley grain level from 30 to 35% of diet DM, barley grain replaced beet pulp. All diets were prepared as a total mixed ration (**TMR**) with 62% concentrate and 38% forage on a DM basis (Tables 2 and 3). Cows were offered TMR twice daily at 0730 and 1600 h, permitting 5-10% orts with unlimited access to fresh water and salty stones. Diets were formulated with the CPM dairy program. The experiments were conducted at the Dairy Facilities of the Lavark Research Station (Isfahan University of Technology, Iran) under the guidelines of the Iranian Council of Animal Care (1995).

Table 1. Ingredient and chemical composition of the diets (DM basis) in the first study

	g/kg of dietary DM
Alfalfa hay	212.7

Maize silage	212.7
Barley grain	255.9
Whole cottonseed	92.3
Cottonseed meal (mechanically processed)	99.8
Soybean meal (solvent extracted)	111.2
Mineral and vitamin supplement[1]	9.9
Sodium chloride	3.0
Sodium bicarbonate	2.5
Chemical composition:	
Crude protein (CP)	161.0
Rumen degradable protein	115.0
Rumen undegradable protein	46.0
ADF	241.0
aNDF	380.0
NFC[2]	351.0
Ca	6.5
P	5.2
Ether extract	43.1
DCAD[3] (meq/kg)	276.0

[1]Contained 196 g Ca, 96 g P, 71 g Na, 19 g Mg, 3 g Fe, 0.3 g Cu,
2 g Mn, 3 g Zn, 100 ppm Co,100 ppm I, 0.1 ppm Se and 50×10^5 IU
of vitamin A ,10×10^5 IU of vitamin D and 0.1 g of vitamin E/kg.
[2]Nonfibre carbohydrates = 1000 g DM – [g/kg DM of (NDF + CP + EE + Ash)].
[3]DCAD = dietary cation anion difference $[(Na^+ + K^+) - (Cl^- + S^{2-})]$.

Table 2. Dry matter-based dietary ingredients at 35% and 30% barley grain in the second study

	Dietary use of barley grain	
Ingredient (DM-based)	35%	30%
Alfalfa hay	15.4	15.4
Corn silage	23.0	23.0
Barley grain (ground or steam-rolled)	35.0	30.0
Beet pulp	4.5	9.4
Soybean meal	16.0	16.0
Whole cottonseed	3.9	4.0
NaCl	0.1	0.1
Calcium carbonate	0.2	0.2
Sodium bicarbonate	1.0	1.0
Minerals and vitamins supplement[1]	0.9	0.9

[1]Contained 250000 IU/kg vitamin A, 50000 IU/kg vitamin D, 1500 IU/kg
vitamin E, 2.25 g/kg manganese, 120 g/kg calcium, 7.7 g/kg zinc, 20 g/kg
phosphorus, 20.5 g/kg magnesium, 186 g/kg sodium, 1.25 g/kg iron, 3 g/kg
sulfur, 1.25 g/kg copper, 14 mg/kg cobalt, 56 mg/kg iodine, and 10 mg/kg selenium.

Table 3. Chemical composition of forages and diets (dry matter basis) in the second study

	Forage		Dietary barley grain[1]	
Nutrient	Alfalfa hay	Corn silage	35%	30%

DM %	95.2	24.1	60.0	61.0
CP %	14.2	8.6	19.0	18.8
NDF %	43.4	52.0	33.6	35.1
ADF %	34.2	37.4	19.0	19.0
Starch %	2.8	26	28.0	25.6
NE_L[2], Mcal/kg	1.11	1.28	1.54	1.52

[1]Dry matter based.

[2]Estimated using NRC (2001).

Sampling and Laboratory Analyses

Feeds and orts. The amount of TMR offered and orts were measured daily from d 15 to 20 of each period to calculate DMI for individual cows. Samples of TMR were taken daily for individual cows during the last 5 days of each period. Feed and ort samples were oven-dried at 60°C for 48-h, ground to pass through 1-mm screen using a Wiley mill (Arthur H. Thomas Co., Philadelphia) and stored at -20°C until analyzed for chemical composition. Feed samples were analyzed for CP (method 984.13; AOAC, 1990), NDF (Van Soest et al., 1991; using heat-stable α-amylase and sodium sulfite) and ADF (method 973.18; AOAC, 1990). Organic matter was determined by ashing feed and fecal samples for 8 h at 550°C. The acid insoluble ash (AIA; Van Keulen and Young, 1977) was used as an internal marker to determine coefficient of total tract apparent digestibility (CTTAD).

Milking and milk sampling. Cows were milked three times daily at 0330, 1300, and 2130 h. Milk production was recorded at each milking during the final 5 d of each period. The amount of milk produced for each cow at each milking was measured using standard graduated jars (Agri & SD Co., Frankfort, Germany). Prior to each milking cows were monitored for udder inflammation and presence of milk clots in the nipples to ensure that milk yield and composition were not affected by mastitis. No cases of mammary infections were observed during sampling weeks. Milk was sampled 3 times daily at each milking in pre-labeled plastic vials, composited for individual cows, preserved with 2-bromo-2-nitropropan and potassium dichromate, and kept at 4°C. Milk samples were analyzed for fat, protein, lactose, and total solids by Milk-O-Scan (134 BN Foss Electric, Hillerød, Denmark).

Rumen fluid and volatile fatty acids analysis. In the first study, on the last day of each period, a sample of rumen fluid was collected from each cow 4 h after the morning feeding using a stomach tube. The initial 100 ml of fluid aspirated was discarded to minimize saliva contamination. The pH of the second portion was measured immediately using a mobile pH meter (HI 8314 membrane pH meter, Villafranca, Italy). Rumen fluid samples were centrifuged at $900 \times g$ for 10 min at 4°C and the supernatants were stored at -20°C for later volatile fatty acids (VFA) analysis. Urine was sampled by manual stimulation of the vulva and its pH was measured immediately. Grab faecal samples were taken from the rectum every day of the collection period and, after pH measurement, all were frozen for later analysis of AIA. To measure faecal pH, a portion of each faecal sample was thoroughly mixed with the same volume of distilled water to obtain sufficient uniformity and, as a result, a representative pH value.

In the second study, using four of cows, rumen fluid from the ventral sac was sampled by rumenocentesis technique (Nordlund and Garrett, 1994) at 4-h after morning feeding on the last day of each period. A 16-cm square area caudoventral to the costochondral junction of the last rib on a line parallel with the top of the patella was clipped and washed with alcohol. The shaved area was scrubbed with Povidone-Iodine and Savlon, and was sedated by injecting 8 ml of 2% lidocaine-epinephrine solution to prevent bleeding. A stainless steel needle was inserted about 4 cm into the ventral sac of the rumen, and a 5-ml syringe was used to aspirate rumen fluid. For 3 days after each sampling, cows were injected intramuscularly with Penicillin-G-Procaine to minimize any chance of infection. The pH of rumen fluid was measured immediately after sampling using a mobile pH-meter (HANNA, instrument, S/N: 137243. Portugal). To cease fermentation, 20 µl of 50% sulfuric acid was added to rumen fluids, and samples were kept at -20°C until analyzed for VFA using gas chromatography (0.25 × 0.32, id 0.3 µ WCOT Fused Silica Capillary, CHROMPACK CP. 9002, Model No. CP- 9002 Serial No. 94 77 B, Vulcanusweg 259 AM DELFT, The Netherlands), as described by Bal et al. (2000).

Eating, ruminating, and chewing activities. Eating and ruminating activities were monitored visually for a 24-h phase on d-17 of each period. The eating and ruminating activities were recorded once every 5 min under the assumption that each activity would persist for the entire 5 min (Yang et al., 2000). Total time spent chewing was calculated as the time spent eating plus the time spent ruminating. All activities were expressed per kg of DM, NDF, and ADF intake (Table 3).

Table 4. Fecal and urine pH, and coefficient of apparent total tract digestibility (CTTAD) for dry matter (DM) and organic matter (OM) in cows fed differently processed barley grains (the first study)

	Treatment[1]					Treatment effect
	GB	SRB	DB720	DB810	SE	P
Fecal DM, g/kg	167.6	163.1	170.0	171.4	3.2	0.08
Fecal pH	6.63	6.67	6.57	6.66	0.04	0.19
Urine pH	7.90	7.90	7.90	7.91	0.02	0.95
CTTAD						
Dry matter	0.68	0.69	0.66	0.68	0.013	0.36
Organic matter	0.69	0.69	0.67	0.69	0.014	0.36
pH	6.65	6.58	6.60	6.71	0.14	0.81
VFA, mol/100 mol						
Acetate (A)	0.651	0.649	0.668	0.669	0.017	0.53
Propionate (P)	0.202	0.207	0.194	0.183	0.016	0.23
Butyrate	0.111	0.111	0.099	0.106	0.071	0.52
Acetate : propionate	3.29	3.24	3.56	3.75	0.29	0.31

[1]GB = finely ground barley, SRB = steam-rolled barley with PI = 68 g/kg, DB720 = dry-rolled barley with PI = 720 g/kg, DB810 = dry-rolled barley with PI = 810 g/kg. Processing index (PI) was the ratio of the processed barley grain density to the whole barley grain density multiplied by 1000 (g).

Barley Processing Techniques

Barley grain (*Hordeum Spp.*) was ground using a conventional hammer mill (Isfahan Dasht, model 5543 GEN, Isfahan, Iran) with a standard screen size of 1 mm. Barley grains were screened during two separate steps and steamed for a minimum of 5 min at 102°C within a 4 m-height stainless steel chamber right above the rollers (Nikkhah et al., 2004). Steamed grains had a moisture content of 18-20% as they were rolled between preheated corrugated rollers (46 × 90 cm, Harris Co., CA, USA). Processing index (PI) was defined as the ratio of rolled grains density to whole grains density times 100 (Yang et al., 2000). For instance, if the density of whole and steam-rolled barley grains were on average 580 and 420 g/L, respectively, the PI for steam-rolled barley was (420/580) × 100 or 72.4. In the first experiment, the steam-rolled barley had PI of 0.68, while finely and coarsely dry-rolled barley had PI of 720 and 810 g/kg, respectively. In the second study, the steam-rolled barley had a PI of 72%.

Statistical Analysis

Data were subjected to Proc Mixed of SAS Institute (2003). The method of estimating least square means was Restricted Maximum Likelihood and the method of calculating denominator degrees of freedom was Kenward-Roger (SAS Institute, 2003). For the first study, effects of treatment (i.e., differently processed barley) and period were fixed and that of cow was random. Period was modeled as a repeated factor with first-order autoregressive (AR (1)) covariance structure to account for the correlation of repeated measures on the same cow (Tempelman, 2004). Tukey's test (SAS, 1999) was used to compare least square means. For the second study, models included fixed effects of processing technique, barley grain inclusion rate, their interaction and square, plus random effects of period and cow within square. The model for rumen data included the fixed effects of processing method, barley grain percentage and their interaction. Normality of distribution and homogeneity of residuals variance were tested using Proc Univariate of SAS Institute (2003). The significant effects were declared at $P < 0.05$ and trends were set at $P \leq 0.10$.

RESULTS AND DISCUSSION

Dry Matter Intake (DMI)

The lack of a difference in DMI (Table 5) indicate no effects of processing technique on diet palatability, and implying a similar impact of differently processed barley grains on short-term feed intake regulation (e.g., VFA; Allen, 2000). Findings are not consistent with the traditional belief that finely ground grains produce dust and depress DM intake (Mathison, 1996; Morrison, 1935), but limited to the conditions of the present study (e.g., dietary level of barley grain). It is generally believed that the larger particles produced by steam-rolling, rather than grinding, may reduce the ruminal degradation rate of barley grain (Mathison, 1996) although, unlike rate, extent of barley DM degradation is probably not altered by steam-rolling. Similarly, DM intake was not affected in feedlot cattle fed dry-rolled, steam-rolled and whole barley grains (Owens et al., 1997), and Bradshaw et al. (1996) also found no impacts of processing method on DM intake in growing and finishing feedlot steers.

172 A. Nikkhah

Table 5. Milk production and composition of cows fed diets with differently processed barley grains (the first study)

	Treatment[1]					Treatment Effect
	GB	SRB	DB720	DB810	SE	P
Milk yield, kg/d	27.8	28.6	29.0	26.7	1.2	0.22
DMI, kg/d	23.47	23.89	23.22	23.96	0.66	0.64
Milk composition						
Fat, g/kg	39.6	37.8	37.6	38.5	1.5	0.55
Fat yield, kg/d	1.07	1.07	1.07	0.99	0.04	0.22
Protein, g/kg	30.2	30.1	30.1	30.2	0.5	0.99
Protein yield, kg/d	0.83	0.85	0.86	0.79	0.03	0.09
Lactose, g/kg	51.0	51.4	51.6	51.6	0.5	0.70
Lactose yield, kg/d	1.42	1.47	1.50	1.37	0.07	0.22
TS, g/kg	122.3	121.0	120.9	121.9	1.4	0.68
TS yield, kg/d	3.37	3.44	3.49	3.22	0.12	0.17

[1]GB = finely ground barley, SRB = steam-rolled barley with PI = 68 g/kg, DB720 = dry-rolled barley with PI = 720 g/kg, DB810 = dry-rolled barley with PI = 810 g/kg. Processing index (PI) was the ratio of the processed barley grain density to the whole barley grain density multiplied by 1000 (g).

Table 6. Treatment effects on nutrient intake and eating, ruminating and chewing times (the second study)

	Processing method (PM)							
	Steam-rolling		Grinding				P-value	
Barley use % (BR)	35%	30%	35%	30%	SEM	PM	BR	PM × BR
DMI, kg/d	24.3	24.5	24.4	25.2	0.42	0.36	0.24	0.50
NDF intake, kg/d	9.8	9.9	9.5	10.5	0.39	0.45	0.04	0.11
ADF intake, kg/d	5.4	5.5	5.5	5.7	0.22	0.46	0.41	0.79
Eating time								
Min/d	350.4	342.4	342.4	344.5	6.8	0.67	0.67	0.46
Min/kg of DMI	14.5	14.0	14.0	13.9	0.4	0.37	0.46	0.61
Min/kg of NDF intake	36.2	34.9	36.0	33.7	1.0	0.52	0.09	0.66
Min/kg of ADF intake	70.0	62.8	62.4	62.7	2.6	0.37	0.44	0.39
Ruminating time								
Min/d	485.6[ab]	428.1[b]	453.6[ab]	498.6[a]	22.3	0.40	0.79	0.04
Min/kg of DMI	20.0	17.5	18.6	19.9	0.9	0.58	0.55	0.06
Min/kg of NDF intake	49.4	43.0	47.9	48.9	2.7	0.43	0.34	0.19
Min/kg of ADF intake	90.6	77.5	83.1	91.8	5.7	0.56	0.70	0.07
Chewing time[1]								
Min/d	835.9[ab]	769.0[b]	795.9[ab]	843.3[a]	23.8	0.48	0.69	0.03
Min/kg of DMI	34.5	31.5	32.6	33.8	1.1	0.83	0.41	0.06
Min/kg of NDF intake	85.6	77.7	83.8	82.5	3.2	0.63	0.17	0.32
Min/kg of ADF intake	157.6	140.0	145.5	153.9	7.4	0.90	0.54	0.10

[a,b]Within each row, means with different superscripts differ at $P < 0.05$.

[1]Sum of eating and ruminating times.

In the second study, neither dietary level (35% vs. 30%, $P = 0.24$) nor processing method (ground vs. steam-rolled, $P = 0.36$) of barley grain affected DMI (Table 6). Cows on GB and SB consumed 37.3 and 37.9 Mcal NE_L daily, respectively. Results suggest that even at high inclusion rates of barley grain, grinding maintains DMI compared to steam-rolling. The conventional belief that fine grinding produces dust, overly increases rumen fermentation rate, and depresses feed intake, has been based mainly upon results from beef studies. Despite greater DMI of lactating cows than of finishing beef cattle, dairy diets have much greater proportions of dry and ensiled roughages, and lower percentages of concentrate. Such differences will alter physical properties of the ration. Recently, McGregor et al. (2007) reported no effects of feeding either finely (PI = 69%) or coarsely (PI = 83%) steam-rolled barley grain on DMI of mid and late lactation cows. The DMI data alongside similar dietary NE_L and CP concentrations for GB vs. SB suggest that the potency of chemical constraints (e.g., rumen VFA; Allen, 2000) on short-term feed intake regulation were likely not different for GB vs. SB.

Eating, Ruminating, and Chewing Times

Eating time whether daily (343.5 vs. 346.4 min) or per kg intake of DM (13.9 vs. 14.2 min), NDF (34.9 vs. 35.5 min), and ADF (62.5 vs. 64.9 min) was not affected by grinding vs. steam-rolling of barley grain ($P > 0.10$, Table 6), which agree with comparable feed intake among treatments. Likewise, across barley grain inclusion rates, daily ruminating time (476.1 vs. 456.8 min) and rumination per kg intake of DM (19.3 vs. 18.7 min), NDF (48.4 vs. 46.2 min), and ADF (87.4 vs. 84.0 min) were not influenced by GB vs. SB either ($P > 0.10$, Table 6). These similarities can be attributed to the comparable DMI and daily ruminating time. The ruminating time of 13-14 min/kg DMI agrees with other reports (Beauchemin et al., 2000; Maekawa et al., 2002). Considering the lower DMI in those studies compared to the present study (e.g., 17-21 kg vs. 24.8 kg), it may be suggested that higher-producing cows can consume more DMI than lower-producing cows mainly by spending more time eating and not necessarily by eating more quickly. Compared to an increased eating time, an increased eating rate is more likely to compromise rumen environment during larger meals particularly shortly after feeding. Owing to the unchanged eating and ruminating times, total chewing time daily and per kg nutrient intake were unaffected by processing methods. However, daily ruminating time was lower ($P = 0.04$) for SB30 than for GB30 (428.1 vs. 498.6 min/d). As a result, ruminating time per kg of DMI tended to be lower ($P = 0.05$) for SB30 than for GB30 (17.5 vs. 20.0 min/d). In view of the extensive fermentation of barley starch and protein (Herrera-Saldana et al., 1990), it is likely that barley grain particle size did not have a large impact on rumen mat properties. This notion is consistent with the comparable daily rumination time between SB and GB. A tendency ($P = 0.10$) for greater rumination time per kg of ADF intake for GB30 (91.8 min/d) than for SB30 (77.5 kg/d) might reflect a prolonged fiber exposure to microbial enzymes for effective digestion. The consistent rumen fiber mat formation took place possibly earlier and was more efficient for SB30 than for GB30. The hypothetical reduction in rumen mat formation efficiency for GB30 might be due to fine barley particles with high degradation and passage rates (Mertens, 1997).

Rumen Fermentation

Steam-rolling is believed to stabilize rumen pH and lower the risk SARA by reducing barley starch degradation rate (Mathison, 1996; Tothi et al., 2003). Rumen pH and molar percentages of major VFA at 4-h after the morning feed delivery were unaffected by treatments in both studies (Tables 4, 7). Steam-rolling rather than grinding of barley grain increased (P < 0.01) isobutyrate and valerate concentrations, and tended to increase (P = 0.06) isovalerate concentration in the second study (Table 7). Branched-chain VFA (BCVFA) are products of AA metabolism in the rumen and play important roles in microbial energetics (Brockman, 2005). Due to small BCVFA contributions to milk secretion (Brockman, 2005), BCVFA effects on milk energy are not as significant as the effects of major rumen VFA. Rumen data in both studies including lower and higher barley inclusion rates provide evidence that grinding of barley grain at up to 35% of diet DM does not compromise rumen pH. Noteworthy, rumen conditions were assessed using samples taken at 4-h after feeding when VFA concentrations were expectedly at peak (Stone, 2004). As such, the lack of a treatment difference in rumen pH when VFA molar percentages suggests that grinding vs. steam-rolling of differently included barley grain had little effect on rumen fermentation in cows with an average DMI of 23-25 kg/d.

Table 7. Treatment effects on rumen fluid pH and volatile fatty acids (VFA) molar percentages at 4-h after the morning feed delivery (TMR was delivered twice daily at 0730 and 1600 h) (the second study)

| Item | Processing method (PM) | | Barley use % (BR) | | SEM | P-value | | |
	Steam-rolling	Grinding	35%	30%		PM	BR	PM × BR
Rumen pH	5.73	5.70	5.74	5.70	0.07	0.73	0.58	0.76
VFA, mol/100 mol								
Acetate	68.3	68.0	67.6	68.6	1.0	0.87	0.44	0.42
Propionate	17.3	19.5	18.9	17.8	0.9	0.13	0.43	0.40
Butyrate	8.8	9.6	9.1	9.3	0.5	0.34	0.86	0.41
Isobutyrate	1.8	0.6	1.0	1.3	0.3	<0.01	0.30	0.70
Valerate	1.5	0.8	1.2	0.14	0.1	<0.001	0.57	0.46
Isovalerate	1.4	0.8	1.1	1.0	0.3	0.06	0.69	0.29
Acetate : propionate	4.0	3.6	3.7	3.9	0.2	0.22	0.45	0.21

Milk Production

Milk yield was not affected by processing method and dietary inclusion rate of barley grain in either study (Table 5, 8). From a rumen health standpoint, these findings suggest that in low and high-producing cows, increasing barley grain use above certain limits (e.g., 30%) does not improve milk production. In the second study, feeding GB30 (37.3 kg/d) led to the same ECM as feeding SB30 (37.8 kg/d) and SB35 (37.5 kg/d), but GB35 (34.9 kg/d) decreased ECM compared to other treatments (Table 6). At both levels of barley grain, SB consistently increased feed efficiency compared to GB (P < 0.01), which was an accumulative effect of the numerical increase and decrease in milk yield and DMI, respectively (Table 6). Overall, therefore, treatments had smaller effects on productivity with more pronounced effect on feed efficiency. The increased feed efficiency by steam-rolling was about 4.7%. Since the average dietary barley grain use was 32.5% on a DM basis, the barley-related improvement in feed efficiency was about 15%. This means that based on the results of the current study, steam-rolling of barley could be affordable would it not cost more than 15% of what would grinding cost. In addition, the greater the difference between milk price and feed cost, the greater economic magnitude of an improved feed efficiency by steam-rolling. The milk fat data seem to rule out the possibility that ruminal digestibility of dietary fibres differed among treatments, supported by the similar concentration of ruminal acetate and equal acetate to propionate ratio. The comparable milk yield of fat, protein, lactose and total solids among GB, DB, and SRB supports Mathison (1996) who found no benefits of SRB over DB for growing-finishing cattle in Western Canada.

Table 8. Treatment effects on milk production and feed efficiency in the second study

| | Processing method (PM) | | | | | P-value | | |
| | Steam-rolling | | Grinding | | | | | |
Barley use % (BR)	35%	30%	35%	30%	SEM	PM	BR	PM × BR
Milk yield, kg/d	38.3	38.8	36.9	38.0	0.8	0.24	0.38	0.72
4% FCM, kg/d	35.4	35.7	33.4	35.3	0.8	0.17	0.19	0.32
ECM, kg/d[1]	37.5[a]	37.8[a]	34.9[b]	37.3[a]	0.9	0.09	0.12	0.23
ECM : DMI	1.54[a]	1.54[a]	1.45[b]	1.47[ab]	0.03	<0.01	0.74	0.70
Milk yield : DMI	1.57[a]	1.58[a]	1.50[b]	1.51[b]	0.02	<0.01	0.89	0.92
Milk fat %	3.36	3.47	3.38	3.6	0.14	0.89	0.42	0.46
Fat yield, kg/d	1.39	1.35	1.24	1.34	0.04	0.25	0.26	0.33
Milk protein %	2.87	2.86	2.85	2.87	0.02	0.88	0.97	0.58
Protein yield, kg/d	1.09	1.11	1.04	1.09	0.03	0.16	0.26	0.53
Milk lactose %	5.47[a]	5.42[ab]	5.45[ab]	5.37[b]	0.03	0.30	0.05	0.78
Lactose yield, kg/d	2.10	2.11	2.02	2.05	0.05	0.21	0.70	0.46
Milk SNF %	8.54	8.47	8.56	8.45	0.07	0.95	0.20	0.75
SNF yield, kg/d	3.27	3.29	3.16	3.22	0.09	0.32	0.66	0.82
Total solids (TS) %	12.00	11.95	11.94	12.05	0.10	0.88	0.79	0.40
TS yield, kg/d	4.59	4.63	4.39	4.57	0.11	0.24	0.33	0.51
Milk fat% : protein%	1.21	1.21	1.18	1.26	0.05	0.86	0.47	0.51

[1]Energy-corrected milk calculated as (kg of milk × 0.3246) + (kg of milk fat × 12.96) + (kg of milk protein × 7.04)(Jenkins et al., 1998).
[a,b]Within each row, values with different superscripts differ at $P \leq 0.05$.

It may be suggested that differently processed barley grains at both 30% and 35% dietary inclusion rates did not influence microbial protein synthesis, intestinal AA delivery, and the mammary AA supply. These data were consistent with the comparable DMI and rumen VFA data. Feeding barley grain at 35% instead of 30% of the diet DM tended to increase milk lactose percentage in the second study ($P = 0.05$). Given the precise statistical analysis and the low standard error, biological interpretation of the 0.04-0.05% unit rise in milk lactose percentage requires caution. Owing to the similar milk SNF percentage and yield, total solids percentage and yield was similar among treatments. Milk production data provide little basis for the primacy of steam-rolling over grinding at up to 30% barley grain in the diet while at 35% dietary barley grain, steam-rolling fairly and positively affected milk energy output.

Conclusions

Grinding has conventionally been considered a risk to DMI, rumen function and the balance between highest starch utilization and increased risk of SARA, whilst being an easy-to-access technique to process barley grain. Steam-rolling is believed to reduce such risks, but it is more expensive than grinding. While heat treatment is thought to alleviate or attenuate these challenges, no conclusive efforts have been made to clarify the issue by comparing GB and SB for lactating cows. Results of the first study suggest no differences in DM intake, rumen conditions, total tract digestibility and milk yield of cows fed finely GB or medium-flat SB (PI = 680 g/kg) in a TMR. In the second study, when DM of TMR contained 30% barley grain, grinding resulted in similar feed intake, rumen fermentation at 4-h post-feeding, and milk production as steam-rolling. Compared to grinding, steam-rolling of barley increased feed efficiency at both barley inclusion rates, and positively fairly affected milk energy output only at 35% dietary barley grain. Increasing barley grain use from 30% to 35% in the diet of high-producing cows did not improve cow performance.

References

Allen, M. S. (2000). Effects of diet on short-term regulation of feed intake by lactating dairy cattle. *J. Dairy Sci.* 83, 1598-1624.

AOAC. (1990). Official methods of analysis. 15th ed. Assoc. Offic. *Anal. Chem.*, Arlington, V. A.

Arieli, A., Bruckental, I., Kedar, O., Sklan, D. (1995). In sacco disappearance of starch nitrogen and fat in processed grains. *Anim. Feed Sci. Technol.* 51, 287-295.

Bal, M. A., Shaver, R. D., Jirovec, A. G., Shinners, K. J. & Coors, J. G. (2000). Crop processing and chop length of corn silage: Effects on intake, digestion, and milk production by dairy cows. *J. Dairy Sci.* 83, 1264-1273.

Beauchemin, K., Rode, L., Maekawa, M., Morgavi, D. P. & Kampen, R. (2000). Evaluation of a non-starch polysaccharidase feed enzyme in dairy cow diets. *J. Dairy Sci.* 83, 543-553.

Bradshaw, W. L., Hinman, D. D., Bull, R. C., Everson, D. O., Sorensen, S. L. (1996). Effects of barley variety and processing methods on feedlot steer performance and carcass characteristics. *J. Anim. Sci.* 74, 18-24.

Brockman, R. P. (2005). Glucose and short chain fatty acid metabolism. In Dijkstra, J., J. M. Forbes and J. France. *Quantitative aspects of ruminant digestion and metabolism.* 11:157-176. 2nd Ed., CABI Publishing. Wallingford, UK.

Cochran, W. G. & Cox, G. M. (1992). *Experimental Designs.* Second Ed., Wiley-Interscience Publications. NY. USA.

Emmanuel, S., Dunn, M. & Ametaj, B. N. (2008). Feeding high proportions of barley grain stimulates an inflammatory response in dairy cows. *J. Dairy Sci.* 91, 606-614.

Fiems, L. O., Cottyn, B. G., Boucque, Ch. V., Vanacker, J. M. & Buysse, F. X. (1990). Effect of grain processing on in sacco digestibility and degradability in the rumen. *Arch. Anim. Nutr.* 40, 713-721.

Hall, M. B. & Huntington, G. B. (2008). Nutrient synchrony: Sound in theory, elusive in practice. *J. Anim. Sci.* 86:E287-292E.

Herrera-Saldana, R. E., Huber, J. T. & Poore, M. H. (1990). Dry Matter, crude protein, and starch degradability of five cereal grains. *J. Dairy Sci.* 73, 2386-2393.

Iranian Council of Animal Care. (1995). *Guide to the Care and Use of Experimental Animals*, vol. 1. Isfahan University of Technology, Isfahan, Iran.

Jenkins, T. C., Bertrand, J. A. & Bridges, W. C. Jr. (1998). Interactions of tallow and hay particle size on yield and composition of milk from lactating Holstein cows. *J. Dairy Sci.* 81, 1396-1402.

Ljøkjel, K., Harstad, O. M., Prestløkken, E. & Skrede, A. (2003a). *In situ* digestibility of protein in barley grain (*Hordeum vulgare*) and peas (*Pisum sativum L.)* in dairy cows: influence of heat treatment and glucose addition. *Anim. Feed Sci. Technol.* 107:87-104.

Ljøkjel, K., Harstad, O. M., Prestløkken, E. & Skrede, A. (2003b). *In situ* digestibility of starch in barley grain (*Hordeum vulgare*) and peas (*Pisum sativum L.)* in dairy cows: influence of heat treatment and glucose addition. *Anim. Feed Sci. Technol.* 107, 105-116.

Maekawa, M., Beauchemin, K. A. & Christensen, D. A. (2002). Effect of concentrate level and feeding management on chewing activities, saliva production, and ruminal pH of lactating dairy cows. *J. Dairy Sci.* 85, 1165-1175.

Mathison, G. W. (1996). Effects of processing on the utilization of grain by cattle. *Anim. Feed Sci. Technol.* 58, 113-125.

McGregor, G., Oba, M., Dehghan-banadaky, M. & Corbett, R. (2007). Extent of processing of barley grain did not affect productivity of lactating dairy cows. *Anim. Feed Sci. Technol.* 138, 272-284.

Mertens, D. R. (1997). Creating a system for meeting the fiber requirements of dairy cows. *J. Dairy Sci.* 80, 1463-1481.

Morrison, F. B. (1935). Feeds and Feeding. Chapter 4: *Factors affecting the value of feeds*. The Morrison Publishing Co. Ithaca, NY, pp, 59-71.

National Research Council. (2001*). Nutrient Requirements of Dairy Cattle*. 7th rev. ed. National Academy Press, Washington, DC.

Nikkhah, A. & Ghorbani, G. R. (2003). Effects of dry and steam processing on in situ ruminal digestion kinetics of barley grain. *J. Anim. Sci.* 81, 338 (Suppl. 1).

Nikkhah, A., Alikhani, M. & Amanlou, H. (2004). Effects of feeding ground or steam-flaked broom sorghum and ground barley on performance of dairy cows in midlactation. *J. Dairy Sci.* 87, 122-130.

Nordlund, K. V. & Garrett, E. F. (1994). Rumenocentesis: a technique for collecting rumen fluid for the diagnosis of subacute rumen acidosis in dairy herds. *Bovine Pract.* 28, 109.

Owens, F. N., Secrist, D. S., Hill, W. J. & Gill, D. R. (1997). The effect of grain source and grain processing on performance of feedlot cattle: a review. *J. Anim. Sci.* 75, 868-879.

Ørskov, E. R. (1986). Starch digestion and utilization in ruminants. *J. Anim Sci.* 63:1624-1633.

SAS User's Guide. (2003). Version 9.1. Edition. SAS Institute Inc., Cary, NC.

Stone, W. C. (2004). Nutritional approaches to minimize subacute ruminal acidosis and laminitis in dairy cattle. *J. Dairy Sci.* 87, E13-26E.

Tempelman, R. J. (2004). Experimental design and statistical methods for classical and bioequivalence hypothesis testing with an application to dairy nutrition studies. *J. Anim. Sci.* 82, E162-E172.

Tothi, R., Lund, P., Weisbjerg, M. R. & Hvelplund, T. (2003). Effect of expander processing on fractional rate of maize and barley starch degradation in the rumen of dairy cows estimated using rumen evacuation and in situ techniques. *Anim. Feed Sci. Technol.* 104, 71-94.

Van Keulen, V. & Young, B. H. (1977). Evaluation of acid-insoluble ash as natural marker in ruminant digestibility studies. *J. Anim. Sci.* 26, 119-135.

Van Soest, P. J., Robertson, J. B. & Lewis, B. A. (1991). Methods for dietary fiber, neutral detergent fiber, and nonstarch polysaccharides in relation to animal nutrition. *J. Dairy Sci.* 74, 3583-3597.

Yang, W. Z., Beauchemin, K. A. & Rode, L. M. (2000). Effects of barley grain processing on extent of digestion and milk production of lactating cows. *J. Dairy Sci.* 83, 554-568.

Zinn, R. A. (1993). Influence of processing on the comparative feeding value of barley for feedlot cattle. *J. Anim. Sci.* 71, 3-10.

In: Progress in Food Science and Technology, Volume 1 ISBN: 978-1-61122-314-9
Editor: Anthony J. Greco © 2012 Nova Science Publishers, Inc.

Chapter 9

BIOACTIVITY OF OLEANANE TRITERPENOIDS FROM SOY (*GLYCINE MAX* MERR) DEPENDS ON THE CHEMICAL STRUCTURE

Wei Zhang and David G. Popovich
National University of Singapore, Singapore

ABSTRACT

Soyasaponins are oleanane triterpenoids molecules that are found in Soy (*Glycine Max* Merr) and other legumes such as green peas (*Pisum sativum* L) and lentils (*Lens culinaris*). Soyasaponins are diverse, structurally complex and possess amphiphilic properties, with polar water soluble sugar moieties attached to a non polar, water insoluble pentacyclic ring structure. In legumes, soyasaponins are present primarily as glycosides which are storage forms of saponins and are secondary plant metabolites. These molecules serve as defensive molecules for the plant to defend against biological predators such as fungi and herbivores. A resurgence of interest into secondary plant metabolites has emerged as many of these compounds possess diverse biological activity which includes the potential to act as chemo-preventive agents. Soyasaponins are classified into five groups based on the chemical structure of their respective aglycones but the two main groups are A and B. Group B soyasaponins are the most abundant group of saponins found in soy and are thought to contribute to bulk of soy's chemo-preventative properties attributed to soyasaponins.

Keywords: soyasaponin; soyasapogenol; HPLC; bioactivity; Hep-G2

INTRODUCTION

Soyasaponins are oleanane triterpenoids found in a wide variety of legumes such as soy (*Glycine Max* Merr), green peas (*Pisum sativum* L) and lentils (*Lens culinaris*). Soyasaponins elong to a group of secondary plant metabolites with a triterpenoid aglycone with one or more carbohydrate moiety attached as a glycoside. The soyasaponins content in soy bean seed is about 0.5% [1] and can reach 2% in the hulls [2], but composition can vary widely depending on cultivation year, location, climatic conditions and degree of maturity [3, 4]. Scientific research of soyasaponins has a long history and they have been studied since at least the 1920s [5]. Specifically, the relationship between soyasaponins structure and potential bioactivity such as chemo-preventative properties has attracted increasing research attention [6, 7]. The complex structure and difficulties in extracting and analyzing intact compounds has slowed the determinants of bioactivity. In this review, the isolation, chemical characterization and detection strategies to characterized soyasaponins will be discussed along with a detailed review of reported bioactive effects of soyasaponins and soyasapogenols.

SOYASAPONINS AND SOYASAPOGENOLS CLASSIFICATION

Soyasaponins are composed of a triterpenoid with water-soluble sugar glycosides moieties. The sugar glycosidic attachment consists of 5 types of monosaccharides, β-D-glucopyranosyl, β-D-galactopyranosyl, α-L-arabinopyranosyl, β-D-glucoronopyranosyl and α-L-rhamnopyranosyl [8]. Soyasaponins can be hydrolyzed to liberate aglycones (soyasapogenols). Five soyasapogenols have been characterized which are referred to as soyasapogenol A, B, C, D and E (**Figure 1**) [9]. Three naturally present soyasapogenols have been suggested which are groups A, B and E [6, 10-12]. Some have argued that soyasapogenols C, D are artifacts of extraction and E may also be an artifact [13, 14]. Generally most soyasaponins can be classified into two groups referred to as group A and B. Group A soyasaponins are bidesmosidic saponins with two glycosylation sites on their aglycone (soyasapogenol A) and can be divided into 2 groups which are deacetylated (A1 and A2, major group A soyasaponins) and acetylated forms [15, 16]. In addition, the acetylated group A soyasaponins are considered to be responsible for undesirable bitter and astringent taste in soybean [17]. Soyasaponin Ab which is completely acetylated on the terminal sugar moiety was reported to have stronger undesirable taste compared to soyasaponin A1 which has no acetyl groups [17]. The undesirable taste became stronger as hydrophobicity increases. Group A soyasaponins are also well known for their physical properties such as the foaming activity, emulsification and surface activity [18-20]. Far less bioactive information is available on group A soyasaponins compared to group B.

Group B soyasaponins are monodesmosidic saponins with one glycosylation site on the aglycone (soyasapogenol B) and can also be classified into 2 groups which are DDMP (2,3-dihydro-2,5-dihydroxy-6-methyl-4-pyrone) conjugated (αg, βa, βg, γa and γg) and non-DDMP conjugated (I, II, III, IV and V) [16, 21, 22]. DDMP group B soyasaponins were proposed as the predominant and genuine soyasaponins in legume [16]. Soyasaponin βg is thought to be the main saponin found in cultivated soy, while uncultivated soy is rich in βg, γg and γa [7].

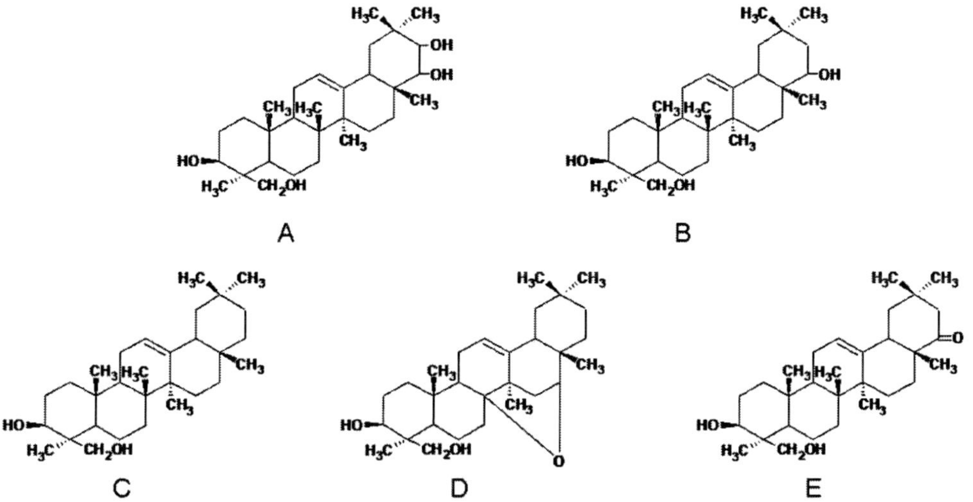

Figure 1. The Chemical Structures of Soyasaponins Classified According to the Aglycone.

Group A Soyasaponins	Structure	MW	Formula
A1 (Ab)	glc(1→2)gal(1→2)glcU(1→3)-A-(22←1)ara(3←1)glc(2,3,4,6-tetra-O-Acetyl)	1463	$C_{67}H_{104}O_{33}$
A2 (Af)	gal(1→2)glcUA(1→3)-A-(22←1)ara(3←1)glc(2,3,4,6-tetra-O-Acetyl)	1274	$C_{61}H_{94}O_{28}$
A3 (Ah)	ara(1→2)glcUA(1→3)-A-(22←1)ara(3←1)glc(2,3,4,6-tetra-O-Acetyl)	1244	$C_{60}H_{92}O_{27}$
A4 (Aa)	glc(1→2)gal(1→2)glcU(1→3)-A-(22←1)ara(3←1)xyl(2,3,4-tri-O-Acetyl)	1364	$C_{64}H_{100}O_{31}$
A5 (Ae)	gal(1→2)glcUA(1→3)-A-(22←1)ara(3←1)xyl(2,3,4-tri-O-Acetyl)	1202	$C_{58}H_{90}O_{26}$
A6 (Ag)	ara(1→2)glcUA(1→3)-A-(22←1)ara(3←1)xyl(2,3,4-tri-O-Acetyl)	1172	$C_{57}H_{88}O_{25}$
Ac	rha(1→2)gal(1→2)glcUA(1→3)-A-(22←1)ara(3←1)glc(2,3,4,6-tetra-O-Acetyl)	1420	$C_{67}H_{104}O_{32}$
Ad	glc(1→2)ara(1→2)glcUA(1→3)-A-(22←1)ara(3←1)glc(2,3,4,6-tetra-O-Acetyl)	1390	$C_{66}H_{102}O_{31}$
Group B Soyasaponins			
I (Bb)	rha(1→2)gal(1→2)glcUA(1→3)-B	942	$C_{48}H_{78}O_{18}$
II (Bc)	rha(1→2)ara(1→2)glcUA(1→3)-B	912	$C_{47}H_{76}O_{18}$
III (Bb')	gal(1→2)glcUA(1→3)-B	796	$C_{42}H_{68}O_{14}$
IV (Bc')	ara(1→2)glcUA(1→3)-B	766	$C_{41}H_{66}O_{13}$
V (Ba)	glc(1→2)gal(1→2)glcUA(1→3)-B	958	$C_{48}H_{78}O_{19}$
DDMP Group Soyasaponins			
β g	rha(1→2)gal(1→2)glcUA(1→3)-B-(22←2')-DDMP	1068	$C_{54}H_{84}O_{21}$
β a	rha(1→2)ara(1→2)glcUA(1→3)-B-(22←2')-DDMP	1038	$C_{53}H_{82}O_{20}$
γ g	gal(1→2)glcUA(1→3)-B-(22←2')-DDMP	922	$C_{48}H_{74}O_{17}$
γ a	ara(1→2)glcUA(1→3)-B-(22←2')-DDMP	892	$C_{47}H_{72}O_{16}$
α g	glc(1→2)gal(1→2)glcUA(1→3)-B-(22←2')-DDMP	1084	$C_{54}H_{84}O_{22}$
Group E Soyasaponins			
Bd	glc(1→2)gal(1→2)glcUA(1→3)-E	956	$C_{48}H_{76}O_{19}$
Be	rha(1→2)gal(1→2)glcUA(1→3)-E	940	$C_{48}H_{76}O_{18}$

Group E soyasaponins Bd and Be were recognized based on soyasapogenol E as the aglycones of the third type of soyasaponins [12]. However, some studies proposed soyasapogenol E as an artifacts formed during extraction [9]. Furthermore, group E soyasaponins Bd and Be were presumed to be transformed into soyasapogenol B during acid hydrolysis [3]. Generally group E soyasaponins are present in very low quantities and are often undetectable in extracts.

EXTRACTION, PURIFICATION AND DETECTION OF SOYASAPONINS

Extraction of soyasaponins is challenging due to the complex purification process, long extraction time and low extraction efficiency. Moreover, there are no complete commercial set of soyasaponins that are available. Organic solvent extraction is a conventional extraction method which has been used to extract soyasaponins from defatted soybean or soy flour. The extraction efficiency of this method is decided by three main factors, time, temperature and solvent type. Aqueous ethanol (70%) is commonly used in many studies [8, 23, 24] with agitation at room temperature to protect and ensure extraction of the intact glycosylated soyasaponins [25]. The optimum extraction time was determined to be 4-6 h for yielding maximum quantities of soyasaponins from soy flour using 100% methanol at 60°C under reflux conditions [26]. Ultrasonic extraction has also been utilized for soyasaponin extraction [27-30] as this extraction technique results in good extraction efficiencies [28], increased speed of extraction [29] and good mass transfer between contents and solvent [31]. Optimized ultrasonic extraction of soyasaponins (Aa, Ab, Ba, Bb, Bb' and βg) with 40% ethanol in water at 25°C has been reported [30]. This method has significantly shortened the extraction time from 2 h to 20 min compared to conventional methods but it varies slightly depending on types of solvents used. A rapid quantification and characterization method of soyasaponins using ultrasound extraction with 80% ethanol at 25°C for soyasaponins has been reported [32]. Ultrasound extraction under low temperature is appropriate to allow DDMP-conjugated soyasaponins B which are unstable at high temperature to be extracted intact.

Current isolation and purification studies are aimed to achieve purified soyasaponins under three criteria which are extracts with relatively high purity, at least in milligram quantities for use in biological evaluation and the minimization of the use of toxic solvents [26].

The extraction of soyasaponins is hampered due to the presence of isoflavones that often share similar polarities as soyasaponins and are often extracted together. Preparation of extracts by removing isoflavones have been proposed and utilized an XAD-2 resin to attract both soyasaponins and isoflavones from crude extraction [8]; this phytochemical extract was then passed through Source 15 RPC column and individual soyasaponin fraction was eluted. This method resulted in a satisfactory isolation of soyasaponins from isoflavones and a good separation between the various soyasaponins. However, the purity of individual soyasaponins Ab, αg and βg needed to be improved (>97%) by a second chromatographic step [8]. Silica gel can be also used for isolation of individual soyasaponins from crude extracts [23]. A semi-preparative LC system has widely been applied to purify and separate individual soyasaponins from extracts. Generally individual soyasaponins are isolated from crude extracts by column chromatography then purified [23, 24]. Preparative HPLC seems to be most effective technology to separate individual soyasaponins but it is limited by a large requirement of solvents and low recovery. A solid phase extraction (SPE) method is a relatively simple and economical alternative for individual soyasaponins purification [26]. However, the SPE methods result in lower purity (about 85-90%) of individual soyasaponins compared to (>97%) preparative HPLC.

The detection and quantification of soyasaponins have generally utilized a reversed-phase high-performance liquid chromatography (HPLC) column coupled to a ultraviolet (UV) or photodiode array detector (PDA) [21, 24, 33], evaporative light scattering detection

(ELSD) [3] or electrospray ionization (ESI) mass spectrometer (MS) detector [23, 32, 34]. The maximum absorption wavelength of most soyasaponins is around 205 nm [23], but the DDMP conjugated soyasaponins can reach 295 nm. Soyasaponins glycosides can vary and it is generally difficult to separate all soyasaponins at their maximum absorption [33] in one analysis employing UV dector. An analytical method was developed using UV detection of all the known group B soyasaponins at 205 nm [24], but this method requires the preparation of non-DDMP and DDMP soyasaponins standards by prep-HPLC do to the lack of a full complement of commercial standards. Ireland and Dziedzic (1985) described the analysis of all four soyasapogenols (A, B, C and E) by HPLC with silica normal phase column [14].

ELSD, which is based on mass detection by light scattering after evaporation of mobile phase, has been successfully used for the detection of soyasapogenols, soyasaponins and ginseng saponins which are dammarane triterpenoids [35]. ELSD has been successfully utilized in detecting soyasapogenol A, B and E [3]. Furthermore a HPLC-ELSD-ESI-MS method for the analysis all groups of soyasaponins including acetyl soyasaponins from group A and DDMP soyasaponins from group B was also reported [35]. However, ELSD like UV suffered from some disadvantages such as extensive sample preparation and interference when analyzing soyasaponins in human serum at low concentration [30].

Using a mass spectrometer to identify group B soyasaponins [32, 36] seems to be most powerful and sensitive method of detection [32]. However, complicated ion fragments patterns can be produced by the ionization process which makes selection of main ions useful for quantification and identification difficult. MS requires an expensive apparatus and is usually not available for daily routine analysis [37].

Soyasapogenols are thermal sensitive compounds. They are difficult to observe during MS-ESI detection. A structurally diagnostic reverse Diels-Alder reaction was described as the source of the main ion patterns in the mass spectra of all four soyasapogenols (A, B, C and E) [13]. We found that the capillary temperature of $220^{\circ}C$ during ESI-MS analysis produced a significantly ($p \leq 0.05$) greater intensity of soyasapogenol B ion ([M-H]$^-$ = 457.4) without the typical fragmentation patterns (produced by a reverse Diels-Alder reaction) reported in the literature [38]. Furthermore capillary temperatures between $200^{\circ}C$ to $250^{\circ}C$ were found to be the optimum temperature range for analyzing soyasapogenol B. At low capillary temperatures ($\leq 200^{\circ}C$), the soyasapogenol B ion ([M-H]$^-$ = 457.4) was converted to ion at m/z at 441,while, at high capillary temperatures ($\geq 250^{\circ}C$), the soyasapogenol B ion produced the ion m/z at 440 and 437 which was confirmed by MS/MS analysis (**Figure 2**).

Thin layer chromatography (TLC) is a simple and rapid method for determination of soyasaponins [39]. Soyasapogenols were initially identified by TLC before LC-MS became widely available [13]. However it is difficult to separate soyasapogenol A and B due to structural similarities soyasapogenol A possesses only an additional hydroxyl group at the C-21 position compared to soyasapogenol B.

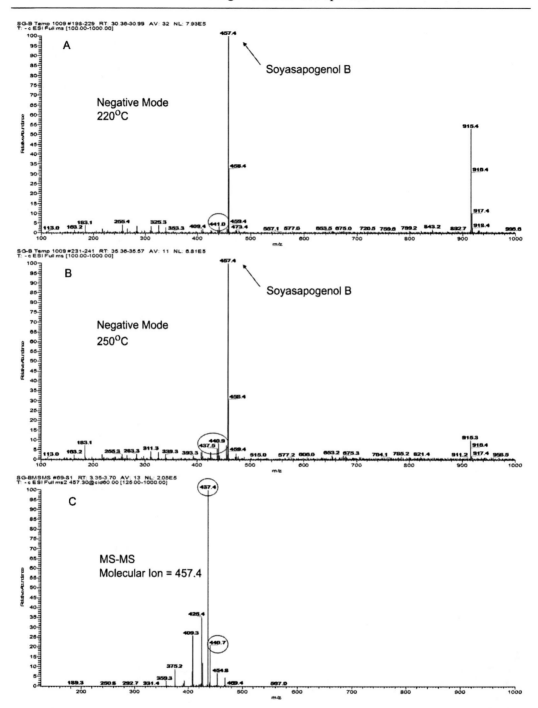

Figure 2. Panels A and B are the mass spectrums of soyasapogenol B at capillary temperatures of 200°C and 250°C, respectively. Panel C is the MS/MS spectrum of soyasapogenol B produced by negative ESI mode at 50% scatter energy.

SOYASAPONINS BIOACTIVITIES

The biological activity of each individual saponin depends upon its polarity, acidity, hydrophobicity [40]. The ability of soyasaponins to lower plasma cholesterol concentration has been widely reported [41-45]. However, the exact mechanism responsible for the hypocholesterolemic effects of soyasaponins needs to be further elucidated. The presence of soyasaponins has also been suggested to affect the hypercholesterolemic action of soy protein [46]. Moreover, Oakenfull and Sidhu (1990) reported that the hypercholesterolemic activity of saponins was a result of saponins forming insoluble complexes with cholesterol due to their particular similarities in chemical structural and this complex could inhibit the intestinal absorption of both endogenous and exogenous cholesterol [42]. Another possible mechanism indicated that soyasaponins could increase fecal bile acid and neutral sterol excretion [42] and this enhanced fecal excretion may lead the cholesterol-lowering effects [41].

Soyasaponins have known antioxidant properties likely preventing lipid per-oxidation [47]. Compared to other antioxidants including L-ascorbic acid, α-tocopherol, all-trans-retinol, and butylated hydroxytoluene (BHT), soyasaponins ranked second in reducing the antimutagenic activity of AFB1 (Aflatoxin B1) and had a slightly lower activity compared to BHT [48].

The anti-carcinogenic activity of soyasaponins in cultured colon carcinoma cells has been well documented [49-52]. A structure- function relationship for soyasaponins have been reported [6]. Soyasaponins have also been reported to inhibit the growth of cervical adenocarcinoma (Hela) cells [53] and hepatocarcinoma (Hep-G2) cells [53, 54] and induce apoptosis. We reported that extracts with soyasapogenol A (SG-A), soyasapogenol B (SG-B) and total or crude soyasaponins (TS), had an effect on Hep-G2 proliferation in a dose-response manner with 72 h LC50 values of 0.594 ± 0.021 mg/mL for TS, 0.052 ± 0.011 mg/mL for SG-A and 0.128 ± 0.005 mg/mL for SG-B [54]. Moreover, cell cycle analysis indicated a significant ($p < 0.05$) greater sub-G1 build up of apoptotic cells at 24h ($25.63 \pm 2.1\%$), and 72h ($47.1 \pm 3.5\%$) for the SG-A extract compared to SG-B and TS [54].

Colonic mucosal ornithine decarboxylase (ODC) activity in rat colon is regarded as a biomarker for short-term screening assay for potential chemopreventive agents of colon cancer [55]. Soyasaponin A_2 is a triterpenoid saponin and has been reported to reduce mucosal ODC activity and polyamine levels in the colon [56]. ODC is the polyamine synthesis enzyme, which is related to cell proliferation and transformation [57]. Presently, there is no evidence to indicate soyasaponin A_2 could inhibit the intestinal tumorigenesis induced by azoxymethane. However, soyasaponin A_2 was considered to have potential anti-promoting effects against tumor formation by altering key enzyme activity through inhibiting ODC activity and polyamine levels [56]. Human colon cancer cells treated with soyasaponins suppressed proliferation, induced differentiation and inhibited protein kinase C activity [58]. *In vivo* group B soyasaponins could metabolize to the aglycone soyasapogenol B by bacterial glycosidases in the colon, while the mixture of soyasaponins and soyasapogenol may interact with the colonic epithelium [6, 37]. Group B soyasaponins played an inhibitory role in the growth of colon cancer (HCT-15) cells resulting in S-phase accumulation and showed a nearly 4.5-fold increase in cell morphologies and related autophagic death [59]. The level of LC-3 which was a specific protein for macroautophagy, increased significantly comparing to non-treated cells. Furthermore, controlling the induction of macroautophagic cell death was reported by two signaling pathway, a reduction of the activating ser^{473}phosphorylation and enhancement of activity of ERK1/2 [60]. These results provided evidence that may help to develop group B soyasaponins as a potential colon cancer chemopreventive agent.

The effect of two methanolic extractions, room temperature (RT) and reflux (RE), on composition and bioactive properties in hepatocarcinoma cells (Hep-G2) were investigated in our laboratory. The RT extract which is rich in DDMP soyasaponins showed a limited ability to inhibit the growth of Hep-G2 cells however it increased the differentiation of Hep-G2 cells measured by forwarding side scatter flow cytometry [61]. Morphologically, RT treatment induced changes in the shape, size as well as growth patterns (**Figure 3**). RT treated cells segregated and formed small clusters or individual cells and were larger than untreated control cells. Untreated Hep-G2 cells grew into colonies and further into a mono-layer [61]. The mechanism is still unclear and research in progressing in our laboratory regarding the significant of differentiation of cultured hepatocarcinoma cells.

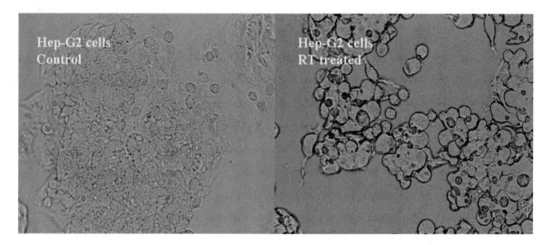

Figure 3. The representative morphological observation of untreated control Hep-G2 cells and a group B soyasaponin room temperature (RT) extract after 72 h of treatment.

Sialytransferase (ST) activity is associated with tumor metastasis and invasion [62, 63]. Finding a specific inhibitor on ST becomes a useful tool for the chemotherapeutic treatment of cancer metastasis [64]. Wu et al. (2001) showed that soyasaponin I, which is an amphiphilic molecule, not only had the ability inhibit sialytransferase activity but also was a highly specific inhibitor [65]. Hsu et al. (2005) confirmed that soyasaponins I was a ST inhibitor in an *in vitro* cell model and could decrease α2,3-sialylations and ST3Gal IV expression to modify the invasive behavior of the tumor cells [66]. Additionally, α2,3-linked sialic acids were suggested to play an important role in metastasis potential of melanoma (B16F10) cells. Chang et al. (2006) utilized soyasaponins I as a ST inhibitor in an experimental metastasis model [67]. However, there is still limited understanding of the anti-metastasis mechanism of soyasaponins I.

A soyasaponin extract has been reported to induce apoptosis in Hela cells [53] and a group B soyasaponins had the pro-apoptotic and anti-invasive activities in SNB 19 human glioblastoma cells [68]. We reported that an extract that contained a majority of the soyasaponins as I (62%) and III (29%) inhibited Hep-G2 cells and induced apoptosis measured with three distinct assays cell cycle, caspase assay and TUNEL assay [69]. The accumulation of sub-G1 cells measured during cell cycle analysis suggest apoptotic cells death, while the caspase and TUNEL measurement confirm apoptotic cell death (**Figure 4**) [69].

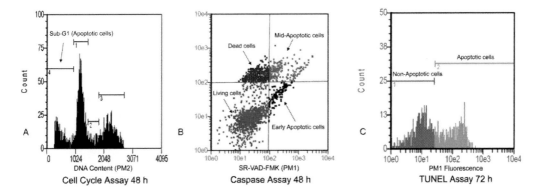

Figure 4. Panel A is the cell-cycle histograms of Hep-G2 cells treated with a soyasaponins I and III extract at the LC50 concentration; Panel B is representative flow cytometer analysis of caspase activity after soyasaponin I and III extract treatment for 48hr; Panel C is TUNEL apoptotic analyses of soyasaponin I and III extract treated Hep-G2 cells for 72 h.

Yoshiki et al (1996) found both soyasaponin Ab and soyasaponins βg had an important role in the scavenging of active oxygen. However, soyasaponin Ab expressed a different effect on Cl reaction against H_2O_2 compared soyasaponin βg [70]. Structural differences between soyasaponin Ab which has a sugar chain attached to C-22 position of the ring structure which is absent in soyasaponin βg [70] may be the difference. Group A soyasaponins have been reported to inhibit the CCl_4-depedent lipid per-oxidation due to their sugar chains at C22 position and is proposed to react directly with peroxy radicals [47, 70].

Group B soyasaponins have been reported to inhibit viruses such as Herpes simplex virus (HSV-1) with soyasaponin II, influenza and human immunodeficiency virus (HIV-1) by soyasaponin I [71]. Group B soyasaponins can be classified according different sugar moieties linked at C-3 (**Figure 1**). The trisaccharides of soyasapogenol B showed different anti-HSV-1 activities in the order of Rha-Glc-GlcA, Rha-Ara-GlcA, Rha-Xyl-GlcA and Rha-Gal-GlcA [72]. The potential anti-HSV-1 activity of the soyasapogenol moieties was also in the order of soyasapogenol E > soyasapogenol B. Group B soyasaponins having a glucosyl residue and the carbonyl group at C22 may be a more effective inhibitor of HSV-1 than the soyasaponin with hydroxyl group at C24 (**Figure 1**) [72]. The hydroxyl group at position C29 might reduce anti-HSV-1 activity but a methoxy carboxyl group at C20 might enhance the activity [73].

SOYASAPONINS CONVERSION

We have reported that DDMP-soyasaponins convert to non-DDMP soyasaponins by controlling temperature, pH and hydrolysis time [74]. We produced an extract that yielded soyasaponins I and III by base hydrolysis of a purified group B soyasaponin extract [74]. This may have an influence in bioactive response. Non-DDMP soyasaponins were more stable than DDMP conjugated soyasaponins during gut passage in Atlantic salmon [75]. Non DDMP saponins would also more abundant in the colon. Utilizing hydrolysis under acidic conditions does not produce the same resultant compounds and induced formation of soyasaponins artifacts and alcohols and can make certain soyasaponins esterify [25] potentially altering bioactivity.

Soyasaponin aglycones have been reported to the final metabolic products that are biotransformed by the human intestinal microflora [37, 41, 76]. Soyasaponins have been shown to be converted from various glycosides forms of group A or B soyasaponins into soyasapogenols by fermentation with human colonic microflora [6]. Intestinal microflora has been reported to be able to transform other soyasaponins. Soyasaponin Ab has been reported to be metabolized to soyasapogenol A and soyasapogenol A 3-β-D-glucuronide and these metabolites likely possess different bioactive properties compared to group A soyasaponins [76]. Soyasaponin I, an abundant group B soyasaponin, has been reported to be transformed into structurally related soyasaponin III after 24 h and soyasapogenol B after 48 h of incubation with human fecal microorganisms [77]. We are currently studying the biotransformation of group B soyasaponins with lactic bacterial enzymes extracts and have noted new metabolite produced from the fermentation process along with the biotransformation of group B soyasaponin into soyasaponin I and soyasapogenol B. Further work is underway to establish if these biotransformed products, those that would likely produced in the human colon during fermentation would confer an increase in bioactivity.

CONCLUSION

Soyasaponins are diverse and structural complex molecules with differing levels of bioactivities. The extraction conditions, purification and separation processing influence the resulted soyasaponins present in extracts. Analysis and determination of soyasaponins are mainly influenced by the stability of soyasaponins and methods of detection. Soyasaponins with different structures possess different bioactive effects depending on the model system employed. Future studies are needed in order to specifically determine an overall molecule mechanism attributed to saponins.

REFERENCES

[1]   Kim, S. L.; Berhow, M. A.; Kim, J. T.; Chi, H. Y.; Lee, S. J.; Chung, I. M. Evaluation Of Soyasaponin, Isoflavone, Protein, Lipid, And Free Sugar Accumulation In Developing Soybean Seeds. *Journal Of Agricultural And Food Chemistry* (2006), *54*, 10003-10010.

[2]   Shi, J.; Arunasalam, K.; Yeung, D.; Kakuda, Y.; Mittal, G.; Jiang, Y. M. Saponins From Edible Legumes: Chemistry, Processing, And Health Benefits. *Journal Of Medicinal Food* (2004), *7*, 67-78.

[3]   Rupasinghe, H. P. V.; Jackson, C. J. C.; Poysa, V.; Di Berardo, C.; Bewley, J. D.; Jenkinson, J. Soyasapogenol A And B Distribution In Soybean (Glycine Max L. Merr.)

In Relation To Seed Physiology, Genetic Variability, And Growing Location. *Journal Of Agricultural And Food Chemistry* (2003), *51*, 5888-5894.

[4] Zhang, W.; Popovich, D. G. Chemical And Biological Characterization Of Oleanane Triterpenoids From Soy. *Molecules* (2009), *14*, 2959-2975.

[5] Sumiki; Y Studies On The Saponin Of Soy-Bean. *Bulletin Of The Agricultural Chemical Society Of Japan* (1929), *5*, 27-32.

[6] Gurfinkel, D. M.; Rao, A. V. Soyasaponins: The Relationship Between Chemical Structure And Colon Anticarcinogenic Activity. *Nutrition And Cancer* (2003), *47*, 24-33.

[7] Yoshiki, Y.; Kudou, S.; Okubo, K. Relationship Between Chemical Structures And Biological Activities Of Triterpenoid Saponins From Soybean. *Bioscience Biotechnology And Biochemistry* (1998), *62*, 2291-2299.

[8] Decroos, K.; Vincken, J. P.; Van Koningsveld, G. A.; Gruppen, H.; Verstraete, W. Preparative Chromatographic Purification And Surfactant Properties Of Individual Soyasaponins From Soy Hypocotyls. *Food Chemistry* (2007), *101*, 324-333.

[9] Ireland, P. A.; Dziedzic, S. Z. Effect Of Hydrolysis On Sapogenin Release In Soya. *Journal Of Agricultural And Food Chemistry* (1986), *34*, 1037-1041.

[10] Macdonald, R. S.; Guo, J. Y.; Copeland, J.; Browning, J. D.; Sleper, D.; Rottinghaus, G. E.; Berhow, M. A. Environmental Influences On Isoflavones And Saponins In Soybeans And Their Role In Colon Cancer. Amer Inst Nutrition: Washington, DC, 2004; Pp 1239-1242.

[11] Kitagawa, I.; Yoshikawa, M.; Wang, H.; Saito, M.; Tosirisuk, V.; Fujiwara, T.; Tomita, K. Revised Structures Of Soyasapogenols A, B, And E, Oleanene-Sapogenols From Soybean. Structures Of Soyasaponins I, II, And III. *Chemical & Pharmaceutical Bulletin* (1982), *30*, 2294-2297.

[12] Tsukamoto, C.; Kikuchi, A.; Kudou, S.; Harada, K.; Kitamura, K.; Okubo, K. Group A Acetyl Saponin-Deficient Mutant From The Wild Soybean. *Phytochemistry* (1992), *31*, 4139-4142.

[13] Heftmann, E.; Luudin, R. E.; Haddon, W. F.; Peri, I.; Mor, U.; Bondi, A. High-Pressure Liquid Chromatography, Nuclear Magnetic Resonance And Mass Spectra Of Biosynthetic Soyasapogenols. *Journal Of Natural Products* (1979), *42*, 410-416.

[14] Ireland, P. A.; Dziedzic, S. Z. Analysis Of Soybean Sapogenins By High-Performance Liquid Chromatography. *Journal Of Chromatography A* (1985), *325*, 275-281.

[15] Kitagawa, I.; Saito, M.; Taniyama, T.; Yoshikawa, M. Saponin And Sapogenol. XXXVIII. Structure Of Soyasaponin A_2,A Bisdesmoside Of Soyasapogenol A, From Soybean, The Seeds Of Glycine Max MERRILL. *Chemical & Pharmaceutical Bulletin* (1985), *33*, 598-608.

[16] Shiraiwa, M.; Harada, K.; Okubo, K. Composition And Structure Of Group-B Saponin In Soybean Seed. *Agricultural And Biological Chemistry* (1991), *55*, 911-917.

[17] Okubo, K.; Iijima, M.; Kobayashi, Y.; Yoshikoshi, M.; Uchida, T.; Kudou, S. Components Responsible For The Undesirable Taste Of Soybean Seeds. *Bioscience Biotechnology And Biochemistry* (1992), *56*, 99-103.

[18] Gohtani, S.; Yamano, Y. Effects Of Nacl And Ph On Emulsion Stability With Soyasaponin-1. *Nippon Nogeikagaku Kaishi-Journal Of The Japan Society For Bioscience Biotechnology And Agrochemistry* (1987), *61*, 1113-1117.

[19] Gohtani, S.; Shinomoto, K.; Honda, Y.; Okubo, K.; Yamano, Y. Surface-Activity Of Soybean Saponin AB (Acetyl-Soyasaponin-A1). *Nippon Nogeikagaku Kaishi-Journal Of The Japan Society For Bioscience Biotechnology And Agrochemistry* (1990), *64*, 901-906.

[20] Gohtani, S.; Yamano, Y. Stability And Zeta Potential Of Soyasaponin-1 Emulsion. *Nippon Nogeikagaku Kaishi-Journal Of The Japan Society For Bioscience Biotechnology And Agrochemistry* (1990), *64*, 139-144.

[21] Kudou, S.; Tonomura, M.; Tsukamoto, C.; Shimoyamada, M.; Uchida, T.; Okubo, K. Isolation And Structural Elucidation Of The Major Genuine Soybean Saponin. *Bioscience, Biotechnology, And Biochemistry* (1992), *56*, 142-143.

[22] Kudou, S.; Tonomura, M.; Tsukamoto, C.; Uchida, T.; Sakabe, T.; Tamura, N.; Okubo, K. Isolation And Structural Elucidation Of DDMP-Conjugated Soyasaponins As Geniune Saponins From Soybean Seeds. *Bioscience Biotechnology And Biochemistry* (1993), *57*, 546-550.

[23] Gu, L. W.; Tao, G. J.; Gu, W. Y.; Prior, R. L. Determination Of Soyasaponins In Soy With LC-MS Following Structural Unification By Partial Alkaline Degradation. *Journal Of Agricultural And Food Chemistry* (2002), *50*, 6951-6959.

[24] Hu, J.; Lee, S. O.; Hendrich, S.; Murphy, P. A. Quantification Of The Group B Soyasaponins By High-Performance Liquid Chromatography. *Journal Of Agricultural And Food Chemistry* (2002), *50*, 2587-2594.

[25] Tava, A.; Mella, M.; Bialy, Z.; Jurzysta, M. Stability Of Saponins In Alcoholic Solutions: Ester Formation As Artifacts. *Journal Of Agricultural And Food Chemistry* (2003), *51*, 1797-1800.

[26] Gurfinkel, D. M.; Reynolds, W. F.; Rao, A. V. The Isolation Of Soyasaponins By Fractional Precipitation, Solid Phase Extraction, And Low Pressure Liquid Chromatography. *International Journal Of Food Sciences And Nutrition* (2005), *56*, 501-519.

[27] Gimeno, R. A.; Marce, R. M.; Borrull, F. Determination Of Organic Contaminants In Coastal Water. *Trac-Trends In Analytical Chemistry* (2004), *23*, 341-350.

[28] Li, H. Z.; Pordesimo, L.; Weiss, J. High Intensity Ultrasound-Assisted Extraction Of Oil From Soybeans. *Food Research International* (2004), *37*, 731-738.

[29] Schmeck, T.; Wenclawiak, B. Sediment Matrix Induced Response Enhancement In The Gas Chromatographic-Mass Spectrometric Quantification Of Insecticides In Four Different Solvent Extracts From Ultrasonic And Soxhlet Extraction. *Chromatographia* (2005), *62*, 159-165.

[30] Yang, Y.; Jin, M.; Huang, M.; Su, B.; Ren, Q. Ultrasound-Assisted Extraction Of Soyasaponins From Hypocotyls, And Analysis By LC-ESI-MS. *Chromatographia* (2007), *65*, 555-560.

[31] Stavarache, C.; Vinatoru, M.; Maeda, Y. Ultrasonic Versus Silent Methylation Of Vegetable Oils. *Ultrasonics Sonochemistry* (2006), *13*, 401-407.

[32] Jin, M.; Yang, Y.; Su, B.; Ren, Q. Rapid Quantification And Characterization Of Soyasaponins By High-Performance Liquid Chromatography Coupled With Electrospray Mass Spectrometry. *Journal Of Chromatography A* (2006), *1108*, 31-37.

[33] Hubert, J.; Berger, M.; Dayde, J. Use Of A Simplified HPLC-UV Analysis For Soyasaponin B Determination: Study Of Saponin And Isoflavone Variability In

Soybean Cultivars And Soy-Based Health Food Products. *Journal Of Agricultural And Food Chemistry* (2005), *53*, 3923-3930.

[34] Dalluge, J. J.; Eliason, E.; Frazer, S. Simultaneous Identification Of Soyasaponins And Isoflavones And Quantification Of Soyasaponin Bb In Soy Products, Using Liquid Chromatography/Electrospray Ionization-Mass Spectrometry. *Journal Of Agricultural And Food Chemistry* (2003), *51*, 3520-3524.

[35] Decroos, K.; Vincken, J. P.; Heng, L.; Bakker, R.; Gruppen, H.; Verstraete, W. Simultaneous Quantification Of Differently Glycosylated, Acetylated, And 2,3-Dihydro-2,5-Dihydroxy-6-Methyl-4H-Pyran-4-One-Conjugated Soyasaponins Using Reversed-Phase High-Performance Liquid Chromatography With Evaporative Light Scattering Detection. *Journal Of Chromatography A* (2005), *1072*, 185-193.

[36] Berhow, M. A.; Cantrell, C. L.; Duval, S. M.; Dobbins, T. A.; Maynes, J.; Vaughn, S. F. Analysis And Quantitative Determination Of Group B Saponins In Processed Soybean Products. *Phytochemical Analysis* (2002), *13*, 343-348.

[37] Hu, J.; Zheng, Y. L.; Hyde, W.; Hendrich, S.; Murphy, P. A. Human Fecal Metabolism Of Soyasaponin I. *Journal Of Agricultural And Food Chemistry* (2004), *52*, 2689-2696.

[38] Zhang, W.; Popovich, D. G. Separation And Purification Of Soyasapogenol B Under Optimized Hydrolysis And Mass Spec Conditions. *Food Chemistry* (2010), *In Press.*

[39] Gurfinkel, D. M.; Rao, A. V. Determination Of Saponins In Legumes By Direct Densitometry. *Journal Of Agricultural And Food Chemistry* (2002), *50*, 426-430.

[40] Baxter, R. L.; Price, K. R.; Fenwick, G. R. Sapogenin Structure-Analysis Of The C-13-NMR And H-1-NMR Spectra Of Soyasapogenol-B. *Journal Of Natural Products* (1990), *53*, 298-302.

[41] Lee, S. O.; Simons, A. L.; Murphy, P. A.; Hendrich, S. Soyasaponins Lowered Plasma Cholesterol And Increased Fecal Bile Acids In Female Golden Syrian Hamsters. *Experimental Biology And Medicine* (2005), *230*, 472-478.

[42] Oakenfull, D.; Sidhu, G. S. Could Saponins Be A Useful Treatment For Hypercholesterolaemia? *European Journal Of Clinical Nutrition* (1990), *44*, 79-88.

[43] Oakenfull, D. Soy Protein, Saponins And Plasma Cholesterol. *Journal Of Nutrition* (2001), *131*, 2971.

[44] Oakenfull, D. G.; Topping, D. L.; Illman, R. J.; Fenwick, D. E. Prevention Of Dietary Hypercholesterolemia In The Rat By Soya Bean And Quillaja Saponins. *Nutrition Reports International* (1984), *29*, 1039-1046.

[45] Potter, S. M. Overview Of Proposed Mechanisms For The Hypocholesterolemic Effect Of Soy. *Journal Of Nutrition* (1995), *125*, 606S-611S.

[46] Potter, J. D.; Illman, R. J.; Calvert, G. D. Soya Saponins, Plasma Lipids, Lipoproteins And Fecal Bile Acids: A Double Blind Cross-Over Study. *Nutrition Reports International* (1980), *22*, 521-528.

[47] Ishii, Y.; Tanizawa, H. Effects Of Soyasaponins On Lipid Peroxidation Through The Secretion Of Thyroid Hormones. *Biological & Pharmaceutical Bulletin* (2006), *29*, 1759-1763.

[48] Jun, H. S.; Kim, S. E.; Sung, M. K. Protective Effect Of Soybean Saponins And Major Antioxidants Against Aflatoxin B1-Induced Mutagenicity And DNA-Adduct Formation. *Journal Of Medicinal Food* (2002), *5*, 235-240.

[49] Koratkar, R.; Rao, A. V. Effect Of Soya Bean Saponins On Azoxymethane-Induced Preneoplastic Lesions In The Colon Of Mice. *Nutrition And Cancer-An International Journal* (1997), *27*, 206-209.

[50] Rao, A. V.; Sung, M. K. Saponins As Anticarcinogens. *Journal Of Nutrition* (1995), *125*, 717S-724S.

[51] Sung, M. K.; Kendall, C. W. C.; Koo, M. M.; Rao, A. V. Effect Of Soybean Saponins And Gypsophilla Saponin On Growth And Viability Of Colon-Carcinoma Cells In Culture. *Nutrition And Cancer-An International Journal* (1995), *23*, 259-270.

[52] Sung, M. K.; Kendall, C. W. C.; Rao, A. V. Effect Of Soybean Saponins And Gypsophila Saponin On Morphology Of Colon-Carcinoma Cells In Culture. *Food And Chemical Toxicology* (1995), *33*, 357-366.

[53]   Xiao, J. X.; Huang, G. Q.; Zhang, S. H. Soyasaponins Inhibit The Proliferation Of Hela Cells By Inducing Apoptosis. *Experimental And Toxicologic Pathology* (2007), *59*, 35-42.

[54] Zhang, W.; Popovich, D. G. Effect Of Soyasapogenol A And Soyasapogenol B Concentrated Extracts On Hep-G2 Cell Proliferation And Apoptosis. *Journal Of Agricultural And Food Chemistry* (2008), *56*, 2603-2608.

[55] Kawamori, T.; Tanaka, T.; Kojima, T.; Suzui, M.; Ohnishi, M.; Mori, H. Suppression Of Azoxymethane-Induced Rat Colon Aberrant Crypt Foci By Dietary Protocatechuic Acid. *Japanese Journal Of Cancer Research* (1994), *85*, 686-691.

[56] Kawamori, T.; Tanaka, T.; Hara, A.; Yamahara, J.; Mori, H. Modifying Effects Of Naturally Occurring Products On The Development Of Colonic Aberrant Crypt Foci Induced By Azoxymethane In F344 Rats. *Cancer Res* (1995), *55*, 1277-1282.

[57] Makitie, L. T.; Kanerva, K.; Anderson, L. C. Ornithine Decarboxylase Regulates The Activity And Localization Of Rhoa Via Polyamination. *Experimental Cell Research* (2009), *315*, 1008-1014.

[58] Oh, Y. J.; Sung, M. K. Soybean Saponins Inhibit Cell Proliferation By Suppressing PKC Activation And Induce Differentiation Of HT-29 Human Colon Adenocarcinoma Cells. *Nutrition And Cancer-An International Journal* (2001), *39*, 132-138.

[59] Ellington, A. A.; Berhow, M.; Singletary, K. W. Induction Of Macroautophagy In Human Colon Cancer Cells By Soybean B-Group Triterpenoid Saponins. *Carcinogenesis* (2005), *26*, 159-167.

[60] Ellington, A. A.; Berhow, M. A.; Singletary, K. W. Inhibition Of Akt Signaling And Enhanced ERK1/2 Activity Are Involved In Induction Of Macroautophagy By Triterpenoid B-Group Soyasaponins In Colon Cancer Cells. *Carcinogenesis* (2006), *27*, 298-306.

[61] Zhang, W.; Yeo, M. C.; Tang, F. Y.; Popovich, D. G. Bioactive Responses Of Hep-G2 Cells To Soyasaponin Extracts Differs With Respect To Extraction Conditions. *Food And Chemical Toxicology* (2009), *47*, 2202-2208.

[62] Gessner, P.; Riedl, S.; Quentmaier, A.; Kemmner, W. Enhanced Activity Of CMP-Neuac:Gal Beta 1-4glcnac:Alpha 2,6-Sialyltransferase In Metastasizing Human Colorectal Tumor Tissue And Serum Of Tumor Patients. *Cancer Letters* (1993), *75*, 143-149.

[63] Majuri, M. L.; Niemela, R.; Tiisala, S.; Renkonen, O.; Renkonen, R. Expression And Function Of Alpha 2,3-Sialyl- And Alpha 1,3/1,4-Fucosyltransferases In Colon

Adenocarcinoma Cell Lines: Role In Synthesis Of E-Selectin Counter-Receptors. *International Journal Of Cancer* (1995), *63*, 551-559.

[64] Wang, X. F.; Zhang, L. H.; Ye, X. S. Recent Development In The Design Of Sialyltransferase Inhibitors. *Medicinal Research Reviews* (2003), *23*, 32-47.

[65] Wu, C. Y.; Hus, C. C.; Chen, S. T.; Tsai, Y. C. Soyasaponin I, A Potent And Specific Sialyltransferase Inhibitor. *Biochemical And Biophysical Research Communications* (2001), *284*, 466-469.

[66] Hsu, C. C.; Lin, T. W.; Chang, W. W.; Wu, C. Y.; Lo, W. H.; Wang, P. H.; Tsai, Y. C. Soyasaponin-I-Modified Invasive Behavior Of Cancer By Changing Cell Surface Sialic Acids. *Gynecologic Oncology* (2005), *96*, 415-422.

[67] Chang, W. W.; Yu, C. Y.; Lin, T. W.; Wang, P. H.; Tsai, Y. C. Soyasaponin I Decreases The Expression Of Alpha 2,3-Linked Sialic Acid On The Cell Surface And Suppresses The Metastatic Potential Of B16F10 Melanoma Cells. *Biochemical And Biophysical Research Communications* (2006), *341*, 614-619.

[68] Yanamandra, N.; Berhow, M. A.; Konduri, S.; Dinh, D. H.; Olivero, W. C.; Nicolson, G. L.; Rao, J. S. Triterpenoids From Glycine Max Decrease Invasiveness And Induce Caspase-Mediated Cell Death In Human SNB19 Glioma Cells. *Clinical & Experimental Metastasis* (2003), *20*, 375-383.

[69] Zhang, W.; Popovich, D. G. Group B Oleanane Triterpenoid Extract Containing Soyasaponins I And III From Soy Flour Induces Apoptosis In Hep-G2 Cells. *Journal Of Agricultural And Food Chemistry* (2010).

[70] Yoshiki, Y.; Kinumi, M.; Kahara, T.; Okubo, K. Chemiluminescence Of Soybean Saponins In The Presence Of Active Oxygen Species. *Plant Science* (1996), *116*, 125-129.

[71] Hayashi, K.; Hayashi, H.; Hiraoka, N.; Ikeshiro, Y. Inhibitory Activity Of Soyasaponin II On Virus Replication In Vitro. *Planta Medica* (1997), *63*, 102-105.

[72] Kinjo, J.; Yokomizo, K.; Hirakawa, T.; Shii, Y.; Nohara, T.; Uyeda, M. Anti-Herpes Virus Activity Of Fabaceous Triterpenoidal Saponins. *Biological & Pharmaceutical Bulletin* (2000), *23*, 887-889.

[73] Ikeda, T.; Yokomizo, K.; Okawa, M.; Tsuchihashi, R.; Kinjo, J.; Nohara, T.; Uyeda, M. Anti-Herpes Virus Type 1 Activity Of Oleanane-Type Triterpenoids. *Biological & Pharmaceutical Bulletin* (2005), *28*, 1779-1781.

[74] Zhang, W.; Teng, S. P.; Popovich, D. G. Generation Of Group B Soyasaponins I And III By Hydrolysis. *Journal Of Agricultural And Food Chemistry* (2009), *57*, 3620-3625.

[75] Knudsen, D.; Ron, O.; Baardsen, G.; Smedsgaard, J.; Koppe, W.; Froklaer, H. Soyasaponins Resist Extrusion Cooking And Are Not Degraded During Gut Passage In Atlantic Salmon (Salmo Salar L.). *Journal Of Agricultural And Food Chemistry* (2006), *54*, 6428-6435.

[76] Chang, S. Y.; Han, M. J.; Joh, E. H.; Kim, D. H. Liquid Chromatography/Mass Spectrometry-Based Structural Analysis Of Soyasaponin Ab Metabolites By Human Fecal Microflora. *Journal Of Pharmaceutical And Biomedical Analysis* (2010), *52*, 752-756.

[77] Hu, J.; Zheng, Y. L.; Hyde, W.; Hendrich, S.; Murphy, P. A. Human Fecal Metabolism Of Soyasaponin I. *Journal Of Agricultural And Food Chemistry* (2004), *52*, 2689-2696.

In: Progress in Food Science and Technology, Volume 1 ISBN: 978-1-61122-314-9
Editor: Anthony J. Greco © 2012 Nova Science Publishers, Inc.

Chapter 10

ENHANCED PRODUCTION OF B – CAROTENE AND L – ASPARAGINASE FROM A TRIBAL FOOD ALGA, *VAUCHERIA UNCINATA*

*Usha Pandey[*1] and J. Pandey[≠2]*

[1]Faculty of Science and Technology, M. G. Kashividyapith, Varanasi, India
[2]Centre of Advanced Study in Botany, Banaras Hindu University, Varanasi, India

ABSTRACT

The growing concern towards the exponential increase in human population, environmental stresses and associated health risks during recent years has enhanced the demand for potential biodiversity resources with high nutritional and therapeutic values. In India, rural and tribal people use a wide variety of wild plants including algae for maintaining vigor and treatment of diseases.

We investigated enhanced production of β – carotene and L – asparaginase in a yellow green alga, *Vaucheria uncinata* in a full factorial design using 1mM each of nitrate (KNO_3^-), ammonium $\{(NH_4)_2SO_4\}$, glutamine and NaCl in free and immobilized cell culture condition. Glutamine was found to be the most suitable N – source for the production of β – carotene and L – asparaginase. Maximum production of β – carotene (7.02 $\mu g\ ml^{-1}$) and about five fold increase in L – asparaginase (5.89 IU mg^{-1} protein) was recorded when glutamine grown cells were exposed to 1mM NaCl under immobilized cell and free cell culture respectively. Exposure to NaCl also leads to a substantial increase in proline and glutathione contents. Our study indicates *V. uncinata* to be a promising bioresource for production of β – carotene and L – asparaginase. Use of immobilized and free cell cultures supplemented with glutamine and NaCl could be an effective approach for scaling – up production of β – carotene and L – asparaginase, respectively for commercial application.

---

* Corresponding author email : usha_pandey28@yahoo.co.in, Faculty of Science and Technology, M. G. Kashividyapith, Varanasi – 221001, India

≠ Centre of Advanced Study in Botany, Banaras Hindu University, Varanasi – 221005, India, Email : jiten_pandey@rediffmail.com

Keywords : Bioresource, L − asparaginase, β − carotene, Glutathione, Algae, *Vaucheria uncinata*

1. INTRODUCTION

Accompanying the changes from nomadic to settled life style, the human being became largely attached to land as there was a need for assured food supply as against uncertain nomadic life of hunters and gatherers. With this change, there was a need felt for enhanced and assured food production. The industrial and agricultural revolution emerged as a mixed blessing in as much as it has helped enhance standard of living, but has generated a number of environmental stresses threatening the very future of human kind. The growing concern towards environmental stresses and associated health risks, during recent years, has enhanced the demand for such biodiversity resources which, in addition to fulfilling nutritional and energy needs, are capable of contributing additional physiological benefits. In India, rural and tribal people use a wide variety of wild plant sources for maintaining vigor and for treatment of diseases. During recent years, the algae have emerged as potential bioresource for food and medicine. A number of algal organisms have been reported to be the rich source of dietary supplements including proteins, carbohydrates, lipids, aminoacids and vitamins (Parikh and Madamwar, 2006; Colla *et al.,* 2007; Pandey and Pandey, 2008a). Some species of algae contain new bioactive compounds, not found in higher plants and traditional therapeutic sources which, in addition to fulfilling nutritional and energy needs, have high therapeutic values and capable of contributing additional physiological benefits (Romay *et al.,* 2003; Pandey and Pandey, 2008b).

Algae have now recognized as a potential bioresource for food and pharmaceutical industries. Similar to higher plants, algal organisms contain a number of antioxidants including enzymatic and non − enzymatic constitution (Pandey and Pandey, 2009). The yellow green alga, *Vaucheria* (Xanthophyceae), is known to be a rich source of carotene and xanthophylls. This alga has emerged as a biotechnological tool for horizontal gene transfer (Rumphoon *et al.,* 2008). Most species of *Vaucheria* grow in saline environment (Christensen, 1988). Even freshwater forms show potential acclimation to high salinity ranges (Schneider *et al.,* 1996). Some species often forms thick mat in littoral regions (Loehvo and Back, 2000). *Vaucheria* contains high concentration of carotenoids (Entwisle, 1988), a class of isoprenoid − derived pigments essential for stability of photosynthetic membranes and photoprotection of light reaction. During recent years, these pigments have gained importance in human health care. Being precursor of vitamin A, these pigments help minimizing incidences of night blindness, xerophthalmia and keratomalacia (Liu *et al.,* 2009). As an antioxidant, these pigments help neutralizing reactive oxygen species (ROS), which are implicated in several human diseases including carcinogenesis induced tumor promotion (Devasagayam, 2004).

Glutathione is an other important antioxidant known for scavenging O_2^- and H_2O_2 thus, preventing the formation of highly toxic OH⁻ radicals (Foyer *et al.,* 1997). This non − protein tripeptide, containing glutamate − cysteine − glycine, is an important acceptor of sulphate group in S − metabolism. The response pattern of glutathione is often used as a stress marker in ecophysiological studies (Tausz *et al.,* 2004). It is an essential component of cellular antioxidative defence system, which keeps reactive oxygen species (ROS) under control (Noctor and Foyer, 1998). Given the multifunction in metabolism, thiol pools and glutathione redox ratios are considered as universal redox sensing and signaling system at the cellular level (Rijstenbil *et al.*, 1998; Noctor *et al.,* 2002; Tausz *et al.*, 2004). Similar to many other stresses, salinity can induce oxidative stress in plant cell including algae (Pandey and Singh; 1995; Tausz *et al.,* 2004). Therefore, for practical purposes, measurement of average tissue values of glutathione concentration is used to mark oxidative stress responses. Similar to glutathione, proline has been shown to play an important role in ameliorating environmental stresses including salinity in algae, microbes and higher plants (Reynoso and Gamboa, 1982; Siripornadulsil *et al.*, 2002; Sairam and Tyagi, 2004).

The enzyme L – asparaginase received increasing attention during recent years for its antineoplastic potential (Selvakumar *et al.*, 1991; Narta et al., 2007). The ezyme has been widely studied for its applications in the treatment of acute lymphoblastic leukemia (Keating et al., 1993; Narta *et al.,* 2007), pancreatic carcinoma (Yunis *et al.,* 1977) and bovine lymphomosarcoma (Mosterson *et al.*, 1988). In normal cells, L – asparagine synthetase catalyzes the synthesis of L – asparagine, an important nutrient for cancer cells. The malignant cells deprived of this enzyme require an exogenous source of L – asparagine for their growth and multiplication. L – asparaginase catalyzes the hydrolysis of L – asparagine into L – aspartic acid and ammonia. Thus, the administration of L – asparaginase depletes the exogenous L – asparagine and cause the death of malignant cells (Selvakumar *et al.*, 1991). This phenomenal behavior of malignant cell is being exploited by scientific community in the treatment of cancers (Narta *et al.*, 2007). L – asparaginase is present in many animal tissues, plants and microorganisms but not in mankind (El – Bessoumy *et al.*, 2004). Ever since L – asparaginase anti – tumor activity was first demonstrated by Broome (1961), its production using microorganisms has attracted considerable attention owing to there being cost – effective and eco – friendly nature. Peterson and Ciegler (1969) reported that all the 44 bacterial species they tested contained measureable quantities of L – asparaginase. However, only a few species could produce the enzyme in larger amounts (Peterson and Ciegler, 1969). L – asparaginase is now produced throughout the world using a large number of microorganisms and under different experimental design to evaluate various nutritional and environmental requirements for microbial growth coupled with enhancement in enzyme production (Wriston and Yellin, 1973; El – Bessoumy *et al.,* 2004; Prakasham *et al.,* 2007). However, there is a dearth of studies explicitly addressing algal organisms and process optimization for enhanced production of this anti – tumor enzyme (Paul and Cooksely, 1981).

Since the first algal L – asparaginase was reported in an euryhaline microalgae (Paul and Cooksey, 1981), we used salinity as a possible regulator of algal – L – asparaginase production in *Vaucheria uncinata*. Salinity, an important stress for freshwater algae, often stimulate synthesis of many biochemical compounds including β – carotene and glutathione as a component of their defense system. Further, since nitrogen has been shown to be the most important nutritional requirement for microbial growth and subsequent biocatalyst production (El – Bessoumy *et al.,* 2004; Prakasham *et al.,* 2007), we considered N – source as a possible co – regulator for enhanced production of L - asparaginase and β – carotene in this freshwater yellow green alga.

2. MATERIALS AND METHODS

2.1. Test Organism

The test organism, *Vaucheria uncinata*, was isolated from lake Baghdara ($24^0 31^/$ N lat; $73^0 48^/$ E long and 577 m above msl), a perennial woodland lake of southern Rajasthan, India. Being situated away from urban settlement (20 Km SE of Udaipur city) in the midth of a dry tropical deciduous forest, this freshwater lake is subjected to least human interference and provide natural habitat for this yellow green alga. *V. uncinata* is a coenocytic, filamentous yellow green alga. The filament is branched and has no septation. Thallus is siphonecious and at the base of the thallus a branched rhizoidal system helps attaching with the substratum.

2.2. Experimental

A unialgal population was raised using axenic culture in Bold's basal medium. *Vaucheria uncinata* cells were immobilized within fibrous network of sponge discs obtained from mature dry fruit of *Luffa cylindrica* (Iqbal and Zafar, 1993). The sponge obtained from mature dry fruits of *L. cylindrica* were cut into pieces of approximately 2.0x1.5 cm, 3 mm thick, soaked in boiling water for about 30 minute and thoroughly washed in distilled water. The sponge discs were oven dried at 70^0C and stored in deciccators before use for immobilization. Batch cultures were maintained at 25 ± 1^0C with a light intensity of 150 µmol m^{-2} s^{-1} and 15 : 9 h light : dark cycle. Cultures were grown in transparent plastic bags (n = 3) with an initial volume of 5 L, air flow rate of 0.36 vvm (LL^{-1} min^{-1}) and initial inoculum of 50 mg dry mass L^{-1}. Similar growth cultures were also prepared for free cells. Cultures were grown in nitrate (1mM, KNO$_3$), ammonium (1mM, (NH$_4$)$_2$SO$_4$) and glutamine (1mM) supplemented medium. Further, the alga was grown in the presence or absence of NaCl. The effect of nitrogen starvation was studied in separate experiments in nitrogen free Bold' basal medium in the presence or absence of NaCl.

A 20 – day – old harvested and freeze – dried biomass was analyzed in triplicate for pigments. Chlorophyll was extracted in N$^/$ N – dimethyl formamide and absorbance recorded at 664 nm. For carotenoid and xanthophylls, acetone extraction procedure of Weybrew (1957) was followed. β – carotene was estimated at regular intervals according to Freed (1966). Acetone and petroleum ether extracted β – carotene was purified through silica gel packed chromatography column using 3 % acetone as eluent. Purity of the sample was checked with standard β – carotene at 452 nm using eluent as blank.

Protein, carbohydrate and lipids were estimated as described in Sadasivam and Manickam (1996). Aspartic acid was estimated following standard ninhydrin method and proline was determined following Bates et al. (1979). For determination of glutathione, algal samples were homogenized in a mixture of 6 % metaphosphoric acid, 1 mM EDTA and 10 % polyvinyl pyrralidon (PVP) at pH 2.8. After centrifugation, total GSH was determined in acid soluble extracts as described by Griffith (1980).

L – asparaginase enzyme was assayed colorimetrically following Wriston and Yellin (1973) by estimating the amount of ammonia produced during L – asparaginase catalysis at 37^0C. The reaction mixture, consisting of 0.5 ml of 0.08 mol L^{-1} L – asparagine, 1.0 ml of

0.05 mol L^{-1} borate buffer (pH 7.5) and 0.5 ml of enzyme solution, was incubated for 10 min at 37 ^{0}C. The reaction was stopped by adding 0.5 ml of 15 % trichloro acetic acid solution. The ammonia liberated was coupled with Nessler's reagent and was quantitatively determined using standard curve. One unit of L – asparaginase (IU) represents the amount of the enzyme capable of producing 1 μmol of ammonia per minute at 37 ^{0}C.

2.3. Statistical

All the results reported are the means of three independent experiments. The precision and accuracy of the data were ensured through repeated analyses of samples. The analytical variances of the data obtained remained below 10 %. Standard error of mean (SEM) and least significant difference (LSD) were computed for expressing data variability. The level of significance of means was compared using analysis of variance (ANOVA). Data were log transformed, whenever necessary, to equalize variances.

3. RESULTS

When *Vaucheria uncinata* was grown under different nitrogen sources, glutamine was found to be the most suitable N – source for production of various biomolecules. Protein and lipid contents increased by more than two folds in the presence of glutamine in the growth medium (Table 1). Chlorophyll, carotenoids, β - carotene and aspartic acid contents increased invariably by different N – sources and the increases were more pronounced in glutamine supplemented growth medium. Exposure to NaCl in the medium, although reduced the protein and carbohydrate contents by 29.1 % and 15.1 % respectively, caused a substantial increase in aspartic acid and lipid contents in *V. uncinata* (Table 1). Glutamine – grown, NaCl – exposed cells showed about 3.4 fold increase in aspartic acid accumulation over the cells grown without N – supplement (Table 1).

Table 1. Changes in biochemical composition (% dry weight) of *V. uncinata* as influenced by NaCl (1mM) and nitrogen sources (1mM)

	Constituent				Treatment				
	-NaCl				+NaCl				
	Control	NO$_3^-$	NH$_4^+$	Glutamine	Control	NO$_3^-$	NH$_4^+$	Glutamine	LSD$_{0.05*}$
Protein	11.50 ± 1.02	16.80 ± 1.13	21.70 ± 1.51	23.21 ± 2.20	8.15 ± 0.67	11.80 ± 1.02	14.00 ± 1.10	18.20 ±1.63	0.152
Carbohydrate	22.30 ± 2.04	30.15 ± 2.35	27.10 ± 2.40	32.51 ± 2.67	18.70 ± 1.45	26.20 ± 2.11	23.40 ± 1.80	28.76 ±2.24	0.146
Lipid	3.32 ± 0.15	4.54 ± 0.33	5.60 ± 0.43	6.83 ± 0.51	3.36 ± 0.20	7.00 ± 0.47	7.21 ± 0.38	9.50 ± 0.73	0.182
Aspartic acid	0.180 ± 0.021	0.242 ±0.018	0.242 ± 0.021	0.245 ± 0.019	0.185 ± 0.013	0.348 ± 0.024	0.296 ± 0.024	0.636 ± 0.042	0.174

Initial inoculum (0 day) was 50 mg dry weight L^{-1}. Values are mean (n = 3) ± 1SE. Effect of time, treatment and their interactions are significant at p < 0.01 (ANOVA). *The least significant difference (LSD) at p = 0.05.

Table 2. Changes in pigment concentrations (mg g^{-1}) of *V. uncinata* as influenced by NaCl (1mM) and nitrogen sources (1mM)

	Constituent				Treatment				
	-NaCl				+NaCl				
	Control	NO$_3^-$	NH$_4^+$	Glutamine	Control	NO$_3^-$	NH$_4^+$	Glutamine	LSD$_{0.05}$*
Chlorophyll	1.12 ± 0.11	1.52 ± 0.12	1.74 ± 0.12	2.86 ± 0.21	0.65 ± 0.03	0.84 ± 0.05	0.87 ± 0.06	1.93 ± 0.15	0.169
Carotenoids	0.72 ± 0.51	1.05 ± 0.10	1.24 ± 0.07	1.45 ± 0.12	0.78 ± 0.02	0.86 ± 0.05	1.25 ± 0.11	1.72 ± 0.13	0.178
Carotenes	0.18 ± 0.01	0.40 ± 0.02	0.52 ± 0.03	0.75 ± 0.04	0.23 ± 0.01	0.35 ± 0.02	0.58 ± 0.04	0.96 ± 0.07	0.187
Xanthophylls	0.52 ± 0.03	0.58 ± 0.04	0.57 ± 0.03	0.86 ± 0.05	0.44 ± 0.02	± 0.42 ± 0.03	0.72 ± 0.04	0.71 ± 0.05	0.156

Values are mean (n = 3) ± 1SE. Effect of time, treatment and their interactions are significant at p < 0.01 (ANOVA). *The least significant difference (LSD) at p = 0.05.

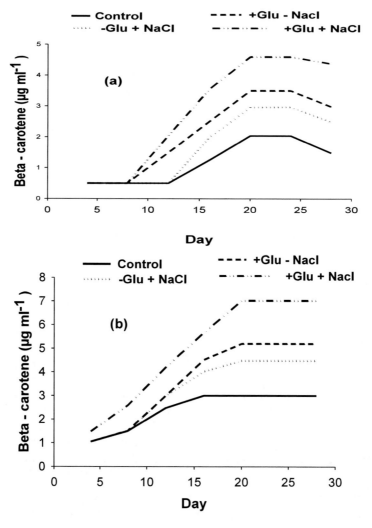

Figure 1. Enhanced production of β – carotene in *V. uncinata* by supplemental glutamine and / or NaCl in (a) free cell and (b) immobilized cell cultures (p < 0.001; ANOVA).

Pigment production was significantly influenced by salt stress and N – supply. When glutamine – grown cells were exposed to NaCl, carotenoid content enhanced by 139 %. However, chlorophyll content reduced significantly under salinity (Table 2). Xanthophylls appeared sensitive to salt stress in comparison to carotene (Table 2). β – carotene production enhanced significantly in glutamine – grown *V. uncinata* in the presence of 1mM NaCl in both free and immobilized cells (Fig. 1) and the later showed superiority in this respect.

The interaction effect of two factors, however, indicated that in the absence of glutamine, NaCl could enhance the production of β – carotene only marginally. Exposure to NaCl enhanced the production of proline and glutathione by more than ten folds and, in this respect, free cells showed superiority over immobilized cells (Fig. 2).

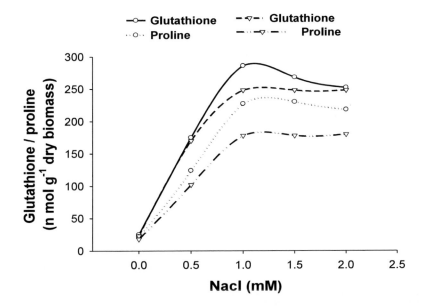

Figure 2. Changes in proline and glutathione contents in glutamine grown cells of *V. uncinata* as influenced by NaCl in free cell (- 0 -) and immobilized cell (- Δ -) cultures. ($p < 0.001$; ANOVA).

L – asparaginase production by *V. uncinata* was studied under free and immobilized cell culture conditions using Bold's basal medium. The results indicate that this alga produced 1.21 and 1.15 IU mg^{-1} protein of L – asparaginase in free and immobilized cell cultures respectively, under normal condition (Fig. 3).

Further experiments were conducted to enhance the enzyme production by varying nitrogen source and NaCl supplemented to the growth medium. In comparison to ammonium and nitrate, glutamine appeared to be a more potential N – source where *V. uncinata* grew most effectively with significant increase in L – asparaginase production. Although the alga grew less effectively in NaCl added medium, the later substantially increased production of L – asparaginase. Maximum enzyme production (5.89 IU mg^{-1} protein) was recorded when glutamine – grown cells were exposed to 1 mM NaCl under free cell culture (Fig. 3), where the enzyme production increased by about five folds over the normal grown cells. On further increase in NaCl concentration, the enzyme production reduced in free cells but remained constant under immobilized cell culture (Fig. 3).

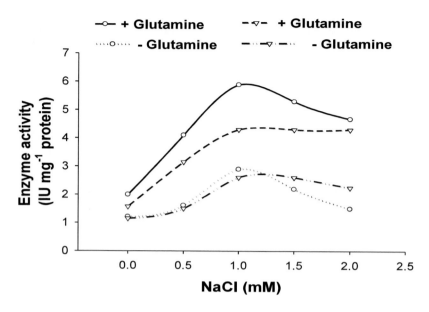

Figure 3. Enhanced production of L – asparaginase in *V. uncinata* under different treatments in free (-0-) and immobilized (- Δ -) cell cultures. (p < 0.001; ANOVA).

4. DISCUSSION

Nutraceuticals and therapeutic enzymes are gaining importance due to their potential role in human health including tumor therapy. Production of such biomolecules, although involves very complex processes and species specific, is regulated by a number environmental and biochemical variables (Colla et al., 2007). During recent years, algae have emerged as potential bioresource for commercially important bioactive compounds (Romay et al., 2003; Pandey and Pandey, 2008a; 2009). In the present investigation, production of β – carotene (pro – vitamin A) increased invariably by many folds when growth medium was supplemented with glutamine and NaCl. These results are in accordance with those reported by Fazeli et al. (2006). Carotenoid content increased in *V. uncinata* under both the treatments in free and immobilized cell cultures and the increase was more pronounced when glutamine – grown cells were exposed to NaCl. Substantially high yield of carotenoids increases the consumptive use of *V. uncinata* due to the potential role of the former in human health as antioxidant and quencher removing free radicals and reactive oxygen species (Liu et al., 2009). Oxygen containing counterpart (xanthophylls) of carotenoids appeared more sensitive to NaCl than carotenes which constitute of pure hydrocarbon. Carotenoids are formed from isopentenyl pyrophosphate (IPP) which is a common precursor of all the isoprenoid compounds and β - carotene is an essential end – product of carotenoid biosynthesis. As reported in higher plants (Halliwell and Gutteridge, 1994), lipid content in *V. uncinata* increased substantially under salt stress. Increased lipid content under NaCl stress may provide necessary carbon pool to isoprene units leading to enhanced synthesis of β – carotene. Substantially enhanced production of β – carotene in *V. uncinata* under immobilized culture supplemented with NaCl and glutamine adds further merit for its consumptive use and on commercial scale.

For fresh water organisms including algae, salinity is an important stress that adversely affects growth and metabolism through osmotic effect, nutritional imbalance and / or toxic ion effects. Algal organisms may counteract the adverse effects of salinity by accumulating a number of biochemical compounds as a component of their defense system (Pandey and Singh, 1995). In the present study, exposure to NaCl, although adversely affected the growth of *V. uncinata* as indicated by a decline in chlorophyll, protein and carbohydrate contents, induced the cell accumulation of proline, lipid, aspartic acid, β – carotene and glutathione both in free and immobilized cell cultures. Such accumulations were more pronounced in case of glutamine supplemented growth media. In organisms ranging from bacteria to higher plants, there exist a strong correlation between increased cellular proline levels and the capacity to survive under salinity stress (Sairam and Tyagi, 2004). Proline also serves as an organic nitrogen reserve that can be utilized during recovery. Although proline can be synthesized from either glutamate or ornithine, glutamate is the primary precursor in osmotically stressed cells. In the present study, proline accumulation was more pronounced in glutamine supplemented growth media. Fats and lipids are commonly stored in specialized bodies in cytoplasm and provide protection to the cells against stresses (Wanner *et al.*, 1981). Henchel *et al.* (1991) have shown significant positive correlation between salinity and three amino acids, alanine, threonine and glutamic acid and suggested it as an acclimatory response of *Vaucheria dichotoma*. Culture media supplemented with NaCl and / or nitrogen sources have been shown to substantially enhance the production of β – carotene in microalgae (Fontana, 2004; Fazeli *et al.*, 2006).

The experiments described in this paper were mainly designed to enhance the production of L – asparaginase, a clinically important enzyme extremely valuable in the chemotherapeutic treatment of human cancers including leukemia. A number of micr-oorganisms including bacteria, fungi and actinomycetes are being used for commercial production of L – asparaginase. Algae are probably the least explored group of biodiversity resources as far as the production of L – asparaginase concerns (Paul and Cooksey, 1981; Hejazi *et al.*, 2002). L – asparaginase production by *V. uncinata* was found to be similar in range reported for most efficient microbial systems (El – Bessoumy *et al.*, 2004; Prakasham *et al.*, 2007). Observations further indicated that environmental and nutritional factors play significant role in the production of L – asparaginase in *V. uncinata*. The yellow green alga, *V. uncinata* when grown in a medium supplemented with glutamine resulted in a substantially high yield of L - asparaginase. Addition of NaCl in the medium although associated with some adverse effects on the organism, resulted in the best yield. The results indicated about 400 % improvement in enzyme production over the normal grown cells.

L – asparaginase production in *V. uncinata* enhanced in response to N – sources added to the growth medium where highest level of enzyme production was observed in cells growing in glutamine supplemental medium. Besides contributing to the synthesis of proline and other amino acids, glutamine can donate it's amino group to aspartic acid to form asparagine, which is strongly activated by Cl⁻ (Rognes, 1980). Continuous supply of aspartic acid is a must for maintaining the level of asparagine and in turn, L – asparaginase activity. In the present study, L – asparaginase production substantially increased in glutamine – grown – NaCl exposed free cells. Roberts *et al.* (1968) observed best yields when *E. coli* was grown in a medium containing glutamic acid, methionine and lactic acid but addition of glucose in the medium depressed the enzyme production. Sun and Setlow (1991) observed enhanced production of L – asparaginase in *Bacillus subtilis* grown in a medium supplemented with aspartate, glutamate, proline or urea. Prakasham *et al.* (2007) reported that supplementing the growth medium with glucose or ammonium chloride could lead to a substantial increase in enzyme production in *Staphylococcus* sp. Our data indicated that, in *V. uncinata* L – asparaginase production was substantially enhanced by glutamine as well as by NaCl. However, in a *Chlamydomonas* species isolated from marine environment, L – asparaginase production was reported to had increased by six fold under N – starvation (Paul and Cooksey, 1981). A comparison of these observations indicates that C – pool and / or N – pool dependence of L – asparaginase production may be species specific.

Green cells accumulate a variety of N – containing compounds (NCC) including aspartic acid and proline under salt stress. Qian *et al.* (1996) reported that aspartic acid protects the active sites of L – asparaginase in *E. coli*. Increased level of aspartic acid in *Vaucheria,* as evidenced in the present study, could protect cells and L – asparaginase against NaCl stress. Proline , in addition to its osmoprotective role (Ali *et al.,* 1997) and an antioxidant to keep reactive oxygen species (ROS) under control, can serve as N – source for the production of L – asparaginase (De Moura Sarquis *et al.,* 2004). *V. uncinata* also showed accumulation of glutathione under salt stress. Glutathione, a major non – protein thiol in plants, plays pivotal role in cellular homeostasis and protects cell against oxidative stress (Foyer *et al.,* 1997). Increased concentration of glutathione would confer better antioxidative protection and would be considered an acclimation under salt stress. Molecular studies of Romero *et al* (2001) have shown that salinity stress also induce synthesis of cysteine, an important precursor of glutathione biosynthesis. In the present study, increased accumulation of glutathione in NaCl exposed *V. uncinata* could protect cells and help maintaining production of β – carotene and L – asparaginase.

CONCLUIONS

The agricultural revolution although has yielded bountiful return to meet growing food demand, our traditional knowledge still forms the base to explore biodiversity resources with high nutritional and therapeutic values which, in addition to fulfilling nutritional and energy needs, contribute to additional physiological benefits. The yellow green alga, *Vaucheria uncinata*, considered in this study, is used as a dietary supplement by local tribes of Udaipur, India. Our study clearly indicates that *V. uncinata* is an attractive source of L – asparaginase and β – carotene. Glutamine and NaCl can be used in the growth medium to obtain high enzyme yield from this yellow green alga. Free cell and immobilized cell culture condition further enhance the production of L – asparaginase and β – carotene respectively, for commercial purpose. Such manipulations require less technological inputs and are cost – effective to scale – up production from a non – toxic biological resource. Further studies of the factors controlling L – asparaginase synthesis in *V. uncinata* and experimental design considering interaction effect for process parameter optimization should help increase the yield still further. These observations make *V. uncinata* an attractive bioresource for its consumptive use from human health perspective and on commercial scale.

ACKNOWLEDGMENT

We are thankful to Prof. H. R, Tyagi, farmer Convener, Department of Environmental Science, MLS University for facilities and help during collection of material.

REFERENCES

Ali, G., Srivastava, P.S. & Iqbal, M. (1999). Proline accumulation, ptotein pattern and photosynthesis in regenerants grown under NaCl stress. *Biol. Plant.* 42, 89 – 95.

Bates, L.S., Waldren, R.P. & Teare, I.D. (1979). Rapid determination of proline for water stress studies. *Plant and Soil* 39, 205 – 207.

Christensen, T. (1988). Salinity preference of twenty species of *Vaucheria* (Tribophyceae). *J. Marine Biologic. Assoc. UK.* 68, 531 – 545.

Colla, L.M., Reinchrt, C.O., Reichert, C. & Costa, J.A.V. (2007). Production of biomass and nutraceutical compounds by *Spirulina platensis* under different temperature and nitrogen regimes. *Biores. Technol.* 98, 1489 – 1493.

De – Moura Sarquis, M.I., Oliveira, E.M.M., Santos, A.S. & da Costa, G.L. (2004). Production of L – asparaginase by filamentous fungi. *Mem Inst. Oswaldo Cruz. Rio de Janeiro* 99, 489 – 492.

Devasagayam, T.P.A., Tilak, J.C., Boloor, K.K., Sane, K.S., Ghaskabdi, S. & Lele, R.D. (2004). Free radical and antioxidants in human health : Current status and future prospects. *J. Assoc. Physician Ind.* 7, 794 – 804.

El – Bessoumy, A.F., Sarhan, M. & Mansour, J. (2004). Production, isolation and purification of L – asparaginase from *Pseudomonas aeruginosa* 50071 using solid – state fermentation. *J. Biochem. Mol. Biol.* 37, 387 – 393.

Entwisle, T.J. (1988). A monograph of *Vaucheria* (Vauchericeae, Chrysophyta) in south – eastern mainland, Australia. *Austral. Syst. Bot..* 1, 1 – 77.

Fazeli, M.R., Tofighi, H., Samadi, N. & Jamalifar, H. (2006). Effects of salinity on β – carotene production by *Dunaliella tertiolecta* DCCBC26 isolated from Urmia salt lake, north of Iran. *Biores. Technol.* 97, 2453 – 2456.

Fontana, J.D. Methods in Biotechnology Vol 14 : Food Microbiology Protocols : Carotenogenic microorganisms : A product based biochemical characterization. In : Spencer, J.F.T. & Ragout de Spencer, A.L. (Editors). Totowa, N. J. : Humana Press Inc (2004); pp. 259 – 271.

Foyer, C.H., Delgado, H., Dat, J.F. & Scott, I.M. (1997). Hydrogen peroxide and glutathione associated mechanisms of acclamatory stress tolerance and signaling. *Physiol. Plant.* 100, 241 – 254.

Freed, M. (1966). Methods of Vitamin Assay. New York : Inter Science Publishers. John Willey and Sons.

Griffith, O.W. (1980). Determination of glutathione and glutamine disulfide using glutathione reductase and 2 – vinyl pyridine. *Anal. Biochem.* 106, 207 – 212.

Halliwell, B. & Gutteridge, J.C. (1994). Lipid peroxidation, oxygen radical, cell damage and antioxidant therapy. *The Lancet* 323, 1396 – 1397.

Hejazi, M., Piotukh, K., Mattow, J., Deutzmann, R., Volkmer- Engert, R. & Lockau, W. (2002). Isoaspartyl dipeptidase activity of plant type asparaginases. *Biochem. J.* 364, 129 – 136.

Henchel, D., Kataoka, H. & Kirst, G.O. (1991).Osmotic acclimation of the brackishwater Xanthophyceae, *Vaucheria dichotoma* (L.) MARTIUS : Inorganic ion composition and amino acids. *Plant Res.* 104, 283 – 295

Keating, M.J., Holmes, R. & Lerner, S.H. (1993). L – asparaginase and PEG asparaginase : past, present and future. *Leuk. Lymphoma* 10, 153 – 157.

Liu, N.G., Zhu, Y. & Jiang, J. (2009). The metabolomics of carotenoids in engineered cell factory. *Appl. Microbiol. Biotechnol.* 83, 989 – 999.

Loehvo, A. & Back, S. (2000). Survey of macro algal mats in the gulf of finland, Baltic Sea. *Auqat. Conser. : Mar. Freshwater Ecosystem.* 11, 11 – 18.

Mansour, M.M.F. (2000). Nitrogen containing compounds and adaptations of plants to salinity stress. *Biol. Plant.* 43, 491 – 500.

Mosterson, M.A., Hull, B.L. & Vollmer, L.A. (1988). Treatment of bovine lymphomosarcoma with L – asparaginase. *J. Am. Vet. Med. Assoc.* 192, 1301 – 1306.

Narta, U.K., Kanwar, S.S. & Azimi, W. (2007). Pharmacological and clinical evalution of L – asparaginase in the treatment of leukemia. *Crit. Rev. Oncol. Hematol.* 61, 208 – 221.

Noctor, G. & Foyer, C.H. (1998). Ascorbate and glutathione : keeping active oxygen under control. *Ann. Rev. Plant Physiol. Plant Mol. Biol.* 49, 249 – 279.

Noctor, G. & Gomez, L., Vanacker, H. & Foyer C.H. (2002). Interaction between biosynthesis, comparmentation in the control of glutamine homeostasis and signaling. *J. Exp. Bot.* 53, 1283 – 1304.

Pandey, U. & Pandey, J. (2008a). Enhanced production of biomass, pigments and antioxidant capacity of nutritionally important cyanobacterium *Nostochopsis lobatus*. *Biores. Technol.* 99, 4520 – 4523.

Pandey, U. & Pandey, J. (2009). Enhanced production of δ – aminolevulinic acid, bilipigments and antioxidants from tropical algae of India. *Biotechnol. Bioprocess Eng.* 14 (3), 316 – 321.

Pandey, U. & Pandey, J., (2008b). Enhanced production of high quality biomass, δ – aminolevulinic acid, bilipigments and antioxidant capacity of a food alga *Nostochopsis lobatus*. *Appl. Biochem. Biotechnol.* 150, 221 – 231.

Pandey, U. & Singh, S. P. (1995). Host – phase interaction under osmtic stress in *Nostoc muscorum J. Basic Microbiol.* 35, 421 – 426.

Parikh, A. & Madamwar, D. (2006). Partial characterization of extra cellular polysaccharides from cyanobacteria. *Biores. Technol.* 99, 4520 – 4523.

Paul, J.H. & Cooksey, K.E. (1981). Regulation of asparaginase, glutamine synthatase and glutamate dehydrogenase in response to medium nitrogen concentration in a euryhaline *Chlamydomonas* species. *Plant Physiol.* 68, 1364 – 1368.

Peterson, R.E. & Ciegler, A. (1969). L – asparaginase production by various bacteria. *Appl. Microbiol.* 17, 929 – 930.

Prakasham, R.S., Rao, C. S., Rao, R. S., Lakshmi, G. S. & Sharma, P. N. (2007). L – asparaginase production by isolated *Staphylococcus* sp. – 6A : design of experiment considering interaction effect for process parameter optimization. *J. Appl. Microbiol.* 102, 1382 – 1391.

Qian, G., Zhou, J., Wang, D. & Hie, B. (1996). The chemical modification of *E. coli*. L – asparaginase by N, O – carboxy methyl chitosanction cells, blood substitutents and immobilization. *Biotechnol.* 24, 567 – 577.

Reynoso, G.T. & de Gamboa, B.A. (1982). Salt tolerance in the freshwater algae *Chlamydomonas reinhardii* : Effect of proline and taurine. *Comp. Biochem Physiol. Part A :Physiol.* 73, 95 – 99.

Rijstenbil, J.W., Haritonidis, S., Malea, P., Seferlis, M. & Wijnholds, J.A. (1998). Thiol pools and glutathione redox ratios as possible indicators of copper toxicity in the green microalgae *Enteromorpha* spp. From the Scheldt Estuary (SW Netherlands, Belgium) and Thermaikos Gulf (Greece, N Aegean Sea). *Hydrobiol.* 385, 171 – 181.

Roberts, J., Burson, G. & Hill, J.M. (1968). New procedure for purification of L – asparaginase with high yield from *Escherichia coli. J. Bacteriol.* 95, 2117 – 2123.

Rognes, S.E. (1980). Anion regulation of lupin asparagine synthetase : chloride activation of the glutamine – utilizing reactions. *Phytochem.*19, 2285 – 2293.

Romay, C., Gonzalez, R., Ledon, N., Remirez, D. & Rimbau, V. (2003). C – phycocyanin : a biliprotein with antioxidant, anti-inflammatory and neuroprotective effects. *Curr. Prot. Pep. Sci.* 4, 207 – 216.

Romero, L.C., Dominguez – Solis, J.R., Gutierrez – Alcala G. & Gotor, C. (2001). Salt regulation of O – acetylserine (thiol) lyase in *Arabidopsis thaliana* and increased tolerance in yeast. *Plant Physiol. Biochem.* 39, 643 – 647.

Rumphoon, E., Worful, J.M., Lee, J., Khannan, K., Taylor, M.S., Bhattacharya, D., Moustafa, A. & Manhart, J.R. (2008). Horizontal gene transfer of the algal nuclear gene psbO to the photosynthetic sea slug *Elysia chlorotica. Proc. Nat. Acad. Sci. USA* 105, 11867 – 11871.

Sadasivam, S. & Manickam, A. (1996). Biochemical Methods. New Delhi : New Age International (P) Ltd.

Sairam, R.K. and Tyagi, A. (2004). Physiology and molecular biology of salinity stress tolerance in plants. *Curr. Sci.* 86, 407 – 421.

Schneider, C.W., Riley, M.K. & Stockton, B.M. (1996). Stability of antheridial morphology in freshwater North American *Vaucheria compacta* var dulcis J. Simons (Tribophyceae, Chrysophyta) grown under different salinities. *Aquatic Bot.* 9964, 301 – 311.

Selvakumar, N., Kumar, V. & Natrajan, R. Bioactive Compounds from Marine Organisms with Emphasis on the Indian Ocean : Partial purification, characterization and anti – tumor properties of L – asparaginase (anti – leukemic agent) from a marine *Vibrio* In : Thompson, M.F., Sarojini, R. & Nagabhushanam, R. (Editors). New Delhi : Oxford and IBM Publishing Co. Pvt. Ltd (1991); pp. 289 – 300.

Siripornadulsil, S., Traina, S., Verma, D.P.S. & Sayre, R.T. (2002). Molecular mechanisms of proline – mediated tolerance to toxic heavy metals in transgenic microalgae. *Plant Cell*, 14, 2837 – 2847.

Sun, D. & Setlow, P. (1991). Cloning nucleotide sequence and expression of the *Bacillus subtilis ans* operon which codes for L – asparaginase and L – aspartase. *J. Bacteriol.* 173, 3831 – 3845.

Tausz, M., Sircelj, H. & Grill, D. (2004). The glutathione system as a stress marker in plant ecophysiology : is a stress response concept valid? J. Exp. Bot. 55, 1955 – 1962.

Wanner, G., Formanek, H. & Theimer, R.R. (1981). The ontogeny of lipid bodies (spherosomes) in plant cells. Planta 151, 109 – 123.

Weybrew, J.A. (1957). Estimation of platid pigments of tobacco. *Tobacco Sci.* 1, 1- 5.

Wriston, I.C. & Yellin, T.O. (1973). L – asparaginase, a review. *Adv. Enzymol.* 39, 185 – 248.

Yunis, A.A., Arimures, G.K. & Russian, D.J. (1977). Human pancreatic carcinoma (MIA Pa Ca - 2) in continuous culture : sensitivity to asparaginase. *Int. J. Cancer* 19, 218 – 235.

In: Progress in Food Science and Technology, Volume 1   ISBN: 978-1-61122-314-9
Editor: Anthony J. Greco   © 2012 Nova Science Publishers, Inc.

Chapter 11

WHEAT GRAIN: A UNIQUE PREPARTAL CHOICE FOR TRANSITION HOLSTEIN HEIFERS

*A. Nikkhah*and F. Ehsanbakhsh*

Department of Animal Sciences, University of Zanjan, Zanjan, Iran

ABSTRACT

Wheat grain is unique in possessing extensive starch and nitrogen fermentation, low cation-anion difference and reasonable palatability, making it a potentially useful prepartal dietary choice. The objective was to determine effects of WG provision to prepartum first-calf heifers on metabolic, health, and productive criteria during the transition period. Wheat grain replaced barley grain as commonly used. Fifteen Holstein heifers at 31 ± 6 days prepartum were blocked based on expected calving date and assigned to three treatments, or feeding totally mixed rations containing either 1) barley grain (13.8%) and wheat bran (6.1%) (BGW), 2) 10% wheat grain (WG10), or 3) 18% WG (WG18) (DM basis) from 31 days prepartum until calving. Prepartal diets contained no supplemental anionic salts. Cows were monitored until 21-day postpartum and received the same early lactation diet. The prepartal provision of WG reduced urine pH at 7-day prepartum, and elevated blood calcium and glucose at 7-day prepartum and at 3-day postpartum. Milk fat and protein yields increased during 21-day postpartum by prepartal WG provision. Blood albumin, globulins, total proteins, and urea concentrations were similar among treatments. Feeding WG did not affect body condition score, calving difficulty, calf weight and health, placenta weight, and the time interval from calving to placenta expulsion. It is suggested that prepartal provision of WG led to simultaneous improvements in energy and calcium status of the heifers experiencing their first periparturient phase without compromising parturition and health.

Keywords: wheat grain; prepartum; lactation; heifer; metabolism

---

* Correspondence: Department of Animal Sciences, University of Zanjan, Zanjan 313-45195 Iran; Tel: +98-241-5283202; Fax: +98-241-5283202; E-mail: anikkha@yahoo.com

INTRODUCTION

Prepartal provision of wheat grain (WG) has recently alleviated periparturient hypocalcaemia and hypoglycemia and improved milk production of multiparous cows (Amanlou *et al.*, 2008). Wheat grain possesses large amounts of easily fermentable starch and protein (Herrera-Saldana *et al.*, 1990), with a low cation–anion difference [CAD; e.g. 5.3 mEq/100 g; NRC, 2001]. A major objective in the management of transition cows (3 weeks pre- and post-calving; Drackley, 1999) has been to hasten adjusting rumen microbes and papillae to high-starch diets, promote feed intake, and reduce the risk of hypocalcaemia and fatty liver (Dirksen *et al.*, 1985; Goff and Horst, 1997; Bobe *et al.*, 2004). Increased prepartal dietary nonfiber carbohydrate (NFC) fermentability stimulates DMI and decreases blood nonesterified fatty acids (Dann *et al.*, 1999; Rabelo *et al.*, 2003). The calcium status of the periparturient cow can be improved by optimizing dietary calcium and potassium and supplementation with active metabolites of vitamin D and anionic salts. These can increase small intestinal calcium assimilation and induce some degree of metabolic acidosis and bone resorption (Oetzel and Barmore, 1993; Goff and Horst, 1997; Horst *et al.*, 1997; Vagnoni and Oetzel, 1998; Charbonneau *et al.*, 2006). However, feeding anionic salts has depressed prepartum DMI in heifers, causing excessive hepatic lipidosis (Moore *et al.*, 2000).

Advancing current insights into transition cow management necessitates developing strategies that simultaneously adapt rumen and intermediary metabolism to high-energy lactation diets and improve calcium and energy balances. The result will be a reduction in the risk of postpartum abnormalities associated with hypocalcaemia and hypophagia such as milk fever, dystocia, retained placenta and displaced abomasum (Goff and Horst, 1997; Overton and Waldron, 2004). The prepartal use of wheat bran and high-potassium forages, with greater CAD and lower NFC than WG, is hypothesized to compromise the timely rumen adjustment to high-starch diets and increases the risk of hypocalcaemia (Amanlou *et al.*, 2008). This hypothesis is of significance to dairy heifers because they do not require, and thus, do not usually receive much non-forage NFC (Hutjens, 2004; NRC, 2001). In addition, first-calf heifers encounter dramatic periparturient elevations in mammary proliferation and calcium demand around parturition, compared to multiparous cows that have undergone such transitions around previous calvings. A major hypothesis was that feeding wheat grain with its uniquely reasonable starch and protein degradation rates (e.g., gluten) to prepartal heifers improves rumen adaptation to postpartal diets, and improves feed intake and Ca dynamics, thereby attenuating hypocalcaemia and hypoglycemia shortly postpartum. These will ease the periparturient metabolic transition and will improve milk production in fresh heifers. The main objective was to determine effects of prepartal WG provision on metabolic and productive indices in transition Holstein heifers.

MATERIALS AND METHODS

Experimental Design, Treatments and Heifer Management

Fifteen Holstein heifers (25.9 ± 1.3 months; body weight, BW = 655.3 ± 28.3, body condition score, BCS = 3.67 ± 0.40, mean ± SD) at 31 ± 6 days prepartum were divided to five blocks based on expected calving date and were randomly assigned to one of three treatments. Treatments were prepartal diets containing 1) ground barley grain and wheat bran (BGW) as commonly fed in many farms (Amanlou et al., 2008; Nikkhah et al., 2004), 2) 10% ground wheat grain (WG10), or 3) 18% ground WG (WG18) (DM basis) (Table 1). In replacing wheat for barley, slight changes were made in the use of wheat bran and cottonseed meal to formulate balanced rations (Table 1). All heifers received the same lactation diet and were monitored for 21 days postpartum. Diets were prepared as a totally mixed ration (TMR) and were delivered once daily at 1000 h for ad libitum intake. The pre-experiment diet had forage to concentrate ratio of 67.8:32.2, with 30.7% alfalfa hay, 21.2% corn silage, 15.9% straw, 11.77% barley grain, 2.85% corn grain, 6.04% wheat bran, 4.12% soybean meal, 3.24% cottonseed meal, 3.61% cottonseeds, 0.4% mineral and vitamin supplement, and 0.17% calcium carbonate. Diets were formulated using the NRC program (2001) to have similar forage percentages i.e., a 2:1 ratio of alfalfa hay and corn silage (Table 1). The prepartal diets contained no anionic-salts. Wheat and barley grains were finely ground using an on-farm hammer mill. All heifers were receiving the same low-energy, dry cow diet before entering the experiment. Heifers had unlimited access to fresh water at all times, and were kept in yard houses (5×21 m), with a roofed area (5×8 m). The bunks and water troughs were designed in the unroofed area. Each heifer had a feed bunk space of 1 m. Prepartum heifers were housed in 3 different pens, one for each treatment. Heifers were moved to a parturition box upon noticing early signs of calving, and were moved to fresh heifer yard after 3 d post-calving monitoring. This experiment was conducted at the dairy facilities of Azar-Negin in the northwestern province of Eastern Azerbayjan from October through December of 2005. The facility had approximately 1100 dairy cattle with an average milk yield of 31.5 kg/cow/d (3.7% milk fat) at the time of the study. Cows were cared for according to the guidelines of the Iranian Council of Animal Care (1995).

Feed, Feces, Urine and Blood Analyses

The prepartal daily amount of TMR and orts were measured for each group to calculate DMI. Feeds and orts were sampled daily, composited for each treatment, coven-dried at 60°C for 48 h, and ground to pass through 1-mm screen. All samples were analyzed for crude protein (Macro Kjeldahl Procedure; method 984.13; AOAC 1990), neutral detergent fiber (NDF, Van Soest et al. 1991; using Na-sulfite) and acid detergent fiber (ADF, method 973.18; AOAC 1990). Blood was sampled via the coccygeal vein at 3-h post-feeding, at 7-day projected prepartum (5.6 ± 2 actual days; mean $\pm$ SE) and at 3-day postpartum, and was placed on ice and centrifuged at 3000 g for 15 min. The serum was stored at -20°C until analysis for glucose (GOD-PAR enzymatic method), calcium (O-Kresolphta leine-Komplexon method), phosphorous (UV method), urea (Berthelot enzymatic method), total protein (Biuret method), albumin (Bromcresol Green method), globulins, and cholesterol (CHOD-PAR enzymatic method, Pars Azmun Laboratory, Tehran, Iran). The absorbance was read using spectrophotometer (Perkin- Elmer, Colemen Instruments Division, Oak Brook, IL, U.S.A.). Body condition score (BCS) was recorded by three skilled people at the beginning of the experiment, at 1-day before parturition, and at 21-day postpartum. A five-point scoring scale, with 1 being an emaciated cow and 5 describing an obese cow, was used (Wildman et al. 1982). Urine was sampled manually by external stimulation of vulva at 7-day before expected calving. The urinary pH was measured immediately after sampling (pH meter; ABB Kent Taylor, Kent EIL, England). Grab fecal samples were taken from the rectum at 12-h intervals for 6 days (from d 4 until d 9 pre-calving) to determine apparent total tract nutrient digestibility. Each day, the fecal samples were obtained 2 h later than the previous day and were stored at −20°C until nutrient analysis. Feed and fecal samples were analyzed for dry matter (DM), crude protein (CP), and acid detergent insoluble ash (AIA) as an internal marker for nutrient digestibility calculation (Van Keulen and Young 1977).

Table 1. Ingredient and chemical composition of treatment diets (% of DM)

Ingredients	Treatment prepartum diet[a]			Postpartum[b]
	BGW	WG10	WG18	
Alfalfa hay	41.12	41.19	41.13	34.18
Corn silage	20.01	19.97	19.98	15.72
Wheat straw	-	-	-	-
Ground corn grain	4.60	4.54	4.58	15.31
Ground wheat grain	-	10.02	18.02	6.07
Ground barley grain	13.81	6.88	-	10.93
Wheat bran	6.14	3.07	-	-
Whole cottonseed	1.53	1.54	1.53	6.92
Cottonseed meal	4.31	4.32	5.52	6.07
Soybean meal	8.18	8.19	8.94	7.57
Rumen-protected fat powder[c]	-	-	-	1.07
Sodium bicarbonate	-	-	-	0.90
Salt	-	-	-	0.56
Vitamin and mineral supplement[d]	0.29	0.29	0.29	0.78
Chemical composition				
NE_L, Mcal/kg	1.53	1.54	1.56	1.70
CP %	15.1	15	15.2	17.1
NFC %	36.4	38.6	39.7	41.9
NDF %	41.2	39.5	38.3	31.1
ADF %	23.2	22.4	22.1	21.2
Ether extract %	2.7	2.5	2.4	
$DCAD^d$ (meq/kg)	225	217	211	301
Ca %	0.6	0.6	0.6	0.8
P %	0.4	0.4	0.4	0.4

[a]BGW, the diet with ground barley grain and wheat bran fed during the last 3 weeks of pregnancy. WG10, diet with 10% wheat grain; WG18, diet with 18% wheat grain on a dry matter basis.
[b]After parturition, all heifers received the same postpartum diet.
[c]GP Feeds Ltd., Byley Industrial Estate, Middlewich, Cheshire, UK.
[d]Contained 8000 Mn g, 2000 mg Cu, 10,000 mg Fe, 8000 mg Zn, 20,000 Mg, 21 mg CO, 55 Mg Se, 55 Mg I, 750,000 IU Vit A, 100,000 IU Vit D3, and 8500 IU Vit E per Kg.
CP, crude protein; NDF, neutral detergent fiber; ADF, acid detergent fiber; NFC, non-fiber carbohydrate; DCAD, dietary cation–anion difference calculated as $[(Na^+ + K^+) − (Cl^- + S^-)]$; NRC (2001).

Heifers were milked four times daily at 0600, 1000, 1400 and 2200 h in a milking parlor. Milk production was recorded at each milking for 21-day postpartum. Individual milk samples were collected weekly from the four consecutive milkings, preserved with potassium dichromate, and kept at 4°C for fat, protein, lactose and SNF analysis by Milk-O-Scan

(Lactostar 3560, Funke Gerber, Berlin, Germany). Before analysis, milk samples from the four daily milkings were pooled proportionally by volume for individual cows.

Calf snd Periparturient Heifer Status

Neonate calves were weighed within 1 h of birth. Parturition difficulty (i.e., dystocia) was determined using a five-point scale with a score of 1 needing no assistance and the score of 5 requiring caesarean or causing calf death. Pregnancy length, calf body weight, placenta weight, and placenta expulsion time were recorded for individual heifers. A case was diagnosed as retained placenta if the placenta was not expelled within 24 h of calving. A case was diagnosed as milk fever if blood calcium dropped to < 6 mg/dl, and rectal temperature decreased drastically. Subclinical hypocalcaemia was defined at blood calcium levels of < 8 mg/dl. Calf status was scored at birth using a four-point scale (4 = healthy, 1 = dead). Udder edema was scored at 1 d prepartum (i.e., upon observing first signs of calving and moving to calving boxes) based on a four-point scale (0 = no observable edema, 3 = severe edema).

Statistical Analysis

Data were analyzed using mixed linear models procedure of SAS Institute (Littell *et al.,* 1998; SAS, 2003; Wang and Goonewarden, 2004). The randomized complete block design model included a fixed effect of treatment and random effects of heifer within block by treatment and residuals. Repeated milk production data were analyzed with additional effects of time and its interaction with treatment. The prepartal and postpartal blood metabolites were analyzed independently. Mean values were estimated with Restricted Maximum Likelihood method, and degrees of freedom were calculated using Containment method (SAS, 2003). The biological polynomial trends of treatment effects were acquired using CONTRAST statement of the SAS program to deduce how feeding the increasing amount of WG affected heifer performance. Orthogonal coefficients for unequally spaced treatments were acquired using PROC IML of SAS (SAS Institute, 2003). Significance was detected at $P<0.05$ and trends were declared at $P<0.10$.

RESULTS

Peripheral Blood Metabolites

Prepartal WG as 10% and 18% of diet DM linearly increased ($P<0.01$) blood glucose at both 7-day prepartum and 3-day postpartum (Table 2). A similar linear increase was observed for blood calcium concentrations before ($P<0.01$) and after ($P=0.01$) calving. Blood phosphorous, urea, total protein, and globulins did not change ($P>0.10$). Provision of WG tended to decrease ($P=0.07$) blood cholesterol linearly at 7-day prepartum, but did not affect it at 3-day postpartum (Table 2).

Table 2. Prepartum and postpartum blood concentrations of metabolites

	7-d prepartum				3-d postpartum			
	BGW[1]	WG10	WG18	SE	BGW	WG10	WG18	SE
Glucose[2], mg/dl	39.4[b]	51.6[a]	53.0[a]	3.1	30.0[b]	39.0[a]	40.0[a]	2.1
Calcium[3], mg/dl	7.5[b]	8.6[ab]	9.1[a]	0.4	7.5[b]	8.0[ab]	8.8[a]	0.3
Phosphorous, mg/dl	7.2	7.0	6.8	0.5	6.3	5.5	6.0	0.9
Urea, mg/dl	18.4	17.6	17.7	1.2	18.0	17.4	17.6	1.9
Cholesterol[4], mg/dl	88.2	55.2	53.2	12	70.0	70.4	60.8	18
Serum total protein, g/dl	9.2	10.0	10.1	0.9	8.6	8.9	10.2	1.2
Albumin, g/dl	2.6	3.4	3.2	0.4	2.6	2.7	3.1	0.2
Globulin, g/dl	6.6	6.5	6.8	0.7	6.1	6.2	7.1	1.3

[a,b]Within prepartum and postpartum periods, values with different superscripts differ at $P<0.05$.

[1]BGW, the prepartum conventional diet with ground barley grain and wheat bran. WG10, the prepartum diet with 10% wheat grain; WG18, the prepartum diet with 18% wheat grain on a dry matter basis. After parturition, all heifers received a same diet.

[2]Glucose, prepartum and postpartum, $P<0.01$ for both linear and quadratic effects.

[3]Calcium: prepartum, $P=0.01$ for linear and quadratic effects; postpartum, $P<0.01$ for the linear effect and $P=0.02$ for the quadratic effect.

[4]Cholesterol: prepartum, $P=0.07$ for the linear effect; $P=0.04$ for the quadratic effect.

Table 3. Urine pH, parturition and calf status, and changes in BCS

	Treatment[a]				P-value	
	BGW	WG10	WG18	SE	Linear	Quadratic
Urine pH[b]	7.02	6.78	6.61	0.06	<0.001	<0.001
Pregnancy length, day	279	276	281	2	0.55	0.91
Dystocia[c]	1.8	1.4	1.4	0.3	0.43	0.36
Calf status[d]	3.4	3.8	3.8	0.3	0.34	0.27
Calf weight, kg	41.4	41.0	42.0	1.2	0.71	0.94
Placenta expulsion time, h	6.8	4.7	3.8	1.6	0.22	0.21
Placenta weight, kg	6.0	5.9	6.2	0.3	0.58	0.82
Udder edema[e]	1.0	1.2	1.0	0.6	1.00	0.89
Beginning BCS	3.5	3.6	3.2	0.09	0.34	0.24
BCS at 7-d pre-calving	3.60	3.72	3.74	0.08	0.24	0.21
BCS at 21-d post-calving	3.04	3.20	3.24	0.09	0.16	0.14
BCS change between 7-day prepartum and 21-d postpartum	-0.56	-0.52	-0.50	0.10	0.69	0.70

[a]BGW, the prepartum conventional diet with ground barley grain and wheat bran. WG10, the prepartum diet with 10% wheat grain; WG18, the prepartum diet with 18% wheat grain, on a dry matter basis. After parturition, all heifers received a same diet.

[b]Urine was sampled manually at 7-day before expected calving.

[c]Determined based on a five-point scale with a score of 1 needing no assistance and the score of 5 requiring caesarean or causing calf death.

[d]Scored based on a four-point scale with 4 = a healthy and 1 = a dead calf.

[e]Scored using a three-point scale with 0 = no observable edema and 3 = severe edema.

Urine pH and Parturition Status snd Health

Urine pH dropped highly significantly ($P<0.001$) by WG (Table 3). Feeding WG did not affect dystocia, calf status and weight, placenta weight and expulsion time, udder edema, and pregnancy length ($P>0.10$, Table 3).

DMI, Nutrient Digestibility, Milk Production snd Composition, and BCS Changes

Milk fat percentage increased ($P<0.05$) when WG was included at 18% but not at 10% of the prepartal diet DM. Yields of milk fat, fat corrected milk, and energy corrected milk during 21-d postpartum increased linearly ($P<0.05$) by WG. Feeding WG increased milk protein percentage and yield ($P<0.05$, Table 4). Milk lactose and SNF percentages and yields and BCS were unaffected ($P>0.10$). Group DMI were 10.1, 10.6, and 10.9 kg/d for BGW, WG10 and WG18, respectively. The prepartal apparent total tract DM digestibility was 54.3%, 59.9% and 56.3% and that of CP digestibility was 60.6%, 67.7% and 62% for BGW, WG10 and WG18, respectively.

Table 4. Total tract apparent nutrient digestibility, dry matter intake, and milk production and composition during first 3 weeks postpartum

	Treatment[a]				*P*-value	
	BGW	WG10	WG18	SE	Linear	Quadratic
Milk yield, kg/d	27.7	30.1	29.3	1.1	0.32	0.16
4% FCM[b], kg/d	25.7	27.5	28.9	0.8	0.02	0.03
ECM[b], kg/d	27.8	30.3	31.4	0.9	0.01	0.02
Fat %	3.5	3.5	3.9	0.12	0.04	0.34
Fat yield, kg/d	0.98	1.03	1.14	0.04	<0.01	0.03
Protein %	3.2	3.4	3.5	0.07	0.02	0.02
Protein yield, kg/d	0.89	1.02	1.02	0.04	0.04	0.02
Lactose %	4.8	4.9	4.9	0.08	0.62	0.57
Lactose yield, kg/d	1.32	1.46	1.42	0.07	0.36	0.20
Solids Non-Fat (SNF) %	8.9	9.2	9.0	0.2	0.80	0.52
SNF yield, kg/d	2.47	2.76	2.64	0.12	0.36	0.17

[a]BGW, the prepartum conventional diet with ground barley grain and wheat bran. WG10, the prepartum diet with 10% wheat grain; WG18, the prepartum diet with 18% wheat grain on a dry matter basis. After parturition, all heifers received a same diet.

[b]Fat corrected milk (FCM) = (0.399 × kg daily milk yield) + (15.02 × kg daily milk fat yield); Energy corrected milk (ECM) = [(kg of milk × 0.3246) + (kg of milk fat × 12.96) + (kg of milk protein × 7.04)]; Kowsar et al. (1998).

DISCUSSION

Heifers on BGW consumed about 1395 g barley grain and 622 g wheat bran, whilst heifers on WG10 consumed 1060 g WG, 731 g barley grain and 322 g wheat bran; and heifers on WG18 ate about 1962 g WG. Wheat grain contains greater starch (77% vs. 58%) with 2-3 times faster rumen degradation than barley grain (Herrera-Saldana *et al.,* 1990; Huntington, 1997; Sniffen *et al.,* 1992). Daily starch intake was estimably about 25% greater for WG18 than for BGW i.e., about 700 g/day. Intuitively, wheat gluten helps to develop a luscious protein-starch complex upon mixing with saliva and initial digestion in the mouth that in turn stimulate food ingestion likely via psychostimuli. Increased rumen CHO fermentibility, while controlling starch intake, may increase DM digestibility and microbial growth (NRC, 2001).

Feeding palatable feedstuffs with low CAD and extensive starch and protein fermentability can help more effectively improve Ca balance and reduce anionic salts use. Accordingly, WG improved blood Ca and glucose, suggesting improved energy status of the heifers. These were consistent with earlier findings in mature cows (Amanlou *et al.,* 2008).

Maintaining normal blood calcium (i.e., 8-12 mg/dl) is a primary goal in any dairy farm. Intuitively, adapting heifers to controlled enhancement in bone resorption and intestinal calcium absorption during their first periparturient phase will increase their future capacity of sustaining calcium homeostasis. Calcium plays critical roles in smooth muscle contraction and immune cell activation (Partiseti *et al.,* 1994). The peripartal hypocalcaemia compromises normal muscle contractions and predisposes cows to dystocia, retained placenta, and metritis (Goff and Horst, 1997; Moore *et al.,* 2000). Calcium is involved in immune cell activation (Kimura *et al.,* 2006). Thus, strategies preventing or attenuating peripartal hypocalcaemia could improve immunity and energy status. As such, prepartal provision of WG blood calcium and glucose in the present study, associated with reduced urine pH. Decreased urine pH suggested a reduced extracellular alkalinity (Tucker *et al.,* 1988; Vagnoni and Oetzel, 1998). More acidic circulating fluids can fortify bone resorption and stimulate 1, 25-dihydroxivitamine-D3-mediated small intestinal calcium absorption, thereby alleviating hypocalcaemia (Amanlou *et al.,* 2008; Horst *et al.,* 1997). Reduced urine pH was consistent with increased blood calcium for WG than for BGW.

Unchanged postpartum blood cholesterol concurs with Amanlou *et al.,* (2008). Interpreting treatment differences or similarities in blood urea concentrations requires caution because of substantive N interchanges amongst splanchnic tissues (Huntington, 1989). The lack of treatment differences in blood urea suggests a lack of major differences in hepatic urea output at 3-h postfeeding.

Increased microbial protein synthesis, propionate production, and small intestinal starch and protein assimilation can increase milk production (Dann *et al.,* 1999). The WG18-fed heifers produced 3.6 and 3.2 kg more energy-corrected and fat-corrected milk yield, respectively, than the BGW-fed heifers, showing that WG increased mammary fat and protein synthesis. Increased milk protein yield was evident even when WG was used as 10% of the prepartal dietary DM. Others have shown higher milk fat percentage and yield when prepartal cows received greater and more fermentable concentrate (Dann *et al.,* 1999; Minor *et al.,* 1998; Moallem *et al.,* 2004). Wheat grain was fed at controlled levels to prevent rumen acidosis that would otherwise reduce microbial efficiency. Increased milk fat and protein yield alongside elevated blood glucose suggest healthy rumen fermentation. Glucose forms glycerol and provides co-factors (e.g., NADPH) for *de novo* mammary fat biosynthesis (Bauman *et al.,* 1970). Increased milk production agrees with earlier findings in multiparous cows (Amanlou *et al.,* 2008).

CONCLUSIONS

Provision of WG to pregnant heifers for about 5 weeks prepartum led to simultaneous improvements in peripheral glucose and calcium supplies. These were attained with no supplemental anionic salts and no effects on parturition and calf health, followed by increased milk production during 21-d postpartum. The data suggest WG as a suitable cereal choice in prepartal diets to facilitate metabolic transition. Future larger herd studies are warranted to enlighten other metabolic and reproductive aspects of the hypotheses.

REFERENCES

Amanlou, H., Zahmatkesh, D, & Nikkhah, A. (2008). Wheat grain as a prepartal cereal choice to ease metabolic transition from gestation into lactation in Holstein cows. *Journal of Animal Physiology and Animal Nutrition* 92, 605-613.

AOAC. (1990). *Official Methods of Analysis*, 15th edn. Association of Official Analytical Chemist, Arlington, VA

Bauman, D. E., Brown, R. E., & Davis, C. L. (1970). Pathways of fatty acid synthesis and reducing equivalent generation in mammary gland of rat, sow, and cow. *Archives of Biochemistry and Biophysics* 140, 237–244.

Bobe, G., Young, J. W., & Beitz, D. C. (2004). Invited review: Pathology, etiology, prevention, and treatment of fatty liver in dairy cows. *Journal of Dairy Science* 87, 3105-3124.

Charbonneau, E., Pellerin, D. & Oetzel, G. R. (2006). Impact of lowering dietary cation-anion difference in nonlactating dairy cows: a meta-analysis. *Journal of Dairy Science* 89, 537-548.

Dann, H. M., Varga, G. A. & Putnam, D. E. (1999). Improving energy supply to late gestation and early postpartum dairy cows. *Journal of Dairy Science* 82, 1765–1778.

Dirksen, G., Liebich, H. & Mayer, K. (1985). Adaptive changes of the ruminal mucosa and functional and clinical significance. *Bovine Practitioner* 20, 116–120.

Drackley, J. K. (1999). Biology of dairy cows during the transition period: the final frontier. *Journal of Dairy Science* 82, 2259–2273.

Goff, J. P. & Horst, R. L. (1997). Physiological changes at parturition and their relationship to metabolic disorders. *Journal of Dairy Science* 80, 1260–1269.

Herrera-Saldana, R. E., Huber, J. T. & Poore, M. H. (1990). Dry Matter, crude protein, and starch degradability of five cereal grains. *Journal of Dairy Science* 73, 2386–2393.

Horst, R. L., Goff, J. P., Reinhardt, T. A. & Buxton, D. R. (1997). Strategies for preventing milk fever in dairy cattle. *Journal of Dairy Science* 80, 1269–1280.

Huntington, G. B. (1989). Hepatic urea synthesis and site and rate of urea removal from blood of beef steers fed alfalfa hay or a high concentrate diet. *Canadian Journal of Animal Sciences* 69, 215–223.

Huntington, G. B. (1997). Starch utilization by ruminants: from basics to the bunk. *Journal of Animal Science* 75, 852-867.

Iranian Council of Animal Care (1995). *Guide to the Care and Use of Experimental Animals*, vol. 1. Isfahan University of Technology, Iran.

Kimura, K., Reinhardt, T. A. & Goff, J. P. (2006). Parturition and hypocalcemia blunts calcium signals in immune cells of dairy cattle. *Journal of Dairy Science* 89, 2588–2595.

Littell, R. C., Henry, P. R. & Ammerman, C. B. (1998). Statistical analysis of repeated measures data using SAS procedures. *Journal of Animal Science* 76, 1216-1231.

Minor, D. J., Trower, S. L., Strang, B. D., Shaver, R. D. & Grummer, R. R. (1998). Effects of nonfiber carbohydrate and niacin on periparturient metabolic status and lactation of dairy cows. *Journal of Dairy Science* 81, 189–200.

Moallem, U., Bruckental, I. & Sklan, D. (2004). Effect of feeding pregnant and nonlactation dairy cows a supplement containing a high proportion of nonstructural carbohydrates on postpartum production and peripartum blood metabolites. *Animal Feed Science and Technology* 116, 185–195.

Moore, S. J., Vandehaar, M. J., Sharma, B. K., Pilbeam, T. E., Beede, D. K., Bucholtz, H. F., Liesman, J. S., Horst, R. L. & Goff, J. P. (2000). Effects of altering dietary cation–anion difference on calcium and energy metabolism in peripartum cows. *Journal of Dairy Science* 83, 2095–2104.

National Research Council (2001*). Nutrient Requirements of Dairy Cattle*, 7th rev. edn. National Academy Press, Washington, DC.

Nikkhah, A., Alikahni, M., Amanlou, H. (2004). Effects of feeding ground or steam-flaked broom sorghum and ground barley grains on performance of dairy cows in midlactation. *Journal of Dairy Science* 87, 122-130.

Oetzel, G. R. & Barmore, J. A. (1993) Intake of a concentrate mixture containing various anionic salts fed to pregnant, nonlactating dairy cows. *Journal of Dairy Science* 76, 1617-1623.

Overton, T. R. & Waldron, M. R. (2004). Nutritional management of transition dairy cows: Strategies to optimize metabolic health. *Journal of Dairy Science* 87 (E suppl.), E105-E119.

Partiseti, M., Deist, F. L., Hivroz, C., Fisher, A., Korn, H. & Choquet, D. (1994). The calcium current activated by T cell receptor and store depletion in human lymphocytes is absent in a primary immunodeficiency. *Journal of Biological Chemistry* 269, 32327–32335.

Rabelo, E., Bertics, S.J., Mackovic, J. & Grummer, R. R. (2001). Strategies for increasing energy density of dry cow diets. *Journal of Dairy Science* 84, 2240-2249.

Rabelo, E., Rezende, R. L., Bertics, S. J. & Grummer, R. R. (2003). Effects of transition diets varying in dietary energy density on lactation performance and ruminal parameters of dairy cows. *Journal of Dairy Science* 86, 916-925.

Rukkwamsuk, T., Wensing, T. & Geelen, M. J. H. (1999). Effect of fatty liver on hepatic gluconeogenesis in periparturient dairy cows. Journal of Dairy Science 82, 500-505.

SAS (2003). *SAS User's Guide: Statistics*, Version 9.1 Edition. SAS Institute, Cary, NC.

Sniffen, C. J., O'Connor, J. D., Van Soest, P. J., Fox, D. J. & Russell, J. B. (1992). A net carbohydrate and protein system for evaluating cattle diets: II. Carbohydrate and protein availability. *Journal of Animal Science* 70, 3562-3577.

Strang, B. D., Bertics, S. J., Grummer, R. R. & Armentano, L. E. (1998). Effect of long-chain fatty acids on triglyceride accumulation, gluconeogenesis, and ureagenesis in bovine hepatocytes. *Journal of Dairy Science* 81, 728–739.

Tucker, W. B., Harrison, G. A. & Hemken, R. W. (1988). Influence of dietary cation-anion balance on milk, blood, urine, and rumen fluid in lactating dairy cattle. *Journal of Dairy Science* 71, 346–354.

Vagnoni, D. B. & Oetzel, G. R. (1998). Effects of dietary cation–anion difference on the acid base status of dry cows. *Journal of Dairy Science* 81, 1643-1652.

Van Keulen, J. & Young, B. A. (1977). Evaluation of acid-insoluble ash as a natural marker in ruminant digestibility studies. *Journal of Animal Science* 44, 282-287.

Van Soest, P. J., Robertson, J. B. & Lewis, B. A. (1991). Methods for dietary fiber, neutral detergent fiber, and nonstarch olysaccharides in relation to animal nutrition. *Journal of Dairy Science* 74, 3583–3597.

Wang, Z. & Goonewardene, L. A. (2004). The use of mixed models in the analysis of animal experiments with repeated measures data. *Canadian Journal of Animal Science* 84, 1-11.

Wildman, E. E., Jones, G. M., Wagner, P. E., Boman, R. L., Troutt, H. F. & Lesch, T. N. (1982). A dairy cow body condition scoring system and its relationship to selected production characteristics. *Journal of Dairy Science* 65, 495–501.

In: Progress in Food Science and Technology, Volume 1
Editor: Anthony J. Greco

ISBN: 978-1-61122-314-9
© 2012 Nova Science Publishers, Inc.

Chapter 12

GERMINATED LEGUMES: A BOON TO HUMAN NUTRITION

Geetanjali Kaushik

Centre for Rural Development & Technology, Indian Institute of Technology Delhi,
Hauz Khas, New Delhi, India

1. INTRODUCTION

Food Legumes: Nutritive Value

Food legumes or pulses are high in protein, carbohydrates, and dietary fibre and are a rich source of other nutritional components viz B group vitamins and minerals (Tharanathan & Mahadevamma, 2003) and their consumption and production extends world-wide. Pulses used for human consumption include peas, beans, lentils, chickpeas, and faba beans (Rochfort & Panozzo, 2007). On account of their high nutritive value they are the main sources for human and animal nutrition especially in the developing countries (Tharanathan & Mahadevamma, 2003). While in the developed countries, legumes are increasingly being used in dietetic formulations in the treatment and prevention of diabetes, cardiovascular diseases and cancer of colon (Brand et al, 1990; Satya et al, 2009a; Satya et al, 2010). Frequent legume consumption (four or more times compared with less than once a week) has been associated with 22% and 11% lower risk of coronary heart disease (CHD) and cardiovascular disease (CVD), respectively (Flight & Clifton, 2006).

However, the consumption of grain legumes is limited due to the presence of several anti nutritional factors, such as α-galactosides, trypsin and chymotrypsin inhibitors, phytates, lectins and polyphenols (Fernandez et al, 1997; Srivastava & Srivastava, 2003) which impede the availability of nutrients (Vidal-Valverde et al, 1992). α- galactosides are known to be responsible for flatulence in humans; trypsin and chymotrypsin inhibitors are able to bind hydrolytic enzymes, such as trypsin and chymotrypsin, inhibiting their activity, where as phytic acid is found to reduce the availability of minerals (Prodanov et al, 2004). Phytic acid forms insoluble complexes with polyvalent cations such as Cu, Zn, Co, Mn, Fe or Ca, thus reducing their availability (Harland & Oberleas, 1987). Phytic acid also forms phytate-protein complexes in peas, which decrease the protein availability of legumes. Polyphenols are commonly known as tannins, and are known to form complexes with proteins. Tannin-protein complexes are responsible for low protein digestibility and decrease amino acid availability (Srivastava & Srivastava, 2003). Therefore the removal of these factors is essential for improving the nutritional quality of legumes and, subsequently to increase their potential as human food (Prodanov et al, 2004, Satya 2009a). A wide range of simple processing techniques such as soaking, boiling, autoclaving, cooking roasting, dehulling, germination, fermentation, supplementation with various chemicals (El-Hady & Habiba, 2003) and in recent years extrusion cooking (El-Hady & Habiba, 2003) have been employed for the reduction of these anti - nutritional factors and for improving legume organoleptic quality.

Germination: A Processing Technique

Processing techniques cause important changes in the biochemical, nutritional and sensory characteristics in legumes. These are known to enhance the nutritional value of legumes, by increasing essential amino acids, protein digestibility, amino acid availability and certain B-vitamins (Khatoon & Prakash, 2006; Fernandez & Berry, 1988). In many parts of the world legumes are consumed after germination during which the nutritional value is increased (Sharma, 2006).

Germination is defined as a process in which the hypocotyl is encouraged to develop and grow in length. During germination, seeds undergo pronounced metabolic changes whereby storage nutrient reserves in the cotyledons are broken down into more usable forms and are translocated to the growing shoots and roots for utilization (Ghazali & Cheng, 1991).

Thus the process causes significant changes in the biochemical characteristics. Germination of legume seeds to produce sprouts is one of the processing methods with the potential to increase nutritive value and health qualities of the seeds (Urbano et al., 2005; Zanabria et al, 2006). This process also raises the levels of free amino acids, available carbohydrates, dietary fibre and other components and augments the content of essential minerals and vitamins. In addition the decrease in antinutrients associated with germination enhances the bioavailability of minerals for the human body (Lee & Karunanaithy, 1991).

Germination: Procedure

The seeds are initially soaked in 4-5 volumes of water and subsequently sprouted on paper towels (Noor et al., 1980), cheese cloth (Labaneiah & Luh, 1981), cotton wool (Giri et al., 1981) and cellulose sponge (Fordham et al., 1975; Farhangi & Valadon, 1981). Seeds have also been germinated on filter paper in trays (Sattar et al, 1989), petridishes (Jood et al, 1988) or on trays lined with moist cotton wool (Lee & Karunanithy, 1990) or on the perforated aluminum trays (Sangronis & Machado, 2007). Automatic seed sprouter (Ghazali & Cheng, 1991) has also been used for germination of the seeds.

2. EFFECT OF GERMINATION ON NUTRIENTS AND ANTINUTRIENTS

Germination produces important biochemical changes in the seeds that have high nutritional significance. So it is necessary to understand these changes in a holistic manner in form of changes in the nutrients and antinutrients. Nutrients include carbohydrates, fats, proteins, minerals and vitamins. While among the antinutrients the effect of germination has been reviewed on phytic acid, tannins, trypsin inhibitor, and the oligosaccharides in legumes.

Nutrients

Carbohydrates- During 7 days of germination of chickpea, starch and polyfructosans were rapidly metabolized leading to the appearance in significant concentration of soluble and easily assimilable sugars (Azhar et al, 1972). Overnight soaked soybeans germinated for 6 days showed slight decreases in the reducing sugars with increasing germination time. This may be due to utilization of simple sugars as a source of energy during the germination process (Mostafa & Rahma, 1987). The chickpea (12 h soaked) upon germination for 60h showed increases of total soluble sugars 2-11%, reducing sugars 2-13% and non-reducing sugars 2-11%. The starch content decreased by 18-31%, while the starch digestibility increased by 152-183%. The increase in sugar contents of soaked seeds during germination may be because of mobilization and hydrolysis of seed polysaccharides, leading to more available sugars. Starch may also be hydrolyzed to oligosaccharides and ultimately to monosaccharides during germination (Jood et al, 1988). Chandrasiri et al, 1990 found that after 5 days of germination there was 80.1% loss of ethanol-soluble carbohydrates as compared to ungerminated seeds. Soaking for 12 h followed by germination for 3 days of Kabuli chickpea seeds resulted in significant decrease in the total carbohydrate contents (5%). These decreases were attributed to their use as a source of energy to start germination. There was a significant reduction in the levels of reducing sugars, sucrose and starch by 25.77%, 28.57% and 6.88% respectively. These reductions could have been due to the hydrolysis of these components by hydrolytic enzymes to monosaccharides, which are used as an energy source during germination (El-Adawy, 2002).

Protein - Overnight soaked soybeans germinated for 6 days showed an apparent increase in the total protein with the increase in time of germination. The increase was due to the oxidation and consumption of the other classes in the germination process. The rate of relative increase in essential amino acids was 8.9% after 3 days of germination, whereas this value reached 22.4% at the end of 6-day germination period. During germination probably there is a turnover of proteins and amino acids with the balance between synthetic and degradative processes determining the resultant pattern. Results show that the greatest increases appeared to be in descending order leucine> tyrosine> phenylalanine and glutamic acid whereas methionine content showed a slight decrease due to the germination process (Mostafa & Rahma, 1987). Soaking followed by 3 day germination of black gram resulted in increase in the protein content (Ghazali & Cheng, 1991).

Similar results on protein content were also obtained by Kylen & McCready (1975) and Lorenz (1980) who studied the effect of germination on alfalfa, green gram, lentils and soybeans, and cereal grains, respectively. It was proposed that the apparent increase in the protein content following germination could probably be due to loss of leachable sugars and seed coats (Kylen & McCready, 1975), protein synthesis (Kylen & McCready, 1975), losses of carbohydrates during respiration and/or an alteration of the nitrogenous substances rather than actual increase in protein (Lorenz, 1980).

In overnight soaked chickpea seeds which were germinated for 24h and 48 h the essential amino acids threonine, lysine, leucine, valine and isoleucine increased after 24h but a slight decrease in especially lysine was also found when germination was extended to 48h (Fernandez & Berry, 1988). Overnight soaked *Glycine* when germinated for 5 days showed an increase of more than 21% in total crude protein in germinated soybean as compared to ungerminated seeds. This increase was accounted to a synthesis of enzyme proteins or a compositional change following the degradation of other constituents (Lee & Karunanithy, 1990). 12 h soaking followed by germination for 48 h increased the protein content from 32.89% to 39.35% extending the period of germination to 96h increased the protein content to 40.48% (Dogra et al, 2001). Soaking for 12 h followed by germination for 3 days of Kabuli chickpea seeds resulted in a significant increase in crude protein (8%) as compared to the raw chickpea seeds. This increase was attributed to the use of seed components and degradation of protein to simple peptides during the germination process (El-Adawy, 2002).

Fat - The fat content of black gram was found to decrease as germination proceeded for 3 days and third day sprouts were found to contain only 50% of the initial fat content, based on dry weight (Ghazali & Cheng, 1991). Other workers (Kylen & McCready, 1975; Noor, *et al.,* 1980) have reported similar decreases in the fat content of green gram during germination. The overall decrease in the fat content of black gram during germination was probably the result of metabolism in order to meet the increased energy requirement of the developing plant tissues.

Minerals- Overnight soaked *Glycine* and *Phaseolus* beans when germinated for 5 days showed loss in the mineral content. The losses amounted to 68% 26% and 17% for K, Mg and Fe in case of *Glycine*. While for *Phaseolus* beans the respective losses were 46%, 14% and 20%. However, the Ca content showed increase of 9% for both the beans (Lee & Karunanithy, 1990). The losses of divalent metals were lower as compared to that of the monovalent metals, which could be explained on account of the following factors. Firstly the higher retention of divalent cations is by their binding to protein (Lee & Karunanithy, 1990) and or due to the formation of phytate-cation-protein complex at a higher pH. Secondly the divalent cation phytates are relatively insoluble at intermediate pH (Lee & Karunanithy, 1990) so the losses are probably minimal. Soaking for 12 h followed by germination for 3 days of Kabuli chickpea seeds resulted in noticeable decreases in the contents of K, Ca, Mg, Mn and Cu (El-Adawy, 2002). The content of mineral elements in the seeds of black beans, white beans and pigeon beans were significantly modified by soaking for 5h followed by germination for 5 days. The germination significantly increased the content of calcium of white beans by 8.4%, 5% for black beans and 17.2% for pigeon beans. Germination significantly reduced the magnesium content by 23.7% in white-casing *P. vulgaris*, while black casing *P. vulgaris* L. and *C. cajan* did not show significant differences between ungerminated and germinated seeds (Sangronis & Machado, 2007). Kumar et al. (1978) indicated that the germination of legumes caused a significant loss in the magnesium content, which was attributed to the lixiviation during the germination process. Moreover, the germination of legumes significantly increased their zinc content to 128.8% in white beans, 200% in black beans and 37.7% in pigeon beans. A decrease in iron content was also observed due to the effects of germination. The white-beans showed a decrease of 78.3%, while the black beans and pigeon beans showed meaningful decreases of 42.6% and 82.6%, respectively, compared to the iron content of ungerminated grains. These results may be a consequence of previous soaking and lixiviation of the grains during germination.

Vitamins- Overnight soaked chickpea seeds when germinated for 24h and 48 h showed increases of 394% and 721% respectively in the ascorbic acid (Fernandez & Berry, 1988). Soaking of black gram led to a dramatic increase in the vitamin C content probably due to *de novo* synthesis since many enzymes are generally activated during this period. Further increases in the vitamin C content were observed as germination proceeded.

Upon germination for 3 days the total carotenoids were found to increase in content. Carotenoids are important to the diet as many members of the family are provitamin A, in that these can be converted to vitamin A by the human body (Ghazali & Cheng, 1991).

8 h soaking followed by 3 day germination increased the ascorbic acid (91.3%), riboflavin (69.35%) drastically and thiamine contents (22.36%) considerably (Ahmad & Pathak, 2000). Soaking for 12 h followed by germination for 3 days of Kabuli chickpea reduced both thiamine (37.5%) and niacin (78.83%) significantly, there was also a significant increase in riboflavin and pyridoxine. The retentions of pyridoxine and riboflavin were 103.58% and 116.15% respectively (El-Adawy, 2002). 8 h soaked 4 day soy sprouts had 218 mg/kg of ascorbic acid content which had varied drastically since it was not detected at 0 days. The increase in ascorbic acid level was considered to be a consequence of the reactivation of ascorbic acid biosynthesis in the seeds during germination (Mao-Jun et al, 2005). Sangronis & Machado, 2007 observed that soaking for 5h followed by germination for 5 days resulted in a significant increase in the content of ascorbic acid of the white beans, black beans and pigeon beans. For the white beans the increase was 300%, 208.4% for pigeon beans and 33.2% for black beans. The germination process is therefore an effective method for the increment of ascorbic acid. During germination, the respiration process is triggered by the ascorbic acid. This could be related to the observed increase as a consequence of germination (Sangronis & Machado, 2007). Germination significantly increased the thiamine content of the white, black and pigeon beans. The increment was 12.8%, 26.5% and 7.4%, respectively.

Digestibility- A marked improvement in digestibility by pepsin (88%) was noticed in overnight soaked soybeans germinated for 6 days. Since the germinated samples undergo protein hydrolysis by proteolytic enzymes this leads to improved digestibility as compared to the dry samples (Mostafa & Rahma, 1987). 8h soaking followed by germination for 96 h increased the in-vitro protein digestibility to 95% in Kabuli chickpea from an initial value of 77%. Similarly an increase of 93% was observed in desi cultivar from 70% initially (Khalil et al, 2007). Sangronis & Machado, 2007 observed that soaking for 5h followed by germination for 5 days resulted in a significant increase in the in vitro protein digestibility of white beans, black beans and pigeon beans as effect of germination. The increase was 2% for the white beans, 3% for the black beans and 4% for the pigeon beans (Sangronis & Machado, 2007). Such an increase was explained due to the partial or complete reduction of different antinutrients. The phytic acid and the condensed tannins interact with protein molecules, forming complexes and therefore reducing the protein susceptibility to enzymatic attacks (Alonso et al, 2000).

Antinutrients

Phytic acid- Presoaked (24 h) pigeon pea seeds germinated for 72 h showed a 30% loss in phytic acid content (Igbedioh et al, 1994). Soaking for 12 h followed by germination for 3 days of Kabuli chickpea seeds reduced the phytic acid by 56.19%. The reduction in phytic acid could have been due to phytase activity during germination (El-Adawy, 2002). 8h soaking followed by germination for 96 h lead to a considerable reduction in the phytic acid content in Kabuli (73%) and desi (32%) type chickpea cultivars (Khalil et al, 2007). Sangronis & Machado, 2007 observed that soaking for 5h followed by germination for 5 days lead to a significant statistical reduction in the phytic acid content of white beans, black beans and pigeon beans due to the effect of germination. The reduction percentages were 44.7% for the white beans, 52.9% for the black beans and 40.7% for pigeon beans. Earlier it has been mentioned by Bau et al, 1997 that the increase in phytase activity during germination is responsible for the decrease in content of phytic acid. 24h soaking of chickpea seeds followed by germination for 48 hr significantly reduced the phytic acid content from 1.01% to 0.6% (Khattak et al, 2007).

TIA- Overnight soaked soybeans germinated for 6 days showed a decrease in TIA with the increasing germination time and a 32% decrease in activity was found after 6 days of germination. The decrease in TIA due to germination could be attributed to leaching out of TI during washing of sprouts on a daily basis (Mostafa & Rahma, 1987). Soaking for 12 h followed by germination for 3 days of Kabuli chickpea seeds reduced the trypsin inhibitor by 33.95% (El-Adawy, 2002). Sangronis & Machado, 2007 observed that soaking for 5h followed by germination for 5 days resulted in a significant decrease in the content of TIA of pigeon beans (41%), white beans (52.5%) and black beans (25%). It was reasoned that the decrease in trypsin inhibitors could be due to their utilization as a source of energy during the early stages of germination (Burbano et al, 1999).

Tannins- Soaking of white beans, black beans and pigeon beans for 5h followed by germination for 5 days lead to reduction in levels of tannins by 36.2%, 19% and 14.3%, respectively for the beans. The loss of tannins could be on account of leaching into soak water (Sangronis & Machado, 2007). Also, during germination, an enzymatic hydrolysis by polyphenolase causes the loss of tannins (Reddy et al, 1985).

Oligosaccharides- East et al 1970 studied the changes in stachyose, raffinose, sucrose and monosaccharides during soybean germination. Sucrose degraded fast and about 75% was degraded after 96h of germination thereafter it remained constant till 144h. The effect was attributed to the supplemental sucrose resulting from the degradation of raffinose and stachyose. Stachyose and raffinose also showed progressive decreases and only 6.4% of the original stachyose content remained after 96h. The cumulative degradation of sucrose, raffinose and stachyose during germination was associated with increase in the level of monosaccharides. Pazur et al 1962, reported that 6 days of germination effectively eliminated α– galactosides of soybean. Germination of *Cicer arietinum* (Var, 'Pant G-114') seeds for 72 h resulted in almost complete removal of galactose containing sugars (alpha-galactooligosaccharides) – raffinose & stachyose along with accumulation of sucrose during the early stages germination (Tewari, 2002).

3. CONCLUSION

Germination is a process for improving the nutritional content of legumes. It significantly reduces the trypsin inhibitors, phytic acid and tannins. This process also increases the crude protein, ascorbic acid and the contents of thiamine and riboflavin while improving the protein digestibility. It decreases the starch content and increases the level of total soluble sugars, reducing and non-reducing sugars. The levels of both essential and non-essential amino acids markedly increase during germination. Losses of divalent metals (Ca, Fe and Zn) are low during germination and these could become bioavailable due to decrease in phytic acid content. Thus the use of germinated seeds could make a significant contribution to human diet when considered in the light of the high nutrient/energy ratio in germinated seeds.

4. FUTURE SCOPE

The use of additives (viz citric acid and sodium bicarbonate) in soak media has been reported in the literature to affect the nutrient content. Addition of citric acid to soak medium has been found to result in minimum vitamin loss and also proven to be optimal for best retentions of other nutrients. However, instead of using chemicals further research on the use of natural sources of citric acid like lemon juice as additives in the soaking medium should be undertaken to study their beneficial effect on nutrient retention which would lead to popularization of natural additives for enhancing the nutrient retention.

Germination is a simple inexpensive process for enhancing the nutritional value of legumes however, there is need to focus efforts on optimizing the duration of germination for the various legumes so that the nutrients are enhanced to an optimal level and also the antinutrients are minimized significantly and the balance between the two processes is maintained (Satya et al, 2010).

The available studies on effect of germination on micronutrients present an ambiguous picture and lack of clarity on status of important micronutrients. Hence, indepth investigation on this important aspect is warranted. Overall, germination should be popularized as a simple, inexpensive process that improves the nutritional value of the legumes.

Over the years the indiscriminate use of pesticides has lead to a widespread contamination of the food commodities. The food legumes are highly susceptible to pest attack right from the crop production to the storage level on account of which there is an excessive application of pesticides on them. This leads to the presence of harmful pesticide residues in them. Hence, it is important to study the bioavailability of nutrients in pesticide contaminated food legumes, i,e the effect of pesticide on the nutrient bioavailability in legumes. This theme would address a very important dimension of food quality and safety in the modern paradigm (Satya et al, 2009b).

REFERENCES

Ahmad, S. & Pathak , D.K, (2000). Nutritional Changes in Soybean during germination. *J food Science Technology* 37(6), 665-666.

Alanso, R., Orue, E. & Marzo, F. (1998). Effects of extrusion and conventional processing methods on protein and antinutritional factor contents in pea seeds. *Food Chemistry*, 63(4), 505-512.

Azhar, S., Srivastava, A.K. & Murti, C.R.K.(1972). Compositional changes during the germination of Cicer Arietinumm. *Phytochemistry*, 11,3173-3179.

Bau, H., Villaume, C., Niwlas, J., & Mcjean, L.(1997). Effect of germination of chemical composition, biochemical constituents and anti nutritional factors of soybean (*Glycine max*) seeds *.J of the Science of food of Agri*,73, 1-9.

Brand, J.C, Snow, D.J, Nabhan, G.P & Truswell, A.S. (1990). Plasma glucose and insulin responses to traditional Pima Indian meals. Am J of Cl Nut, 51, 416-420.

Burbano, C., Muzquiz, M., Ayet, G., Cuadrado,C., & Pedrosa, M.(1999). Evaluation of anti nutritional factors of selected varieties of P. Rilgaris. *J of the science of food of Agri*, 79, 1468-1472.

El-Adawy, T.A. (2002). Nutritional composition and antinutritional factors of chickpeas (Cicer arietinum L.) undergoing different cooking methods and germination. *Plant Foods for Human Nutrition*, 57(1), 83-97.

El-Hady, E.A & Habiba, R.A. (2003). *Effect of soaking and extrusion conditions on antinutrients and protein digestibility of legume seeds*. Lebensm-Wiss U-Technol. 36, 285-293.

Farhangi, M. & Valadon, L. R. G. (1981). Effect of acidified processing and storage on carotenoids (provitamin A) and vitamin C in mung bean sprouts. *J. Food Sci.*, *46*, 1464-6.

Fernandez, M, Aranda, P Lopez-Jurado, M., Garcia-Fuentes, M.A & Urbano, G. (1997). Bioavailabilty of phytic acid phosphorous in processed *Vicia faba L var. Major*, *Journal of Agric & Food Chemistry*, 45, 4367-4371.

Fernandez. M.L. & Berry, J. W. (1988). The effect of germination on chickpea starch. *Starch/Starke*, 41(1), 17-21.

Flight, I., & Clifton, P. (2006). Cereal grains and legumes in the prevention of coronary heart disease and stroke: A review of the literature. *European Journal of Clinical Nutrition*, 60, 1145–1159.

Fordham, J. R., Wells, C. E. & Chert, L. A. (1975). Sprouting of seeds and nutrient composition of seeds and sprouts. *J. Food Sci.*, 40, 552-6.

Ghazali, H. M. & Cheng, S. C. (1991). The effect of germination on the physico-chemical properties of black gram (*Vigna mungo* L.). *Food Chemistry* 41, 99-106.

Giri, J., Parvatham, R. & Santhini, K. (1981). Effect of germination on the levels of pectins. *Food Chemistry* 117 (2009) 599–607 607

Harland, B.F. & Oberleas, D.(1987). Phytate in foods. *World Review of Nutrition and Dietetics*, 52,235-259.

Igbedlioh, S.O., Olugbemi, K.T. & Akpapunam, M.A.(1994) effects of processing methods on phytic acid level and some constituents in bambara groundnut (Vigia Subterranea) and pigeon pea (cajanus cajan).*Food chemistry*, 50, 147-151.

Jood, S., Chauhan, B.M. & Kapoor, A.C.(1988). *Contents and Digestibility of carbohydrates of chickpea and black gram as affected by domestic processing and cooking food chemistry*, 30, 113-127.

Khatoon, N. & Prakash, J.(2006). Nutrient retention in microwave cooked germinated legumes. *Food Chumistry*, 97,115-121.

Khattak, A.B., Zeb, A., Bibi, N., Khalil, S.A., & Khattak, M.S. (20070. Influence of germination techniques on phytic acid any poly phenols content of chickpea (*Cicer Arietinum*) sprouts, *Food chemistry*, 104, 1074-1079.

Kumar, K., Venkataraman, L., Jaya, T., & Krishnamurthy, K.(1978). Cooking of some germinated legumes ; changes in phytins, Ca++, Mg++ and pectins. *J of food science* 4385-88.

Kylen, A. M. & McCready, R. M. (1975). Nutrient in seeds and sprouts of alfalfa, lentils, mung beans and soybeans. *J. Food Sci.,* 40, 1000-9.

Labaneiah, M. E. O. & Luh, B.S. (1981). Changes of starch, crude fibre and oligosaccharides in germinating dry beans. *Cereal Chem.,* 58(2), 135-8.

Lee, C.K. & Karunanithy, R. (1990). Effects of germination on the chemical composition of *Glycine* and *Phaseolus* beans. *J. Science Food Agri* 51,437-445.

Lorenz, K. (1980). Cereal sprouts: Composition, nutritive value and food applications. *Critical Rev. Food Sci. and Nutr.,* 13, 353-5.

Mao-Jun, X., Ju-fang, D. & Mu-Yuan, Z. (2005). Effects of germination conditions on ascorbic acid level and yield of soybean sprouts. *J of Sci. food Agri*, 85,943-947.

Mostafa MM & Rahma, EH. (1987). Chemical and nutritional changes in soybean during germination. *Food Chemistry* 23, 257-275.

Noor., M.I, Bressani, R. & Luiz, G. E. (1980). Changes in chemical and selected biochemical components, protein quality and digestibility of mung bean *(Phaseolus aureus)* during germination and cooking process. In *Proceedings of Legumes in the Tropics.* Malaysia. pp. 479-88.

Prodanov, M., Sierra, I. & Vidal-Valverde, C. (2004). Influence of soaking and cooking on the thiamin, riboflavin and niacin contents of legumes. *Food Chemistry*, 84(2), 271-277.

Rochfort, S., & Panozzo, J. (2007). Phytochemicals for health, the role of pulses. *Journal of Agricultural and Food Chemistry*, 55, 7981–7994.

Sangronis, E. & Machado, C.J. (2007) *influence of germination on the nutritional quality of Phaseolus vulgaris* and *Cajanus cajan*, LWT 40, 126-120.

Sattar, A., Durrani, S.K., Mahmood, F. Ahmad, A. & Khan, I. (1989). Effect of soaking and germination temperatures on selected nutrient and anti nutrients of mung bean *Food Chemistry* 34, 111-120.

Satya, S; Kaushik, G. & Naik, S. N. (2009a). *Enhancing Food Quality and Safety*- A Holistic Perspective. Chapter to be published in book on Foods and Nutrition by Studium Press, Houston (USA).

Satya, S; Kaushik, G. & Naik, S. N. (2010).Processing of Food Legumes – A Boon to Human Nutrition. *Mediterranean Journal of Nutrition and Metabolism*- springer verlaug (article in press).

Satya, S; Kaushik, G., Naik, S. N. & Tripathi, B. (2009b).*Effect of pesticide Chlopyrifos residues on micronutrient bioavailability in chickpea* – A preliminary investigation. 4th International Conference on Food Quality & Safety and ISO22000 Training November 3-4, and November 5-6, 2009. Redondo Beach, CA. USA. (Abstract selected for oral presentation).

Sharma, J. (2006). *Effect of processing methods on the quality of cereal and legume products.* PhD Thesis, Centre for Rural Development & Technology, Indian Institute of Technology Delhi, New Delhi, India.

Srivastava, R.P. & Srivastava, G.K. (2003). Nutritional value of Pulses. Indian J Agric Biochem 16(2), 57-65.

Tewari, L.(2002). Removal of the flatulence factors(alpha-gala tooligo saccharides) from chickpea(cicer arietinum) by germination and mold fermentation *J Food Sci technol*, 39(5), 458-462.

Tharanathan, R.N. & Mahadevamma, S. (2003). Grain legumes a boon to human nutrition. *Trends in Food Science & Technology* 14, 507-518.

Urbano, G., Aranda, P., Vilchez, A., Aranda, C., Cabrera, L., & Porres, J. S. M. (2005). Effects of germination on the composition and nutritive value of proteins in Pisum sativum, L.. *Food Chemistry*, 93, 671–679.

Vidal-valverde, C., Frias, J. & Valverde, S. (1992). Effect of processing on the soluble carbohydrate content of lentils. *J of Food Protection*, 55, 301-306.

Zanabria, E. R., Katarzyna, N., De Jong, L. E. Q., Birgit, H. B. E., & Robert, M. J. N. (2006). Effect of food processing of pearl millet (Pennisetum glaucum) IKMP-5 on the level of phenolic, phytate, iron and zinc. *Journal of the Science of Food and Agriculture,* 86, 1391–1398.

In: Progress in Food Science and Technology, Volume 1   ISBN: 978-1-61122-314-9
Editor: Anthony J. Greco   © 2012 Nova Science Publishers, Inc.

Chapter 13

SUCROSE METABOLISM AND GRAIN DEVELOPMENT OF CONTRASTING SPIKELETS LOCATED AT DIFFERENT POSITIONS OF THE RICE PANICLE

Pravat Kumar Mohapatra[*]

CSIR Emeritus Scientist, School of life Science, Sambalpur University,
Jyoti vihar, Sambalpur, India

ABSTRACT

Individual grain weight varies in rice panicle, because some of the caryopses located mostly on the basal part of the panicle fill poorly or partially. Physiological mechanism responsible for premature cessation of grain growth in an inferior caryopsis, thereby resulting in a heterogeneous architecture of panicle morphology, is elusive. During grain filling sucrose is the major phloem solute, which moves symplastically in the sieve tube from the source to the pericarp of grain caryopsis by an osmotically driven pressure gradient. Pericarp comprising cells of maternal tissue is however, not connected symplastically to the filial endosperm cells. Carrier mediated efflux of sucrose from the inner most nucellar tissue of pericarp into an apoplasm and its subsequent influx there from into the outermost aleurone tissue of the endosperm are energy dependent. Entry of sucrose into endospermal cells primes the starch biosynthesis pathway for grain filling; the activity is controlled by a number of enzymes. Empirical evidences ruled out deficiency of sucrose as a possible cause of decreased growth of the inferior caryopsis and instead identified sink limitation responsible for the down regulation of starch filling capacity of grains. The review examines the functions of the pericarp, apoplasmic solutes and enzymes of the starch biosynthesis pathway in sink control of grain filling in rice caryopsis for the purpose of improving yield.

Keywords: rice, sucrose, spikelet, assimilate partitioing

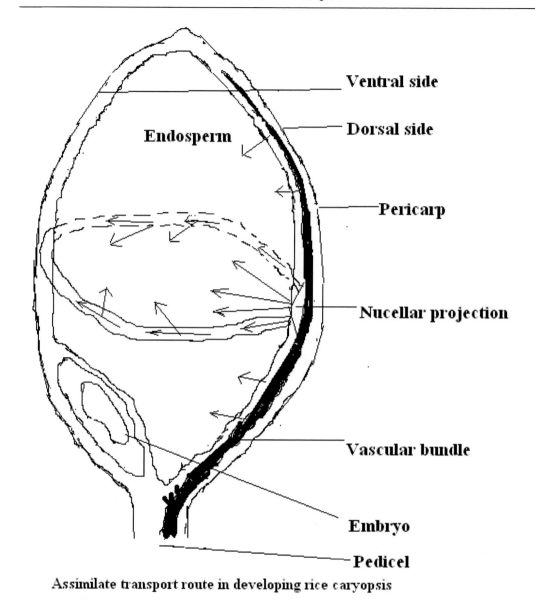

Figure 1. Assimilate transport route in developing rice caryopsis.

1. PANICLE ARCHITECTURE OF RICE

Grain weight is a stable varietal character in rice (Yoshida 1972). Grain development is genetically programmed, but several intrinsic factors often disturb gene expression and limit weight and quality of some of the grains in the panicle. Strong apical dominance within the panicle structure is one of such factors, which precludes adequate filling of grains in the basal

---

* CSIR Emeritus Scientist, School of life Science, Sambalpur University, Jyoti vihar, Sambalpur 768019, India. Tel: 91-9937103747, Email: pravat1948@rediffmail.com

part resulting in considerable variation in single grain weight (Sahu, 1988). The spikelets on the apical primary branches of the panicle fill faster and produce good quality grains, whereas, spikelets on the basal primary branches remain mostly unfilled or partially filled (Venkateswarlu et al 1986a, b; Mohapatra et al 1993). In addition to the spatial disadvantage, temporal difference in spikelet development also affects grain filling of inferior spikelets of rice panicle. The time taken to anthesis is negatively correlated with final size or weight of a mature grain as well as grain quality (Mohapatra et al. 1993), because late-flowering affects the rate and duration of grain growth. Both the components of grain growth vary within the panicle because of the difference in spatial and temporal location of spikelets. The margin of variation widened and number of poorly developed grains increased when either branch or spikelet number increased in the panicle structure (Fig. 1). During the last decade, rice breeders attempted to compromise with the space for freedom of spikelet development, and accommodated large number of spikelets in the panicle branches of the dense or compact panicled rice cultivars. Although grain number increased, the margin of variation in grain weight widened (Wang et al. 2008), and the number of poor quality grains increased in the panicle. Such increases in panicle grain number marginalized the efforts of plant breeders for raising rice yield potential (Khush and Peng 1996; Peng et al. 1999). Because rice is bisexual and exclusively self-pollinated (De Datta 1981; Yoshida 1981), grain development is initiated immediately after anthesis. Fertilisation occurs five hours after anthesis (Hoshikawa 1967; 1970) and completely cellular endosperm is established by day 5. In contrast to apical spikelets, the event of anthesis is delayed in basal spikelets and occurrence of grain filling in them is poor (Mohapatra and Sahu 1991). The partitioning of assimilate to the basal spikelets is poor on account of low sink strength of the developing caryopses.

2. PATHWAY OF ASSIMILATE TRANSPORT IN RICE CARYOPSIS

The physiology of variation in grain filling on rice panicle has been the theme of my research in the School of Life Science, Sambalpur University, India during the past three decades. During grain development assimilates are mostly supplied in the form of soluble sugars to the developing seeds (Thorne 1985). Sucrose is the major phloem solute and it is transported symplastically in the vascular bundle of the pericarp on the dorsal side of rice caryopsis up to the upper terminal part of the grain structure (Figs. 1 & 2). On the dorsal side of the developing caryopsis, phloem releases sucrose symplastically into the maternal tissues of the pericarp, such as sieve element-companion cell complex, vascular parenchyma, pigment strand and nucellus prior to its entry into the embryonic apoplast and ultimate storage in the filial tissues in endosperm (Oparka and Gates 1981 a, b; Furbank et al 2001). Because rice grain possesses no endosperm cavity, sucrose movement from maternal to filial tissue is circumferential. The nucellar tissue (inner layer of the pericarp) extends more than half way through the seed length. Sucrose delivered from the vascular bundle on the dorsal side of the pericarp move sympalstically in the nucellar cells before entering into the apoplasm between nucellus and the aleurone layer of the endosperm (Oparka and Gates 1981a, b). In the dorsal vascular bundle region of caryopsis, nucellar projection/nucellus bears small, densely cytoplasmic cells adjacent to the multilayered aleurone/sub-aleuorne of the filial endosperm. Both nucellus and aleurone become single layered in the non-vascular part of the ventral side of the pericarp. The sieve elements are connected to the companion cells with plasmodesmata and similar connections exist between companion cells and vascular parenchyma. Assimilates moving out of the sieve element companion cell complex laterally traverse cells of the pigment strand and vascular parenchyma. The pigment strand in turn is connected symplastically with the nucellus only at the dorsal side of the caryopsis, just below the vascular bundle. In this part, the cuticle surrounding the endosperm is absent and the pericarp maintains contact with the nucellus. The residual part of the nucellus remains separated from the pericarp by the cuticle. Similar to nucellus, cells of the aleurone layers are symplastically connected, but there is no symplastic connection between the former and the latter. It is believed that the movement of solutes in the phloem from the source to sink end occurs by an osmotically driven pressure gradient (Taiz and Zeiger 2002). In contrast loading of sucrose into the sieve cell-companion cells complex at the source and its unloading from the complex into grain apoplast are energetically controlled. However, the physiological factors responsible for such movement remains obscure and the role of pericarp in assimilates unloading into the developing kernel are not understood. In contrast to wheat or barley (Bagnall et al 2000; Weschke et al 2000) unloading of assimilates in rice caryposis is not confined to a specific part of the maternal filial interface. The transport of sucrose into rice endosperm has been proposed to occur across the nucellus/endosperm interface of the entire upper part of the caryopsis, rather than in the tissues in close proximity to vascular bundle/nucellar projection (Oparka and Gates 1981 a, b; Furbank et al 2001).

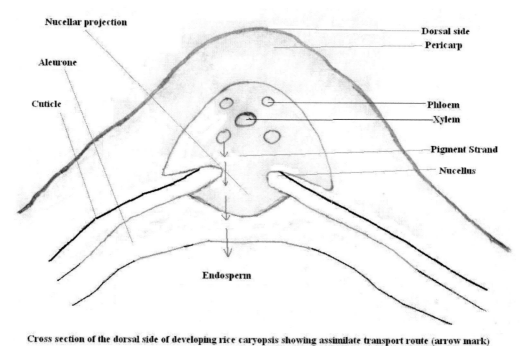

Figure 2. Cross section of dorsal side of developing rice caryopsis showing assimilate transport route.

3. VARIATION IN ALLOCATION AND PARTITIONING OF ASSIMILATES AMONG SPIKELETS OF RICE PANICLE

The total pool of carbon fixed in a leaf is determined by its photosynthetic rate. Depending on metabolic demand from the sink organs, a part of the carbon pool is translocated out of the leaf to the developing sink organs. Physiological regulations determine the share of carbon assimilates available to an individual sink in the process of allocation (Taiz and Zeiger 2002). The carbon assimilates move from the source leaf to various sink organs through a complex network of vascular bundles. Organogenesis not being equal among developing sinks, there is difference in development of vascular tissues, which ultimately fixes the quota of assimilates actually reaching a targeted sink. Such differential distribution of assimilates within a plant or a plant organ is termed partitioning (Taiz and Zeiger 2002). In addition to the vascular structure, partitioning of assimilates is also influenced by position of the sink in relation to the source and competition exacerbated by the growth of nearby sinks. In our laboratory, assessment of growth, development and soluble and insoluble carbohydrate contents precluded the assumption of an unsuccessful competition for assimilates on the part of the inferior spikelets responsible for disparity in growth and assimilate partitioning. During development, an inferior kernel receives enough assimilates from the source, but poor sink strength makes it less competent to use them in the starch biosynthesis pathway resulting in accumulation of unused assimilates (Mohapatra et al. 1993, Patel and Mohapatra 1996). Sink strength is a product of sink size and sink activity (Taiz and Zeiger 2002). In the rice panicle, the apical spikelets are released from the flag leaf enclosure nearly seven days ahead of the basal spikelets. A dominant apical spikelet also possesses higher sink activity to claim larger quota of assimilates from the pool of assimilates moving out of the source organ at the cost of the growth of inferior basal spikelets. When a part of the panicle was removed with surgical manipulations (Mohapatra et al 2004) increased ^{13}C partitioning occured to spikelets of the residual part. In the process grain filling was encouraged in some spikelets potentially incapable of bearing grains. The potentiality for grain filling of the inferior spikelets can be expressed by exogenous application of chemicals that inhibit either action or synthesis of ethylene in the panicle (Naik and Mohapatra 1999; Mohapatra et al. 2000) as well. These experiments reveal the existence of intrinsic difference in sink activity between spikelets of the panicle. The difference in sink activities between spikelets can be quantified in different manners, such as, in terms of activities sucrose splitting and starch synthesizing enzymes, uptake of assimilates from the embryonic apoplast, physiological functions of pericarp and endosperm and functions of sucrose transporters involved in phloem loading and unloading. Some of the evidences showing differences in sink activities between the contrasting spikelets of rice panicle have been discussed below.

3.1. Pericarp Functions

Rice pericarp development is closely knitted with endosperm development. Endosperm cellularisation begins as soon as cell division stops in the inner integument of the pericarp. Prior to grain filling, pericarp functions as a transient storage tissue. During grain elongation, the functions of the pericarp ceases prematurely in an inferior caryopsis (Ishimaru et al 2003) potentially impeding the delivery of assimilates for endosperm growth. In our laboratory, pericarp functions of apical and basal caryopses of rice panicle were compared in two indica rice cultivars by measuring difference in concentrations of chlorophylls and carotenoids, activities of peroxidase and lipid peroxidation (Mohapatra and Mohapatra 2006). The data indicated that rate of development of pericarp was slower during the juvenile phase of seed growth, but degradation was faster during the senescent phase in an inferior spikelet compared to a superior spikelet. A poorly developed pericarp can subsequently impact grain filling by disturbing delivery of assimilates passing through it to the developing endosperm.

3.2. Endosperm Enzyme Activities

Starch constitutes nearly 78% of dry weight of the rice endosperm (Singh and Juliano 1977). The incoming sucrose of the phloem solutes is converted to amylose or amylopectin in the starch biosynthesis pathway (Fig. 3). In the panicle, amylose content of the superior grains is higher than that of the inferior grains (Matsue et al. 1995). Several authors measured difference in sink strength of the apical and basal spikelets in the terms of activities of the enzymes responsible for sucrose splitting and the follow up steps of starch biosynthesis (Umemoto et al. 1994; Smidansky et al. 2003; Ishimaru et al. 2005; Kato et al. 2007). In our work, sucrose was identified as the major translocatable phloem solute entering the endosperm of fertile apical and basal spikelets of rice panicle (Patel and Mohapatra 1996). In the basal spikelet we found no resistance in the delivery of sucrose or deficiency of the material causing any constraint for endosperm starch biosynthesis. Sucrose synthase was the enzyme mostly responsible for priming sucrose into the starch biosynthesis pathway, and poor activity of this enzyme limited starch biosynthesis in the inferior basal grains of the panicle. Sucrose splitting by the other pathway through invertase enzyme action was mostly utilized during the first four days of grain filling (Hirose et al. 2002). In our subsequent work (Naik and Mohapatra 2000), this proposition was confirmed. The dominant apical spikelets possessed higher activities of both sucrose synthase and invertase. It was concluded that sucrose synthase activity could be a potential indicator of sink strength of developing rice kernels (Counce and Gravois 2006).

Altogether, there are 33 enzymes involved in the starch biosynthesis of rice endosperm (Nakamura et al 1989). However, four enzymes including sucrose synthase play key role in starch synthesis (Kato 1995; Yang et al. 2001). The other three enzymes are adenosinediphosphoglucose pyrophopshorylase (AGPase), starch synthase (StSase) and starch branching enzyme (BE). Apart from the original research of Perez et al (1975) and Nakamura et al (1989), several authors have shown evidences in favour of AGPase, starch synthase and branching enzyme controlling starch biosynthesis and sink strength of developing rice endosperm (Nakamura and Yuki 1992; Umemoto et al. 1994; Smidansky et al. 2003; Kato et al. 2007; Mohapatra et al 2009). Ishimaru et al (2005) studied the expression pattern of the genes encoding the carbohydrate metabolizing enzymes discussed above in both superior and inferior spikelets and reiterated differential action of the enzymes responsible for the determination of sink strength and starch synthesis of rice caryopsis during grain filling

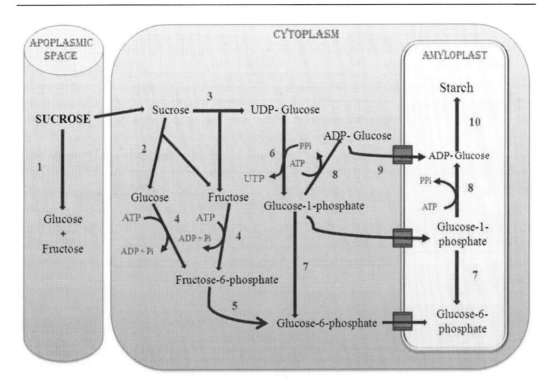

Figure 3. Sucrose metabolic pathway in the endosperm of developing rice caryopsis. Enzymes and transporter controlling the pathway are 1. Cell wall invertase, 2. Cytoplasmic invertase, 3. Sucrose synthase, 4. Hexokinase, 5. Phosphoglucoisomerase 6. UDPglucose pyrophosphorylase, 7. Phosphoglucomutase, 8. ADPglucose pyrophosphorylase, 9. ADPglucose transporter, 10. Starch synthase.

3.3. Functions of Sucrose Transporters at the Maternal and Filial Interface of the Developing Caryopsis

It is suggested that the transfer of sucrose between nucellus and endosperm is energy dependent in rice (Oparka and Gates 1981a,b). Abundance of SUT1 (sucrose-proton symporter) and H^+-ATPase transporters in the inner layer of nucellus and outer layer of aleurone justify the energy dependant unloading of sucrose into the apoplasm and loading there from into the endosperm aleurone (Furbank et al 2001). The H^+-ATPase generates the proton motive force with hydrolysis of ATP for driving the SUT1 symporter of the plasma-membrane. It results in ferrying of one proton along with a sucrose molecule across the membrane. The importance of OsSUT1 transporters in grain filling is highlighted further because antisense suppression of the *OsSUT1* gene leads to impaired grain filling in rice (Scoefield et al. 2002). This suggestion is corroborated in a recent review (Lim et al. 2006). The early part of seed development is a pre-storage phase comprising pericarp growth and cellularisation of endosperm. The maturation or storage phase is determined by initiation of starch accumulation and increase of dry weight. During storage phase, the rise in filial demand for assimilates in sink activity is integrated with an increase of phloem import. As assimilates are consumed in endosperm growth, a turgor homeostat of the maternal tissue perceives the change in solute concentration of the apoplasmic fluid (Zhang et al 2007).

Activity of the turgor homeostat encourages further symplastic flow assimilates from the source into the sink apoplasm. Therefore, turgor motivated mass flow of assimilates to the rice endosperm is dependent on co-ordinate action of the sucrose transporters (Lim et al. 2006), which in turn determines sink strength of the developing caryopsis. Poor activity of the transporters can reduce assimilate unloading and decrease mass flow of solutes to the caryopsis. In fact, assimilate concentration in the apoplast of the slow developing inferior caryposes is much lower that that of the fast developing apical caryopses (Kuanar et al. 2010). Hence, poor activity of the transporters or solute concentration of the apoplasmic fluid can be reliable indicator of sink strength of a developing caryopsis. Our work (Kuanar et al. 2010) is the first of its kind to report of correlation between sink strength and assimilates concentration of apoplasmic fluid in rice caryopsis. In addition to OsSUT1 transporter, some suggestions have been made to include the role of cell wall invertase and monosccharide transporters in endosperm development and longitudinal growth of pericarp (Lim et al. 2006). However, empirical evidences showing importance of these proteins in sucrose unloading and loading are scant.

CONCLUSION

A holistic approach is necessary to amalgamate all attributes of the developing caryopsis that influence sink strength for grain filling, such as the contribution of pericarp, apoplasmic space solutes, assimilates unloading from nucellar cells of pericarp, uptake of sucrose from the apoplasm by aleurone layer of endosperm and activities of sucrose transporters and enzymes of starch biosynthesis pathway of the endosperm for improving grain filling in the inferior spikelets of rice panicle. These activities are closely interlinked with each other. Further, expression analysis profile of the genes responsible for synthesis and action of the starch synthesizing enzymes and the sucrose transporters should be compared between the superior and inferior caryopses to illustrate the factors controlling grain development. Additionally more information is required on the chemical signals that drive the genes into action in a superior caryopsis and absence of which impairs filling in an inferior caryopsis. A good beginning has been made in understanding the temporal and spatial expression profile of sucrose synthase gene during grain filling of spikelets of rice panicle with the support of Emeritus Scientist project to the undersigned by CSIR, New Delhi at School of Life Science, Sambalpur University. This project will help to identify, if sucrose synthase gene regulation or manipulation is necessary for raising sink strength of developing caryopsis for the purpose of increasing the number of filled grains of the rice panicle and yield potential of the crop.

ACKNOWLEDGMENT

The author thanks Council of Scientific and Industrial Research, New Delhi for the award of Emeritus scientist fellowship.

REFERENCES

Bagnall, N., Wang, X-D., Scoefield, G. N., Furbank, R.T., Offler, C.E. & Patrick, J.W. (2000). Sucrose transport related genes are expressed in both maternal and filial tissues of developing wheat grains. *Australian Journal of Plant Physiology* 27, 1009-1020.

Counce, P.A. & Gravois, K.A. (2006). Sucrose synthase activity as a potential indicator of high rice grain yield. *Crop Science* 46, 1501-1507.

De Datta, S.K. (1981). *Principles and practices of rice cultivation*. John Wiley and Sons, New York.

Furbank, R.T., Scoefield, G.N., Hirose, T., Wang, X-D., Patrick, J.W. & Offler, C.E. (2001). Cellular localization and function of a sucrose transporter OsSUT1 in developing rice grains. *Australian Journal of Plant Physiology* 28, 1187-1196.

Hirose, T., Takano, M. & Terao, T. (2002). Cell wall invertase in developing rice caryopsis: molecular cloning of *OsCIN1* and analysis of its expression in relation to its role in grain filling. *Plant and Cell Physiology* 43, 452-459.

Hoshikawa, K. (1967). Studies on the development of endosperm in rice.1. Process of endosperm tissue formation. *Japanese Journal of Crop science* 36, 151-161.

Hoshikawa, K. (1970). Studies on the development of endosperm in rice.2. Development of protein forming plastids. *Japanese Journal of Crop science* 37, 97-106.

Ishimaru, T., Hirose, T., Matsuda, T., Goto, A., Takahashi, K., Sasaki, H., Terao, T., Ishii, R., Ohsugi, R. & Yamagishi, T. (2005). Expression patterns of genes encoding carbohydrate metabolising enzymes and their relationship to grain filling in rice (*Oryza sativa* L.): comparison of caryopses located at different positions in a panicle. *Plant Cell Physiology* 46, 620-628.

Ishimaru, T., Matsuda, T., Ohsugi, R. & Yamagishi, T. (2003). Morphological development of rice caryopsis located at the different positions in a panicle from early to middle stage of grain filling. *Functional Plant Biology* 30, 1139-1149.

Kato, T. (1995). Change of sucrose synthase activity in developing endosperm of rice cultivars. *Crop Science* 35, 827-831.

Kato, T., Shinmura, D. & Taniguchi, A. (2007). Activities of enzymes for sucrose-starch conversion in developing endosperm of rice and their association with grain filling in extra-heavy panicle types. *Plant Production Science* 10, 442-450.

Khush, G.S. & Peng, S. (1996). Breaking the yield frontier of rice. In: Reynolds MP, Rajaram S and McNab A (Eds*). Increasing yield potential of wheat: breaking the barriers*. Proceedings of a workshop held on 26-28 March, 1996 in Ciudad Obregon, Sonora. International Maize and Wheat improvement Centre, Mexico, pp. 36-51.

Kuanar, S.R., Panigrahi, R., Kariali, E. & Mohapatra, P.K. (2010). Apoplasmic assimilates and grain growth of contrasting rice cultivars differing in grain dry mass and size. *Plant Growth Regulation* 61, 135-151.

Lim, J.D., Cho, J-I., Park, Y-I., Hahn, T-R., Choi, S-B. & Jeon, J-S. (2006). Sucrose transport from source to sink seeds in rice. *Physiologia Plantarum 126*, 572-584.

Matsue, Y., Odahara, K. & Hiramatsu, M. (1995). Difference in amylase content, amylographic charactersistics and storage proteins of grains on primary and secondary branches of in rice. *Japanese Journal of Crop Science* 64, 601-606.

Mohapatra, R. & Mohapatra, P.K. (2006). Ethylene control of seed coat development in low and high sterile semidwarf indica rice cultivars. *Plant Growth Regulation* 50, 47-55.

Mohapatra, P.K., Sarkar, R.K. & Kuanar, S.R. (2009). Starch synthesizing enzymes and sink strength of grains of contrasting rice cultivars. *Plant Science* 176, 256-263.

Mohapatra, P.K. & Sahu, S.K. (1991). Heterogeneity of primary branch development and spikelet survival in rice in relation to assimilate contents of primary branches. *Journal of Experimental Botany* 42, 871-879.

Mohapatra, P.K., Patel, R. & Sahu, S.K. (1993). Time of flowering affects grain quality and spikelet partitioning within rice panicle. *Australian Journal of Plant Physiology* 20, 231-241.

Mohapatra, P.K., Naik, P.K.& Patel, R. (2000). Ethylene inhibitors improve dry matter partitioning and development of late flowering spikelets of rice panicles. *Australian Journal of Plant Physiology* 27, 311-323.

Mohapatra, P.K., Masamoto, Y., Morita, S., Takanashi, J., Kato, T., Itani, T., Adu-Gyamfi, J.J., Shunmugasundaram, M., Nguyen, N.T. & Fujita, K. (2004). Partitioning of ^{13}C-labelled photosynthate varies with growth stage and panicle size in high yielding rice. *Functional Plant Biology* 31, 131-139.

Naik, P.K. & Mohapatra, P.K. (1999). Ethylene inhibitors promote male gametophyte survival in rice. *Plant Growth Regulation* 28, 29-39.

Naik, P.K. & Mohapatra, P.K. (2000). Ethylene inhibitors enhanced sucrose synthase activity and promoted grain filling of basal rice kernels. *Australian Journal of Plant Physiology* 27, 997-1008.

Nakamura, Y. & Yuki, K. (1992). Changes in enzyme activities associated with carbohydrate metabolism during the development of rice endosperm. Plant Science 82, 15-20.

Nakamura, Y., Yuki, K. & Park, S.Y. (1989). Carbohydrate metabolism in the developing endosperm of rice grains. *Plant Cell Physiology* 30, 833-839.

Oparka, K.J. & Gates, P. (1981a). Transport of assimilates in the developing caryopsis of rice (*Oryza sativa* L.). Ultra-structure of the pericarp vascular bundle and connection with the aleurone layer. *Planta* 151, 561-573.

Oparka, K.J. & Gates, P. (1981b). Transport of assimilates in the developing caryopsis of rice (*Oryza sativa* L.). The pathways of water and assimilated carbon. *Planta* 152, 388-396.

Patel, R. & Mohapatra, P.K. (1996) Assimilate partitioning within floret components of contrasting rice spikelets producing qualitatively different types of grains. *Australian Journal of Plant Physiology* 23, 86-92.

Peng, S., Cassman, K.G., Virmani, S.S., Sheehy, J. & Khush, G.S. (1999). Yield potential trends of tropical rice since release of IR8 and the challenge of increasing yield potential. *Crop Science* 39, 1552-1559.

Perez, C.M., Perdon, A.A., Resurreccion, A.P., Villareal, R.M. & Juliano, B.O. (1975). Enzymes of carbohydrate metabolism in the developing rice grain. *Plant Physiology* 56, 579-583.

Sahu, S.K. (1988). *Physiology of inflorescence growth and development in rice varieties differing in duration of growth.* PhD thesis in Life Science, Sambalpur University, India.

Scoefield, G.N., Hirose, T., Gaudron, J.A., Upadhyaya, N., Ohsugi, R. & Furbank, R.T. (2002). Antisense suppression of the rice sucrose transporter gene, OsSUT1, leads to impaired grain filling and germination, but does not affect photosynthesis. *Functional Plant Biology* 29, 815-826.

Singh, R. & Juliano, B.O. (1977). Free sugars in relation to starch accumulation in developing rice grain. *Plant Physiology* 59, 417-421.

Smidansky, E.D., Martin, J.M., Curtis-Hannah, L., Fischer, A.M. & Giroux, M.J. (2003). Seed yield and plant biomass increases are conferred by deregulation of endosperm ADP-glucose pyrophosphorylase. *Planta* 216, 656-664.

Taiz, L. & Zeiger, E. (2002). *Plant Physiology*, 3rd Edition, Sinaur Associates Inc. Publishers, Sunderland, USA.

Thorne, J.H. (1985). Phloem unloading of C and N assimilates in developing seeds. *Annual Review of Plant Physiology* 36, 317-343.

Umemoto, T., Nakamura, Y. & Ishikura, N. (1994). Effect of grain location on the panicle on activities involved in starch synthesis in rice endosperm. *Phytochemistry* 36, 843-847.

Venkateswarlu, B., Vergara, B.S., Parao, F.T. &Visperas, R.M. (1986a) Enhancing grain yield potential by increasing number of high density grains. *Philippines Journal of Crop Science* 11, 145- 152.

Venkateswarlu, B., Parao, F.T., Visperas, R.M. & Vergara, B.S. (1986b) Screening quality grains of rice with a seed blower. *Sabrao Journal* 18, 19-24.

Wang, F., Chang, F. & Zhang, G. (2008). Impact of cultivar variation in density of rice panicle on grain weight and quality. *Journal of Science of Food and Agriculture* 88, 897-903.

Weschke, W., Panitz, R., Sauer, N., Wang, Q., Neubohn, B., Weber, H. & Wobus, U. (2000). Sucrose transport into barley seeds: molecular characterization of two transporters and implications for seed development and starch accumulation. *The Plant Journal* 21, 455-469.

Yang, J., Peng, S., Gu, S., Visperas, R.M. & Zhu, Q. (2001). Changes in activities of theree enzyme associated with starch synthesis in rice grains during grain filling. *Acta Agronomica Sinica* 27, 157-164.

Yoshida, S. (1972). Physiological aspects of grain yield. *Annual Review of Plant Physiology* 23, 437-464.

Yoshida, S. (1981). *Fundamentals of rice crop science*. International Rice Research Institute, Philippines.

Zhang, W-H., Zhou, Y., Dibley, K.E., Tyerman, S.D., Furbank, R.T. & Patrick, J.W. (2007). Nutrient loading of developing seeds. *Functional Plant Biology* 34, 314-331.

In: Progress in Food Science and Technology, Volume 1
Editor: Anthony J. Greco

ISBN: 978-1-61122-314-9
© 2012 Nova Science Publishers, Inc.

Chapter 14

WHEAT GRAIN IN DAIRY RATIONS: A MAJOR FEASIBLE COMMODITY OVERLOOKED

A. Nikkhah[*]*, F. Amiri, H. Khanaki and H. Amanlou*
Department of Animal Sciences, Faculty of Agriculture,
University of Zanjan, Zanjan, Iran

ABSTRACT

Wheat grain (WG) is in many regions less common than corn and barley grains as dietary starch sources for dairy cows, due to climatic, nutritional and economical constraints. Wheat starch and proteins usually possess high rumen degradation rates that, if used at high levels, could predispose cows to subacute rumen acidosis, laminitis and milk fat depression. Beliefs in such risks as well as competition for human uses of WG have restricted major WG use in dairy diets. Nevertheless, when and where is WG widely accessible, optimizing its dietary use and processing becomes economically and nutritionally critical. A paucity of in vivo data exists on feeding lactating cows different levels of differently processed WG. The objective was to determine ground WG (GW) level and particle size (PS) effects on blood metabolites, nutrient digestibility, and production. Eight midlactation cows (176 ± 8 days in milk; 554 ± 13 kg body weight (BW), 3.12 ± 0.14 body condition score (BCS); mean ± SE) were used in a 4 × 4 replicated Latin square design study with 21-d periods. Treatments were feeding either 20% or 10% WG, ground either finely or coarsely. Alfalfa hay-based total mixed rations with forage to concentrate ratio of 47.5:52.5 were offered individually at 0900, 1600 and 2300 h. Tail veins blood was sampled at 0 and 2 h relative to feeding. WG at 10% vs. 20% of diet DM tended to decrease blood BHBA (0.64 vs. 0.54 mmol/L), increased blood total proteins (8.28 vs. 8.46), and tended to increase blood albumin (3.72 vs. 3.83) levels. Dry matter intake increased (19.9 vs. 19.4 kg/d) when WG replaced half of dietary barley grain (i.e., 10% WG). Milk NE_L yield and the ratio of milk NE_L to NE_L intake, milk solids content and yield as well as urine pH were unaffected. Fecal pH tended to increase (6.9 vs. 6.7, P=0.10) by increasing WG from 10% to 20%. Total tract apparent DM digestibility was greater for coarse than for fine WG (70 vs. 65, P<0.01). Findings suggest feasible major uses of ground wheat grain in midlactation mixed rations.

---

[*] Correspondence: anikkha@yahoo.com or nikkhah@znu.ac.ir, Department of Animal Sciences, Faculty of Agriculture, University of Zanjan, Zanjan 313-45195 Iran

Keywords: Wheat grain, Midlacttaion, Holstein cow, Economics.

INTRODUCTION

Wheat grain (WG) is an invaluable nutrient source for humans and animals. Usually due to its fast rumen degradation, and enormous intake by humans, WG has less commonly been fed to ruminants compared with corn and barley grains. Subsequent to oats, WG possesses the highest starch and protein degradation extent and rate amongst cereals (Herrera-Saldana et al., 1990; Huntington, 1997). As such, higher WG levels would be more likely to cause subacute rumen acidosis (SARA) than would corn and even barley (BG) grains (Stone, 2004). However, such a SARA risk would exist rather shortly postfeeding (e.g., 44% vs. 60% starch degradability at 4 h; Toankova and Homolka, 2004), as WG and BG would probably be similar in degradation patterns later on postfeeding (Herrera-Saldana et al., 1990).

Therefore, as WG becomes more accessible and cheaper, its dietary use for lactating cows shall be feasible. To be well utilized along the gut (87-100% vs. 16-32%; Fulkerson and Michell, 1985), whole WG must be processed e.g., crushed or milled. Prepartal ground wheat grain (GW) has recently been shown to increase blood glucose and calcium, and improve milk production of periparturient cows and heifers (Amanlou et al., 2008; Nikkhah et al., 2010). Nevertheless, optimum inclusion rates in lactation diets in relation to processing methods and particle size of WG are not conclusively known. Recently, Doepel et al. (2009) found that steam-rolled WG may replace steam-rolled barley grain at 20% of diet DM in second-calf midlactation cows without compromising healthy rumen conditions and nutrient digestibility. Whilst is grinding an easy-to-access and widely available technique to process WG, optimal inclusion rates and particle size of GW in mixed rations are yet to be explored. We hypothesized that unique starch and protein (e.g., gluten) configuration of GW at controlled levels will benefit chronological patterns of rumen energy and protein fermentation, thereby enabling GW to replace BG without depressing milk fat, especially when and where is WG competitive in price and abundance. The objective was to determine effects of feeding a higher versus a lower level of either coarsely or finely ground WG on feed intake, blood metabolites, and production parameters of lactating Holstein cows.

MATERIALS AND METHODS

Cows, Diets, and Management

Eight midlactation Holstein dairy cows [176 ± 8 days in milk; 554 ± 13 kg body weight (BW), 3.12 ± 0.14 body condition score (BCS); mean ± SE] were randomly assigned to one of four treatments in a 4 × 4 replicated Latin square design with four 21-day periods. Each period had 14 days of adaptation and 7 days of sampling and data collection. Treatments included four isoenergetic and isonitrogenous diets containing either 1) coarsely or 2) finely ground wheat grain (GW) at either 1) 10% or 2) 20% of diet DM. Thus, GW level and particle size (PS) effects were compared in a 2 × 2 factorial arrangement. Alfalfa hay-based total mixed rations with forage to concentrate ratio of 47.5:52.5 were offered individually in equal portions 3 times daily at 0900, 1600 and 2300 h to permit about 5-10% orts. Diets were formulated using NRC (2001) program and are shown in Table 1. Cows were housed indoor in individual 3 × 4 m boxes equipped with concrete feed bunkers and automatic metal waterers. Cows were allowed 1 h of daily exercise three times daily prior to each milking. The experiment was conducted at the Dairy Facilities of the University of Zanjan (Zanjan, Iran) under the guidelines of the Iranian Council of Animal Care (1995).

Table 1. Dry matter-based dietary ingredients at 35% and 30% barley grain in the second study

Ingredient (DM-based)	Dietary use of barley grain	
	35%	30%
Alfalfa hay	42.8	42.8
Wheat straw	4.8	4.8
Ground barley grain	10.5	1.1
Ground wheat grain	9.7	19.1
Ground corn grain	5.8	5.8
Soybean meal	12.0	12.0
Whole cottonseed	6.8	6.8
Fish meal	3.6	3.6
Fat supplement	0.8	0.8
NaCl	0.42	0.42
Calcium carbonate	1.05	1.05
Sodium bicarbonate	1.05	1.05
Minerals and vitamins supplement[1]	0.63	0.63

[1]Contained 250000 IU/kg vitamin A, 50000 IU/kg vitamin D, 1500 IU/kg vitamin E, 2.25 g/kg manganese, 120 g/kg calcium, 7.7 g/kg zinc, 20 g/kg phosphorus, 20.5 g/kg magnesium, 186 g/kg sodium, 1.25 g/kg iron, 3 g/kg sulfur, 1.25 g/kg copper, 14 mg/kg cobalt, 56 mg/kg iodine, and 10 mg/kg selenium.

DM Intake, Feed Analyses, and Total Tract Nutrient Digestibility

The amount of TMR offered and orts were measured daily from d 15 to 21 of each period to calculate DMI for individual cows. Samples of TMR were taken daily for individual cows during the last 7 days of each period. Feed and ort samples were oven-dried at 100°C for 24-h, ground to pass through 1-mm screen using a Wiley mill (Arthur H. Thomas Co., Philadelphia) and stored at -20°C until analyzed for chemical composition. Feed samples were analyzed for CP (method 984.13; AOAC, 1990), NDF (Van Soest et al., 1991; using heat-stable α-amylase and sodium sulfite) and ADF (method 973.18; AOAC, 1990). Organic matter was determined by ashing feed and fecal samples for 8 h at 550°C. The acid insoluble ash (AIA; Nikkhah et al., 2004; Van Keulen and Young, 1977) was used as an internal marker to determine coefficient of total tract apparent digestibility. Particle size distribution of TMR samples was measured using Penn State Particle Separator (PSPS) (Lammers et al., 1996). The daily samples were composited to obtain 2 feed samples of both TMR and forage per period for the PSPS analysis. Physically effective NDF was estimated by multiplying dietary NDF percentage by 1) the proportion of DM retained on the 19.0- and 8.0-mm sieves of the PSPS (peNDF$_{>8}$, Lammers et al., 1996) or 2) the proportion of DM retained on the 19.0-, 8.0-, and 1.18-mm sieves of the PSPS (peNDF$_{>1.18}$; Mertens, 1997; Kononoff et al., 2003). After sieving, materials from each sieve were removed and dried at 55°C to determine DM content.

Table 2. Chemical composition of forages and diets (dry matter basis) in the second study

Forage		Diets (wheat grain%)[1]	
Nutrient	Alfalfa hay	10%	20%
DM %	88.5	92.5	93.5
CP %	13.5	18.5	18.6
NDF %	41.6	39.5	40.5
peNDF$_{>8}$[2]	-	18.4	18.0
peNDF$_{>1.18}$[2]	-	37.3	34.8
NE$_L$[3], Mcal/kg	1.2	1.60	1.60

[1]Dry matter based.

[2]NDF content of TMR or forage multiplied by their physical effective factor (pef) or the proportion of particles retained on the 19-, 8-, and 1.18-mm sieves of the PSPS (Mertens, 1997; Kononoff et al., 2003). 3NDF content of TMR or forage multiplied by its pef or the proportion of particles retained on the 19- and 8-mm sieves of the PSPS (Lammers et al., 1996).

[3]Estimated from NRC (2001).

Milk Production and Composition

Cows were milked three times daily at 0800, 1500 and 2200 h in a milking parlor. Milk weights were recorded during the 7-day data collection period. Milk production was also monitored daily during adaptation period to have under control possible unusual fluctuations in production. Before milking, cows were monitored for udder inflammation and presence of milk clots to ensure that they were not infected by mastitis. Milk samples were collected during 2 consecutive days from 6 milking times into plastic vials containing potassium dichromate. Samples in each plastic vials were analyzed separately for protein, fat, solid-non-fats and lactose using Milk-O-Scan (134 BN Foss Electric, Hillerd, Denmark).

Blood Sampling and Analysis

Blood samples from individual cows were taken from tail veins at 0 and 2 h relative to feeding on the last day of each period. Blood was collected into coagulant-containing tubes and centrifuged at 1000 g for 15 min to harvest plasma for storage at -20°C until analyzed for BHBA, albumin (Bromcresol Green method), globulin and glucose (GOD-PAR enzymatic method) concentrations using commercial kits (Pars Azmun Laboratory, Tehran, Iran). The absorbance was read using spectrophotometer (Perkin- Elmer, Colemen Instruments Division, Oak Brook, IL, USA).

Fecal and Urine Parameters

During the 4 days of sampling period, urine and fecal samples were collected twice in the morning and evening. Urine was collected by manual stimulation of the lower vulva. Urine samples in plastic vials were stirred well and analyzed for pH immediately after sampling using a portable pH meter (HI 8314 membrane pH meter, Hanna Instruments, Villafranca, Italy). Fecal samples were taken from rectum and stored at -20°C for later digestibility measurements (Nikkhah et al., 2004). To determine fecal pH, a portion of it was mixed with an equal portion of distilled water and stirred for sufficient uniformity and thus for representative pH values.

Statistical Analysis

Data were analyzed using the MIXED MODEL procedure (SAS Institute, 2003). The REML method was used to estimate least squares means, and the containment method was used to calculate denominator degrees of freedom. The fixed effects in the model included WG level, PS, and their interaction plus random effects of cow, period and residuals. For blood parameters, effects of time and its interactions with treatments were also considered fixed. Normality of distribution and homogeneity of variance for residuals were tested using PROC UNIVARIATE (SAS Institute, 2003). The P-values < 0.05 were declared as significant and those ≤ 0.10 were declared as trends.

RESULTS

Feed Intake, and Milk Properties

Ground wheat grain (GW) inclusion rate effect was significant on DMI and NE_L intake, so that cows fed 10% GW had greater DM and NE_L intakes than cows fed 20% GW (Table 3, $P < 0.01$). However, grinding extent (GE) and its interaction with GW inclusion rate did not affect feed intake. Treatments did not affect any of milk production and composition parameters, including feed efficiency, milk solids yield, milk energy content, and milk fat to protein ratio (Table 3, $P > 0.10$).

Nutrient Digestibility and Fecal and Urine pH

Cows fed coarser GW had greater apparent total tract DM digestibility ($P < 0.001$) and tended to have greater total tract NDF digestibility ($P = 0.09$) than cows fed finer GW (Table 4). Total tract NDF digestibility tended to be greater ($P = 0.07$) with 10% than with 20% GW in the diet (Table 4). Whilst urine pH was unaffected by treatments, fecal pH tended to be higher in cows fed coarse GW particles than in cows fed finer GW particles ($P = 0.10$, Table 4).

Table 3. Treatment effects on milk production and feed efficiency of cows fed either finely or coarsely ground wheat grain (W) at 20% or 10% of diet dry matter

	Coarse		Fine			P-value		
Wheat grain % (W)	20%	10%	20%	10%	SEM	GE	W	GE × W
DMI, kg/d	19.4	19.7	19.4	20.1	0.15	0.22	0.002	0.25
NE_L intake, Mcal/d	31.2	31.7	31.2	32.4	0.25	0.22	0.002	0.25
Milk yield, kg/d	25.1	25.4	25.5	25.1	0.31	0.83	0.99	0.28
4% FCM, kg/d	22.5	23.1	23.0	23.2	0.34	0.42	0.29	0.52
Milk energy[1], Mcal/kg	0.695	0.700	0.699	0.695	0.006	0.93	0.93	0.55
Milk energy yield[1], Mcal/d	17.30	17.69	17.68	17.53	0.29	0.70	0.68	0.36
Milk energy yield : DMI	0.90	0.90	0.92	0.89	0.02	0.72	0.20	0.12
Milk fat %	3.36	3.41	3.39	3.52	0.08	0.39	0.26	0.60
Fat yield, kg/d	0.83	0.86	0.86	0.87	0.02	0.35	0.18	0.76
Milk protein %	3.06	3.07	3.06	3.15	0.05	0.43	0.37	0.45
Protein yield, kg/d	0.77	0.78	0.78	0.79	0.01	0.35	0.41	0.98
Milk lactose %	5.55	5.54	5.58	5.50	0.06	0.85	0.43	0.60
Lactose yield, kg/d	1.39	1.40	1.42	1.39	0.03	0.75	0.86	0.54
Milk SNF %	9.71	9.77	9.74	9.65	0.07	0.56	0.83	0.29
SNF yield, kg/d	242.7	247.5	247.5	244.0	3.57	0.85	0.87	0.27
Total solids (TS) %	13.02	13.39	13.33	13.25	0.16	0.61	0.40	0.19
TS yield, kg/d	3.24	3.38	3.37	3.34	0.05	0.43	0.34	0.15
Milk fat % : protein %	1.10	1.11	1.11	1.11	0.01	0.81	0.41	0.77

[1]Calculated based on milk components (NRC, 2001).

Table 4. Treatment effects on blood metabolites and fecal and urine pH in cows fed either finely or coarsely ground wheat grain (W) at 20% or 10% of diet dry matter

	Coarse		Fine			P-value		
Wheat grain % (W)	20%	10%	20%	10%	SE	GE	W	GE × W
Glucose, mg/dL	51.6	51.1	51.1	50.8	1.7	0.81	0.84	0.95
Total protein, mg/dL	8.2	8.5	8.4	8.5	0.08	0.38	0.03	0.42
Albumin, g/L	3.75	3.83	3.68	3.82	0.05	0.46	0.06	0.60
Globulin, µg/L	4.46	4.62	4.67	4.64	0.09	0.51	0.25	0.34
BHBA, mol/L	0.64	0.53	0.64	0.55	0.05	0.83	0.06	0.81
Fecal pH	6.76	6.72	6.77	6.70	0.03	0.74	0.10	0.66
Urine pH	8.07	8.11	8.02	8.08	0.04	0.47	0.34	0.71
Apparent total tract digestibility %								
Dry matter	69.1	71.0	64.4	64.8	2.6	<0.001	0.41	0.57
NDF	32.0	34.8	31.9	32.1	0.78	0.09	0.07	0.11

Blood Metabolites

Blood albumin and total protein concentrations were greater (P = 0.03) and tended to be greater (P = 0.06) in cows fed 10% GW than in cows fed 20% GW (Table 4). Blood BHBA levels tended to increase with increased GW in the diet (P = 0.06). Neither wheat grain grinding extent (i.e., GW particle size) nor its interaction with GW dietary inclusion rate affected blood metabolites (Table 4).

DISCUSSION

This study provides unique applied insights into dietary use of ground wheat grain (GW) in relation to its particle size. Wheat grain is considered a nutritionally overlooked cereal for feeding dairy cows, as it has traditionally been considered a risk to rumen acidosis, feed intake and milk production. Such traditional over-carefulness in feeding GW to dairy cows has mainly been due to its competitive use by humans and its high starch and protein degradation rate in the rumen that could easily predispose cows to subacute rumen acidosis (SARA) (Stone, 2004; NRC, 2001). However, optimum dietary inclusion rate and processing type of WG has never been emphasized to avoid such potential challenges. The current data demonstrate that GW, regardless of its particle size, can feasibly be included in alfalfa hay based dairy ration at up to 20% of diet DM without affecting milk production. Even, with 20% compared with 10% GW in the diet, DMI was moderately increased, thus contradicting the belief that GW may unfavorably affect ration palatability. Recently, Doepel et al. (2009) found no effects on DMI of feeding up to 20% steam-rolled WG to dairy cows in barley silage based rations.

The present study is the first to find no independent and interactive effects of GW inclusion rate and its particle size on milk production and composition of lactating cows. This along with DMI data rule out a possibility of reduced diet palatability by GW (Morrison, 1935; Mathison, 1996). As a matter of fact, reasonable statistical power of the experiment and thus precise testing of the hypotheses were revealed by a significantly greater DMI when GW increased. It has been recently shown that GW can positively affect prepartal diet palatability for cows and heifers (Amanlou et al., 2008; Nikkhah et al., 2010).

Adequately high milk fat and solids contents as well as normal milk fat to protein ratios suggest no major treatment effects on normal rumen function and its conditions (e.g., pH and VFA concentrations) fluctuations (Mertens, 1997; Nikkhah et al., 2004). Most recent data (Nikkhah et al., 2010; Soltani et al., 2009) demonstrated that steam-rolling was no superior to fine grinding of barley grain at up to 30% barley grain – with usually high degradation rate - in the diet DM. In the present study, GW replaced either half or almost all of barley grain in alfalfa hay, soybean meal based rations, with slight use of corn grain (see Table 1). Fecal pH data suggest that under such dietary circumstances with about 19-20 kg DMI, including GW at 20% of diet DM did not compromise rumen and small intestinal starch assimilation, which could otherwise result in reduced milk fat due to compromised healthy rumen environment, and reduced fecal pH due to more extensive hindgut fermentation of undigested or partially assimilated starch and glucose (see Table 4). Similar urine pH suggests no or little impact of treatments on extracellular fluids anion-cation balance and acidity (Nikkhah et al., 2004). The lower total tract DM digestibility with finer compared to coarser GW particles could likely be due to increased digesta passage rate along the gut of particularly reticulo-rumen (NRC, 2001). Feeding lactating cows 20% steam-rolled WG instead of steam-rolled barley grain in barley silage-alfalfa hay based rations with about 17.1% CP and 33.2% NDF, Doepel et al. (2009) reported no major effects on total tract nutrient digestibility. Our results suggest that grinding compared to steam-rolling creates more responsive total tract digestive capacity, and thus, requires more accurate management to yield optimal outcomes. However, compartmental more and more quantitative interpretation of the total tract DM and NDF digestibility (as affected by particle size) warrants further research. In light of the present data and the literature (Falkerson and Michell, 1985; Faldet et al., 1989; Doepel et al., 2009), ground WG may also be feasibly included in dairy rations at reasonable levels should its price and availability be competitive to that of corn and barley grains. Nevertheless, for higher inclusion rates (e.g., > 20%) appropriate feeding frequency and systems must be exercised (NRC, 2001), as feeding too high GW levels can depress milk solids production (Faldet et al., 1989). The latter concern has in fact been such over-emphasized that optimum levels of WG in dairy diets have rarely been sought and determined. The finding that particle size of GW at 20% of diet DM is not affecting DMI and milk production encourages its commercial use.

Given that cows were in positive energy balance, increased blood BHBA with increased GW had most likely an alimentary origin (Reynolds, 2002; Nikkhah et al., 2008). As such, increased peripheral blood BHBA concentrations could indicate a more extensive rumen epithelial metabolism and output of BHBA (Nikkhah et al., 2008), possibly due to some relatively increased fermentation by 20% GW. The higher blood total protein concentrations with 10% rather than 20% GW suggests that even in lower producing cows under no major metabolic challenges, increasing GW in the diet affects hepatic synthesis or clearance of blood proteins, especially albumin. Albumin is required for peripheral nutrient (e.g., fatty acids and Ca) transfer. Lower blood albumin could be indicative of either reduced peripheral nutrient needs or depressed liver function in synthesizing sufficient albumin to cope with increased nutrient demands. Thus, definitive interpretation of blood protein data should be made cautiously. Overall, increased DMI and thus nitrogen by feeding 10% instead of 20% GW was consistent with increased blood concentrations of albumin and total proteins.

CONCLUSIONS

Including 10% and up to 20% of either coarsely or finely ground wheat grain (replacing dietary barley grain) in alfalfa hay based mixed rations of midlactation dairy cows well maintained milk production and composition. Cows on 10% vs. 20% ground wheat grain (DM based) had moderately improved energy intake, increased blood albumin and total proteins concentrations, a tendency for increased apparent total tract NDF digestibility, and reduced blood BHBA levels. Coarsely rather than finely ground wheat increased total tract DM digestibility, tended to increase NDF digestibility, and resulted in higher fecal pH. Wheat grain inclusion rate did not interact with its particle size on any of production, digestibility and blood parameters.

REFERENCES

Amanlou, H., Zahmatkesh, D, & Nikkhah, A. (2008). Wheat grain as a prepartal cereal choice to ease metabolic transition from gestation into lactation in Holstein cows. *Journal of Animal Physiology and Animal Nutrition* 92, 605-613.

AOAC. (1990). *Official methods of analysis*. 15th ed. Assoc. Offic. Anal. Chem., Arlington, V. A.

Doepel, L., Cox, A., & Hayirli, A. (2009). Effects of increasing amounts of dietary wheat on performance and ruminal fermentation of Holstein cows. *J. Dairy Sci.* 92, 3825-3832.

Faldet, M. A., Nalsen, T., Bush, L. J., & Adams. G. D. (1989). Utilization of wheat in complete rations for lactating cows. *J. Dairy Sci.* 72, 1243-1251.

Falkerson, W. J. & Michell, P. J. (1985). Production response to feeding wheat grain to milking cows. *Aust. J. Exp. Agric.* 25, 253-256.

Herrera-Saldana, R. E., Huber, J. T. & Poore, M. H. (1990). Dry Matter, crude protein, and starch degradability of five cereal grains. *J. Dairy Sci.* 73, 2386-2393.

Huntington, G. B. (1997). Starch utilization by ruminants: from basics to the bunk. *J. Anim. Sci.* 75, 852-867.

Iranian Council of Animal Care. (1995). *Guide to the Care and Use of Experimental Animals,* vol. 1. Isfahan University of Technology, Isfahan, Iran.

Kononoff, P. J., A. J. Heinrichs, and D. R. Buckmaster. 2003. Modification of the Penn State forage and total mixed ration particle separator and the effects of moisture content on its measurements. *J. Dairy Sci.* 86:1858–1863.

Lammers, B. P., D. R. Buckmaster, and A. J. Heinrichs. 1996. A simple method for the analysis of particle sizes of forage and total mixed rations. *J. Dairy Sci.* 79:922–928.

Mathison, G. W. (1996). Effects of processing on the utilization of grain by cattle. *Anim. Feed Sci. Technol.* 58, 113-125.

Mertens, D. R. (1997). Creating a system for meeting the fiber requirements of dairy cows. *J. Dairy Sci.* 80, 1463-1481.

Morrison, F. B. (1935). Chapter 4: Factors affecting the value of feeds. Pages 59–71 in *Feeds and Feeding*. The Morrison Publishing Co. Ithaca, NY.

National Research Council. (2001). *Nutrient Requirements of Dairy Cattle*. 7th rev. ed. National Academy Press, Washington, DC.

Nikkhah, A., Alikhani, M. & Amanlou, H. (2004). Effects of feeding ground or steam-flaked broom sorghum and ground barley on performance of dairy cows in midlactation. *J. Dairy Sci.* 87, 122-130.

Nikkhah, A., Furedi, C., Kennedy, A., Crow, G., & Plaizier, J. (2008). Effects of feed delivery time on feed intake, rumen fermentation, blood metabolites and productivity of lactating cows. *J. Dairy Sci.* 91, 1-12.

Nikkhah, A., Ehsanbakhsh, F., Amanlou, H. & Zahmatkesh, D. (2010). Wheat Grain a Unique Prepartal Choice for Transition Holstein Heifers. *In Grain Production*. Nova Science Publishers, Inc, NY, USA.

Nikkhah, A., Soltani, A., Sadri, H., Alikhani, M., Babaei, M., & Samie, A., & Ghorbani, G. R. (2010). *Optimizing barley grain use by dairy cows: a betterment of current perceptions. In Grain Production*. Nova Science Publishers, Inc, NY, USA.

O'Mara, F. P., Murphy, J. J. & Rath, M. (1997). The effect of replacing dietary beet pulp with wheat treated with sodium hydroxide, ground wheat, or ground corn in lactating cows. *J. Dairy Sci.* 80, 530-540.

Ørskov, E. R. (1986). Starch digestion and utilization in ruminants. *J. Anim Sci.* 63, 1624-1633.

Reynolds, C. K. (2002). Economics of visceral energy metabolism in ruminants: Toll keeping or internal revenue service? *J. Anim. Sci.* 80, E74-84E.

SAS User's Guide. (2003). Version 9.1. Edition. SAS Institute Inc., Cary, NC.

Soltani, A., Ghorbani, G. R., Alikhani, M., Samie, A.H., Nikkhah, A. (2009). Ground versus steam-rolled barley grain for lactating cows: A clarification into conventional beliefs. *J. Dairy Sci.* 92, 3299-3305.

Stone, W. C. (2004). Nutritional approaches to minimize subacute ruminal acidosis and laminitis in dairy cattle. *J. Dairy Sci.* 87, E13-26E.

In: Progress in Food Science and Technology, Volume 1 ISBN: 978-1-61122-314-9
Editor: Anthony J. Greco © 2012 Nova Science Publishers, Inc.

Chapter 15

FOOD DEMAND: COPING WITH ITS GROWTH

Anil Hira, Hayley Jones and James Morfopoulos
Simon Fraser University,
Burnaby, BC, Canada

ABSTRACT

This chapter will examine the growing concern about possible future crises in global food production, from sources such as population growth, climate change, and biofuels production. Some estimates indicate that global farm output might need to double to meet projected demand to avoid price increases (Tweeten and Thompson 2008, 21). We demonstrate that there is a large degree of latent productivity increase possible, well beyond what is needed to alleviate the aforementioned threats, if agricultural productivity in the developing world is improved. We then examine the various impediments to achieving such targets, both on an international and domestic level. We close with some lessons about the role of government institutions in both Brazil and Argentina in terms of major productivity improvements in agriculture over the past two decades.

INTRODUCTION

In 2007-8, a number of news media, international organizations, and activists declared a global food crisis. Unlike previous famine scares, which have been associated with particular countries, such as Ethiopia, going through a combination of drought and civil strife, this was the first case in recent memory of the declaration of a worldwide, long-term crisis that put all of us at risk. The FAO, among others, declared in 2007 that the sudden rise in food prices across the board, including double and triple digit percentage increases in the prices of wheat, soy, and rice put an additional 75 million people into hunger (BBC 2008). In short, this moment was depicted as a turning point for "food security," ie the assurance that there will be adequate supplies of food for the world.

WHY PETROLEUM PRICES ARE THE MOST LIKELY CAUSE

A wide range of causal factors have been posited in regard to the food crisis. These include population growth, expected to reach 9 billion by mid-century; desertification related to climate change; economic growth in India and China leading to greater consumption of meat; and biofuels (BBC 2008). Biofuels in particular were singled out as a policy-driven and therefore avoidable driver of the food crisis. The Jan. 27, 2008 issues of Time magazine featured a number of studies laying the blame on biofuels, which the title called "The Clean Energy Myth."

While all of these factors may play some role, besides a poor wheat harvest due to weather in Russia and Australia (Alston, Beddow, and Pardey 2009, 1) the most plausible explanation for the crisis and its decline after 2008 is the spike in petroleum prices during that period. Clearly, the role of transportation costs, and secondarily, the petroleum-based inputs into agriculture, ranging from fertilizer to running tractors, has been underestimated. Moreover, such price spikes are exacerbated by the cascading nature of financial speculation, which can exacerbate the reactions in the short-run (Cooke and Robles 2009). The following graphs shows the rise of food prices over the last 2 decades coincides with the price of petroleum prices.

Figure 1 shows the historical development of world oil prices. It is quite interesting to note that in the decades prior to this century, prices were on a par with those during the recent price spikes in the 1970s, 80s, and the last few years of the 2000s.

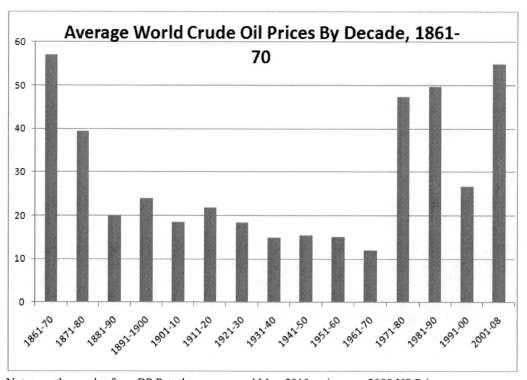

Notes: authors calcs from BP Petroluem, accessed Mar. 2010; prices are 2008 US Prices.

Figure 1. Historical World Oil Prices.

Figure 2 shows the annual average weekly spot price for petroleum exports clearly spikes in 2007-8, directly in line with the timing of the food crisis, and declines temporarily in 2009, with the onset of the current world recession.

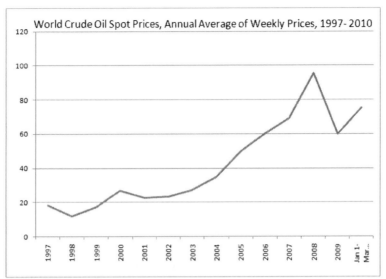

Source: Author calcs from USEIA stats, accessed March 2010, $/barrel.

Figure 2. Petroleum Prices' Upward Trend, Spiking in 2007-8.

If we now examine global food price behavior over the last 2 decades, we see that the spike in food prices coincides almost precisely with the spike in petroleum prices, as demonstrated in Figure 3.

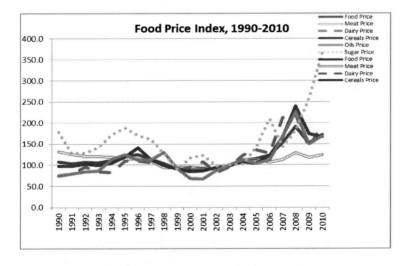

Notes: Stats Source FAO, accessed Mar. 2010, index based on 2002-4 – 100.

Figure 3. Food Prices Spike at the Same Time as Petroleum.

We do see a spike in the price of sugar in Figure 3, that continues through the early part of 2010. We might be tempted to suggest that this shows the responsibility of sugarcane-based biofuels, however, trade and production in sugarcane biofuels is extremely limited, with Brazil being the only significant producer at this time (Hira, forthcoming). More importantly, India, one of the largest producers of sugar in the world, had severe shortages related to drought from 2007-9. Unlike other crops, sugar has a longer harvesting period, thus the supply response may take a few years to kick in. Thus, we need to wait a few more years before guessing about the long-term directions of the sugar price, which likely will settle down considerably.

Figure 3 also allows us to examine the hypothesis that Chinese and Indian increases in meat consumption or climate change are causing the food crisis. While there may be some cause for long-term concern, our data suggest that only petroleum (and drought in the case of sugar) can explain the timing of the sudden spike and decline of the prices of food. While these factors should not be dismissed as having the potential to affect food prices in the long-term, there is no evidence that they are responsible for the most immediate crisis. The only plausible explanation for the decline in oil prices in 2009 is the worldwide recession, which meant a slump of demand.

While we have isolated petroleum as the immediate cause of the food crisis of 2007-8, we can not dismiss the fact that the rise of China and India, the eventual recovery from recession, and the lack of commensurate increases in oil supply mean that the long-term trend is almost assuredly upward, and with it, the rise in the price of food will resume unless serious adjustments are made to global production.

ARE RISING FOOD PRICES NECESSARILY BAD?

The food crisis creates an interesting paradox for development specialists. On the one hand, the study of development has including an ongoing concern for the terms of trade. Classic dependency theory suggests that "neo-colonial" relations exist between the North and the South, whereby the South is exploited in good part through the fact that it exports commodities in exchange for manufactures and advanced services. Thus, there is a problem in the sense that value and technology are leading over time to more sophisticated manufactures, while commodities remain the same. This is the basis of the Singer-Prebisch hypothesis that suggests that there is an overall decline in commodity prices over time vis a vis manufactures. On top of this, we add the fact that commodity prices are by nature more volatile. By nature we mean that agricultural goods are subject to weather, pestilences and other widely varying circumstances beyond the control of farmers. On top of this, agriculture tends to create a strongly concentrated ownership matched with very low level unskilled labour (eg pickers). With the growing capitalization and technology advances of agriculture, such as increasingly costly equipment, fertilizers, seeds, and the desirability of exporting to global markets, there is a general movement towards agribusiness, with a high concentration of land and capitalization, leading to marginalized and underpaid labour in the sector. In addition, it is considerably more costly to organize collectively in widespread farms then in a factory. As a result, rural areas tend to have much more patrimonial-type politics, as seen in the differences between northeast and southeastern Brazil, with the former being dominated by agriculture. It is no surprise, then, that rural areas of the South have the highest concentrations of poverty and marginalisaiton. Therefore, the general view going back to the 1960s was that the South needed to diversify away from agriculture in order to have a more stable economy that would provide greater employment, particularly considering the ongoing global phenomenon of increasing urbanization.

Thus, the increase in commodity prices towards new levels is not by any stretch of the imagination, necessarily a bad thing. An increase in commodity prices could potentially bolster the Southern economies with higher revenue streams while they continue to diversify, provide additional boosts for employment, productivity, and increasing processing in rural areas, ease pressures of migration and on urban services, and reduce urban bias in political decisions. Moreover, increases in prices should naturally lead to increases in supply, which will in turn bring down prices again, unless the overall price of inputs, namely petroleum and by-products, continues to increase. If other factors, such as climate change, come to fruition, the pressure will be even greater on food prices over the coming decades.

Since an increase input prices is likely to be the case, the only solution for reducing food prices is to increase productivity. We tackle that question in the rest of the article, adding in concerns about the ways that productivity increases can also take into account equity and sustainability.

AGRICULTURAL PRODUCTIVITY- WHY SPREADING THE GREEN REVOLUTION MAY NOT BE ENOUGH

Introduction- A New Era for Food Prices?

There is a strong consensus among agricultural economists that agricultural productivity has improved steadily over the course of the 20th century. This includes all food products, ranging from rice to soy to corn, all of which declined in price by at least 1.5% per year from 1950-2008 (Alston, Beddow, and Pardey 2009, 5).

However, most economists also see a deceleration of the price decline in more recent years, some spotting the trend as starting around 1990 (Alston, Beddow, and Pardey 2009, 6 & 11). Global yield growth has declined dramatically from the 1980s as compared to previous decades across all food staples, with the exception of corn (Cline 2007). Annual yield growth dropped from 3.2% per year in 1960 to 1.5% in 2000 (FAO 2009, 2). Analysts expect a change in the types of crops grown over time as consumers in the fast growing countries of the South (India and China) change consumption habits towards more meat and dairy and away from cereals (wheat and rice). Increasing use of biofuels may eventually also increase demand for feedstock crops, such as sugar and oil seed crops. Meanwhile, there is growing concern for the exhaustion of marine stocks (OECD-FAO 2009, 23-4).

Figure 4 below demonstrates the slow down in agricultural productivity in recent years. It also shows that most of the productivity gains of the past 2 decades centred around Asia, South America, and Africa. Observers note the particularly important contributions of China, India, and Brazil in increasing their productivity.

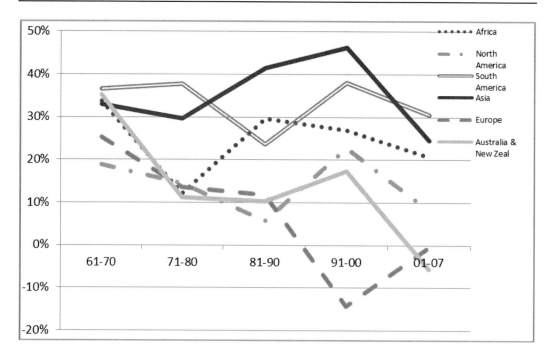

Notes: Author calcs based on FAOStat, accessed March 2010, based on food production indices.

Figure 4. Food Production Rate of Increase by Decade, 1961-2007.

The Green Revolution beginning in the 1960s was behind much of the increase in productivity, but it may have reached some limitations. The Green Revolution depended upon increasing conversion of new cropland, and increasing use of irrigation and fertilizer. It focused on increasing the yields of cereal crops.

While there are some possibilities around the world for expanding cropland, most economists believe that increases in global food production must come from productivity increases. Not only is the potential for expansion limited, but expansion tends to come at a higher cost, for example requiring new infrastructure and roads, and with greater environmental destruction as conversion to cropland may entail deforestation and other mechanisms resulting in reduced carbon sequestration. In addition, increasing urbanization is bound to further reduce natural areas and increase the price of land (Tweeten and Thompson 2008).

Though Figure 4 shows promise in terms of increases in productivity globally, the fact remains that levels of productivity vary considerably from region to region. Most analyses conclude that the most gains are to be found in sub-Saharan Africa (SSA) and parts of Latin America. Irrigation is a major bottleneck. Analysts report that irrigation from river or groundwater is being used at unsustainable levels in large parts of China, South Asia, the Middle East, and North Africa. In SSA, only 4% of the cultivated land was irrigated, and there is a severe lack of basic infrastructure, including roads and institutional support for further expansion Tweeten and Thompson 2008, 6) According to the latest figures from FAOStat (accessed Mar. 2010), the % of irrigated land considering the total potential that could be irrigated varies from 99% for South Africa, 90% for Jordan, and 84% for China to 2.2% for Angola, 9.8% for Brazil, and 25% for Argentina. Meanwhile, population growth is expected to centre in sub-Saharan Africa, where the FAO estimates 2.1% growth. As it states, "By 2030, every third person added to the world's population will be a sub-Saharan African. By 2050, this will rise to every second person (FAO 2002, 16).

A similar concern arises in regard to the ability to increase fertilizer use. Many fertilizers are tied to the petroleum industry. The price of the most important component of fertilizer, nitrogen, is widely expected to rise as yields decline and increasing demands are placed on agriculture. Moreover, there is growing concern about the effects of fertilizer use on the environment.

Finally, the Green Revolution also raised serious questions about equity. Irrigation systems were often hijacked or dominated by larger producers who, over time, were able to consolidate land through their relative increases in income. These trends went hand-in-hand with the increasing capitalization and globalization of agriculture to promote agribusiness replacing family farming. Indeed, the number of undernourished people in developing countries remained at 777 million, or 1/6 of the world's population in 1999 (FAO 2002, 15).

What is Behind Agricultural Productivity

The keys to agricultural productivity are no secret. Large efforts have been made over the past 2 centuries in the North towards agricultural research and development and extension services. In fact, the US higher education system was set up with this motivation in mind. The development of new fertilizers, better crops, new and better equipment, and, most recently, the use of genetic modification and markers have all increased yields and decreased the amount of labour needed per yield. Agricultural R&D has estimated rates of return of between 30 and 75% (FAO 2009, 2).

If R&D and modern techniques of agriculture are already widespread in the North, with considerably higher yields, then obviously spreading such techniques to the developing world is the only way to improve global yields. In general, Southern countries have been increasing efforts over the last decade, but high income countries still accounted for 44.3% of global public agricultural R&D in 2000, with Asia-Pacific countries accounting for another 32.7%. Asia-Pacific increased their share in 1981-2000 from 20-32.7, and the Middle East and North Africa from 5-6, but Latin America and the Caribbean (12.5-10.7) and Sub-Saharan Africa (7.9-6.3) decreased over the same period. In fact, agricultural R&D is concentrated in a few countries. In 2000, China accounted for 13.7%, India, 8.1%, Brazil 4.4%, Japan 7.2%, and the U.S. 16.6%, a combined total of half of all spending! Even in the North, spending has slowed down, with an accompanying greater reliance on the private sector. In the South, the move to private sector R&D has not happened, with the public sector still dominating, particularly in Latin America and sub-Saharan Africa (Alston, Beddow, and Pardey 2009, 15-16 and Pardey, Beintema, Dehmer, and Wood 2006, 4 & 6-7). The move to the private sector as a source for new global R&D will evidently be oriented towards those who can pay, namely farmers in the North. Yet, R&D needs extension and modification for local conditions to be effective in agriculture. In short, the areas where the world has the easiest gains from agricultural productivity are those most neglected.

Ironically, the long-term volatility and declining terms of trade for agriculture have led to disincentives for increased investment in agricultural productivity in the South. Therefore, perhaps the most important step to accelerating productivity gains would be to reduce agricultural subsidies and protectionism in the North. Such subsidies costs Northern taxpayers and consumers billions of dollars, make a mockery of their claims that the South should follow market principles, and generally reduce global efficiency. Northern protectionism takes the form of tariff and non-tariff barriers (such as US quotas on sugar imports), export subsidies, price supports, and indirect subsidization of the sector (such as tax breaks for the oil sector). In 2000, such subsidies amounted to $327 billion. The EU alone spent $5.8 billion in export subsidies in 1998 (FAO 2002, 27). Such Northern taxpayer subsidization of large agribusiness is not only ethically questionable, but counter-productive as we have seen. Yet, as the paralysis of the Doha global trading round demonstrates, Northern farmers are very well-organized, creating a strong lobby group that prevents progress on the issue.

A third set of problems revolves around institutions in the South. There is a raft of well-known problems that prevent markets from functioning. These include weak or non-existent infrastructure, lack of property rights, and a lack of information. The unavailability of finance dooms all but the largest producers to an inability to access the obvious sources of productivity improvements, such as mechanization, improved seeds, etc. Concerns about intellectual property rights have further created a hesitation by Northern private firms to invest in the South or develop or transfer appropriate technology there (James, Pardey, and Alston 2008, 3).

Even if agricultural productivity increases through R&D, trade liberalization, and institution-building, we still need to address how those gains will be distributed. The current trends towards large agribusiness do not portend well for the equally important issues of equity and sustainability. For example, organic farming requires considerably more labour-intensive efforts as the benefits of fertilizer and pesticide are eliminated. Such a shift suggests a move back towards lower economies of scale in agriculture may be commensurate with an improvement in productivity and a healthier food and environment.

HOW TO INCREASE AGRICULTURAL PRODUCTIVITY IN THE DEVELOPING WORLD

Part of the problem with raising agricultural productivity, while improving equity and sustainability in the developing world, is that the structures of agricultural production have developed in unproductive fashion since colonialism and been exacerbated by the Green Revolution. In most developing countries, the agrarian sector is divided into a healthy agribusiness sector that receives subsidies and government supports for the exports it provides, and shrinking subsistence-based sector that is the centre of poverty. The desperately poor, then, are responsible for deforestation, which is used for lack of affordable land and firewood, and the lack of access to food and mechanisms of increasing productivity at reasonable cost. Knox, et. al (2008, 258) report that between 1940-65 half the credit in Peru went to the wealthiest cotton producers. In Mexico in 1970, a quarter of the farms over 5 hectares had mechanization, while smaller ones had a rate of only 4.3%.

Favouring agribusiness in the developing world often makes sense as the strength of the urban poor and large landowners has been a winning formula in Southern politics. During the 1950s-60s, state agricultural policies took on the form of state promotional institutions, such as marketing boards and new agencies, to promote Green Revolution techniques. Under neoliberalism from the 1980s, the fiscal crisis and prevailing wisdom pushed the promotion towards agribusiness exports. Common sense critiques that markets are more efficient and stable and that previous institutions such as marketing boards are corrupt and inefficient paint the previous period with a broad and inaccurate brush, as country and institutional experience varied considerably (Knox, Agnew, and McCarthy 2008, 261). In fact, a more accurate understanding of the state interventionist period for both agriculture and manufacturing is that the interventions themselves set up successful exporting businesses (Hira 2007). The Green Revolution undoubtedly led to vast improvements in productivity in Latin America, China, and South Asia, reducing the specter of massive starvation related to population growth in those areas (Knox, Agnew, and McCarthy 2008, 275).

The success of large-scale agribusiness exports also takes wind out of the sails of the argument that land reform is necessary to improve agricultural productivity. However, there are other problems with the nature of the agribusiness sector in the South. Much of the agribusiness sector is tied to multinational corporations, both for inputs and marketing channels. This reduces the possibility for national linkages and the value-added returns from selling the final product. Moreover, Southern countries can find themselves in the situation of being food importers or even suffering foods shortages at the same time that they are exporting agricultural products! For example, it is a well-known fact that 4 coffee roasters dominate the bulk coffee production chain, using oligopsony power to capture much of the returns on coffee. The oligopolistic nature of agricultural markets refutes the claim that markets exist. Two companies control 65% and 44% of the world seed markets for maize and soya. Six corporations control 75-80% of global pesticides. Five companies control over 80% of the banana trade, and the 30 largest food retailing corporations control about 1/3 of global grocery sales (UNCTAD 2008, 9). Collusion under such circumstances of vertical production control is not only expected but inevitable. Absent the possibility of new international norms requiring alternative trading channels such as fair trade, it is hard to foresee a change in the overall trading and investment relations of agricultural production in the global economy (Hira and Ferrie 2006). The forms of capital-intensive agribusiness that have taken over the South are tied to deforestation, rural poverty and deprivation (as land is increasing concentrated and mechanized), concentration of income and reduction of social mobility, and a loss of national control (Knox, Agnew, and McCarthy2008, 273). A recent spate of stories regarding Chinese, Gulf State, and Indian purchases of large tracts of land in Africa may eventually bring a nationalist backlash if local owners are agribusiness are pushed out. Moreover, such trends will continue to marginalize family farming and increase the push towards subsistence farming among dispossessed populations.

Yet, subsistence farming evidently offers few possibilities for social mobility, as children are needed for labour-intensive techniques, thus by geography and necessity, access to education is limited. Given the natural cycle of weather and pestilence, a classic cycle of indebtedness takes hold, as subsistence farmers take out loans to get through bad years. The stagnation of large parts of Southern populations has been so far ignored, but it has real implications beyond productivity. Rural deprivation is the source a host of increasing ills in the South and globally, including environmental degradation, migration to cities and abroad, soaring crime rates, drug production and trafficking, terrorism, and civil unrest and guerrilla wars.

New techniques offer promise in terms of resolving the multiple strands of the agricultural problematique. As incomes have grown in the North, the market for organics has as well. In other words, as incomes increase, the emphasis shifts from quantity of food production to quality. If consumers can be sold on the idea that sustainable and equitable processes of farming are to their benefit, there could be another revolution in agriculture.

Besides organics, there are several other emerging ideas for increasing sustainable agriculture. One is no-till/conservation agriculture, which attempts to reduce soil erosion through reduced tillage and plant cover and use crop rotation, with an estimated 20-50% improvement in yield. Integrated pest management aims to reduce the use of pesticides through a variety of techniques, including introducing pest-resistant varieties, crop rotations, new irrigation and fertilizer usage techniques, the use of bio-insecticides and traps, and more strategic use of chemical pesticides (FAO 2002, 54).

Despite these recent pushes for sustainable and organic production methods, these products may directly conflict with productivity gains made in developing countries over the past four decades. The mechanization of agriculture employs less workers but produces more food, while organic and sustainable production often produces less food, though of higher quality, but employs more workers. Thus, with the global population expanding, sustainable and organic production increases would decrease the amount of available food, which, coupled with increasing demand would suggest that food prices would be driven higher, a situation which would be good for developing countries' fiscal accounts, but not so good for their domestic consumers. This method will decrease agriculture as a percentage of GDP, but improve prices for the domestic consumer. Despite massive increases in productivity over the course of the previous two decades, particularly as a result of the expanded use of mechanization and technological inputs, the gains have not all been positive; small farmers have founded themselves increasingly indebted and agricultural policy has hardly been synonymous with sustainable development (Lattuada & Estrada 2001). This would suggest that neither route provides a clear solution and that combinations will undoubtedly be adapted to best suit the domestic and fiscal needs of developing countries. Tailored domestic policy must be applied to agriculture in developing countries in order to maximize their fiscal and social benefits, using both sustainable and organic, and highly technological production.

EXAMPLES OF INCREASING AGRICULTURAL PRODUCTIVITY IN THE DEVELOPING WORLD- THE CASES OF BRAZIL AND ARGENTINA

Introduction

Agriculture remains a major strength of both the Brazilian and Argentine economies, despite dramatic increases in manufacturing and technology industries. The development of the agricultural sector in both countries, particularly with regard to productivity, has been hamstrung by currency fluctuation, external debt, inflation, unfavorable interest rates, infrastructure, and a generally poor investment climate within the agriculture industry (Schnepf, Dohlman, Bolling, 2001; Economist Intelligence Unit, 2008). Despite these domestic hindrances, agriculture has remained a key element in economic growth and stability, underwriting both economies against fluctuations in other fledgling industries (Economist Intelligence Unit, 2008). For example, in Brazil agriculture makes up 30 percent of the country's GDP and accounts for 35 percent of unemployment (Chaddad & Jank 2006). However problems have been further compounded by complex and restrictive international and domestic tariffs, which have led to reduced incentive and ability to increase agricultural productivity (Schnepf, Dohlman, Bolling, 2001). Both countries have developed a variety of policy initiatives aimed at increasing agricultural productivity, and in this section we review both the Brazilian and Argentine cases and their experiences with increasing domestic agricultural productivity and the barriers that remain. Despite these initiatives, both have struggled to move up the value-added chain, which remains one of their principal challenges.

Brazil

Brazil has a long history in the international public eye of clearing vast swaths of Amazonian forest in its seemingly ruthless expansion of agriculture and timber. The reality is very different. Brazil is a leading world producer and exporter of a wide variety of agricultural products such as organics, soybeans, sugar, beef, and chicken despite the fact that its farmlands are underexploited and underdeveloped (Economist Intelligence Unit, 2008). Despite its agricultural dominance, much of Brazil's existing and cleared farmland remains either entirely undeveloped or poorly utilized, leaving Brazil with plenty of room for continued productivity increases and expansion. Productivity has been a focus of government policies since the 1970s and Brazil's agricultural sector has become a major contender in world markets in addition to generating a large agricultural surplus (Chaddad & Jank 2006).

The country has seen huge increases in agricultural production since it began to focus on boosting agricultural productivity in the 1970s, which have led to expanded export markets, and to agriculture retaining a position of prominence in the Brazilian economy despite the rise of the technology and service sectors. With a focus on increasing agricultural productivity, agricultural exports from Brazil increased from 13 billion to 32 billion from 1990 to 2005, making Brazil the world's third largest agricultural exporter after the United States and the European Union (Chaddad & Jank 2006). Indeed, since the 1970s Brazil has continued to seen a steady rise in total factor productivity in its agricultural sector, surpassing that of developed countries such as the United States (Chaddad & Jank 2006). This rise in productivity and efficiency came from both technological advancements stemming from research and support policies, as well as the basic evolution of Brazilian agriculture, namely, the discovery of new productive regions, the optimization of land use in new regions, increasing investment and orientation towards exports, and, behind all this, the crucial role of the government R&D and extension agency, EMBRAPA.

Brazil's highly competitive export oriented agricultural production has been developed through a combination of investments in domestic agricultural production, somewhat more available credit, a more diverse range of products, and a shift in government policy (Chaddad & Jank 2006; Helfand & Castro de Rezende 2001). The initial growth in productivity was led by increased investment in domestic agricultural production, primarily throughout the 1970s and 1980s. During this period the agricultural sector was heavily taxed to subsidize other industries and urban development, leading to direct export taxes and higher prices on agricultural supplies such as a fertilizers and machinery and making the sector less internationally competitive due to high prices (Schnepf, Dohlman, Bolling, 2001).

The late 1980s, with the end of the military dictatorship, saw a shift from taxing exports to ignoring them all together and a refocusing of the Brazilian economy on domestic production and supply. This led to greater resources being poured into research and development in support of domestic food crops and the removal of any remaining support for exports. By the early 1990s, export tariffs reached an average of 30 percent, with revenue partially being poured into domestic agricultural solutions (Schnepf, Dohlman, Bolling, 2001). Despite severe limitations on export growth, this period strengthened domestic Brazilian producers, greatly increased agricultural efficiency, and led to a stream of agricultural research focused on Brazilian needs. This was followed by a dramatic change in government policies in the 1990s and 2000s which, coupled with tropical agricultural research, has made Brazil an international agricultural powerhouse (Helfand & Castro de Rezende 2001).

One of the principal drivers of increased productivity in Brazil has been the adoption of 'Zero Tillage' technology over the past five decades. In the 1970s zero tillage was a non-existent technology in Brazilian agriculture, yet by 2000 it was used on 14 million ha of farmland (Ekboir 2003). Zero tillage is an example of an agricultural technology that is not at odds with sustainability policies, as it requires sustainable adaptation to local conditions to function efficiently. The technology has been reliably shown to increase the life of agricultural equipment, increase planting and harvesting efficiency, reduce costs, and marginally increase total yield (Ekboir 2003). While zero tillage has not had the impact on agricultural productivity that fertilizers and other internationally used technologies have, its success in Brazil in the efficiency of its results is worth noting. However, unlike many innovations in agricultural production in Brazil, zero tillage was not largely researched and disseminated by national public research and extension institutes, which played a rather minimal role until fairly widespread adoption of the practice had already occurred (Ekboir 2003). At the same time, it is clear that there is an important role for public research and extension services in maintaining the viability of zero tillage methods, which, given their high dependence on a single herbicide, require a sustained research effort in order to maintain or increase the methods' sustainability (Ekboir 2003).

The last two decades have seen a distinct shift in Brazilian government policy with the Brazilian state focusing on reigning in rampant subsidies and tariffs to create a more export friendly economy. These policies have focused on trade liberalization specifically, deregulation, credit reform, and tariff reduction. The liberalization program increased access to high-quality inputs and exposed domestic producers, who had been benefiting from a long stint of research and support funding but limited export possibilities, to international competition (Helfand & Castro de Rezende 2004). Thus domestic producers, building on that research and support period and newly hardened by international competition, surged forward in global agricultural production, leading the sector to outperform both service and industrial sectors in the 1990s.

Brazil's EMBRAPA (Brazilian Agency for Research on Agriculture and Animal Husbandry) has continued to be a crucial factor in the expansion of new techniques in crop and livestock production. This combination of policies and government-directed support has led to Brazil's dominance and growth in international markets. Perhaps the greatest failure of these policies has been a decrease in available credit, despite policies aimed to counter precisely that, which, combined with high interest rates, has dramatically slowed increases in productivity through new technologies (Schnepf, Dohlman, Bolling, 2001). Rapid economic liberalization after a prolonged period of intensive domestic support and protection, has created strong, internationally competitive domestic producers in Brazil, while leaving a legacy of government policy support and extension programs which are still growing today

(Chaddad & Jank 2006). Furthermore, widespread economic liberalization across the economy beginning in the 1990s has led to a huge influx of foreign direct investment in agriculture, much of which through transnational companies as well as some regionally-based companies (CEPAL 2005). The increasing presence of multinational companies and their expansion into marketing and distribution activities has also led to increasing standardization of products and exports (CEPAL 2005).

Brazil currently has approximately 330 million hectares of farmland, much of which much is left completely unused due to government policies on agricultural development, and the vast majority of land controlled by only a few companies (Economist Intelligence Unit, 2008) despite massive increases in the area of cultivated land in Brazil since the 1990s (CEPAL 2005). While there have been policy initiatives in place since the early 1990s aimed at allowing more efficient allocation and use of agricultural land, they have been largely ineffective in their stated goals, though they have helped strengthen economic stability. Despite its global competitiveness, domestically, Brazilian agricultural policies are still focused on export and price controls, largely remnants of the 1980s, lowering agricultural commodity prices and reducing incentives for increased production and export. In addition, Brazil continues to suffer from poor linkages between government institutions and agricultural producers, which have caused added delays and costs in productivity increases and dissemination of research and technologies (Schnepf, Dohlman, Bolling, 2001). Brazil could greatly increase its agricultural productivity both through increases in agricultural efficiency and better use of arable land.

Argentina

Argentina has long been dependant on its agricultural sector, and it still very much is today, though it has undergone a radical transformation in the past three decades (Trigo et. al. 2009). Though important, agriculture in Argentina only accounted for 10 percent of GDP for the majority of the past thirty years, and it has only been recently, with the decline of the once internationally competitive service sector and the rise of international commodity prices (namely soybeans and beef), that it has regained ground (Economist Intelligence Unit, 2008). Surging food prices revitalized Argentine agriculture and today, through export tariffs, it is one of the principal sources of revenue for the beleaguered Argentine state. Most recently these same tariffs have hindered the growth of Argentine agriculture under prosperous conditions and stifled productivity and investment (Economist Intelligence Unit, 2008), causing strikes and threatening political turmoil. These domestic political conflicts have brought to light the importance of the agriculture sector to the Argentine economy and state. Removing these tariffs and quotas, as well a revision of import and value-added taxes to eliminate distortions that could work against the development of local production, will be a key element in maintaining and revitalizing Argentinean agricultural growth (Bisang et. al. 2009).

Argentine agriculture took major leaps forward during the 1990s, with rapid conversion of cultivated lands to several export crops (CEPAL 2005) and, more importantly, the introduction of imported agricultural technology and production methods, including the dramatic expansion of the use of genetically modified seeds and chemical fertilizers and the implementation of zero tillage methods of production (Lattuada & Estrada 2001; Trigo et. al. 2009). This is distinctly different from the Brazilian case, where agricultural extension and productivity were primarily addressed through domestic channels and protection, though much earlier. Similarly to the Brazilian case, however, the recent rise in commodity prices lifted profits and encouraged investment in Argentine agriculture. Despite the fact that the agricultural sector remains only approximately 10 percent of GDP, it accounts for 69 percent of all exports, with soy alone accounting for over 26 percent of exports and cereals accounting for 9 percent (Economist Intelligence Unit 2008). As a result of this agricultural export dependence, high agriculture commodity prices have propped up the Argentine economy in the post 2002 crisis period.

The past three decades have seen the rapid growth of cereals and soy production led primarily by zero tillage and the introduction of genetically modified techniques (Trigo et. al. 2009; Lattuada & Estrada 2001). These advancements moved Argentina on to the global agricultural stage, principally through soy production, and cemented agriculture as a major revenue source for the Argentine state. Zero tillage, in contrast to Brazil, was widely spread through public agricultural extension policies spearheaded by the country's major agricultural R&D agency, INTA (Naitonal Institute of Agricultural Technology). The combination of government policy, researchers, corporations, and farmers working for the expansion of zero tillage methods and the use of genetically modified seeds built a robust agricultural sector (Trigo et. al. 2009). Genetically modified crops have become a staple for Argentine farmers and, while some smaller farmers have struggled with them, they have led to a vast increase in overall productivity and today make up nearly 100 percent of Argentine soybean crops (Reboratti 2004; Trigo et. al. 2009). Both genetically modified crops and zero tillage policies were propped up by large injections of capital from the state in the 1990s and these technologies accounted for over 40 percent of agricultural growth during this period (Reca & Parellada 2001).

In addition to the implementation of zero tillage and genetic technology, the Argentine agricultural sector also experienced productivity gains through the significant expansion of the use of chemical fertilizers (Reca & Parellada 2001). The addition of chemical fertilizers required further imports and drove up the cost of agricultural production, raising the barriers to entry for small farmers. In response the Argentine state implemented more than 50 national projects offering different types of agricultural credit targeted at small farmers (Lattuada & Estrada 2001). Further credit was provided by the Inter-American Development Bank, along with supplementary informational services, technical support, and research and development. Nevertheless, these programs, both public and private, were limited by several factors, including ineffective financing systems and lack of access to programs for the smallest farmers. Thus, despite the considerable investments and government and private programs, a major disconnect remained between policy and action that failed to prevent the negative social impact of unstable macroeconomic conditions on the agricultural sector (Lattuada & Estrada 2001). Despite problems, this combination of fiscal and technological extension supported massive increases in agricultural productivity among both small and medium sized farmers.

Argentina continues to invest heavily in agricultural research, with research and development spending more than doubling from 2002 to 2006, largely due to growth at INTA (Stads & Beintema 2009). Argentine productivity increases have come at a price. Various factors, including increasing mechanization, increasing debt loads, and the growing use of technological inputs have led many Argentine small farmers into bankruptcy (Lattuada & Estrada 2001). As small farmers have fallen away, the Argentine state, resurging in recent years, has relied increasingly on large agribusiness for revenue via export taxes. In addition, unregulated agricultural practices have resulted in extensive soil degradation, damaging many Argentine agricultural regions (Economist Intelligence Unit, 2008). Despite these negative spillovers, Bisang et. al. (2009) highlights that biotechnology appears to be tool to addresses continuing demand for agricultural products while coping with problems with factors of production. As productivity has increased, revenue has not always followed suit, and increased agricultural yield in Argentina has been.

Conclusion

The Argentine and Brazilian cases hold important lessons for developing agricultural producers as well as policy makers abroad. Both countries, though Brazil in particular, have made large productivity gains in recent years which have increased their agricultural yield and led to increased exports. Much of these productivity increases have been the result of a confluence of factors, namely considerable public and private investment in research and extension services; rapid expansion of agricultural technologies, such as genetically modified seeds and chemical fertilizers; and the adoption of new, more environmentally sustainable, production methods. Questions remain, however, as to the social sustainability of such methods. Moreover, lower agricultural prices as a result of increased productivity would undoubtedly have a devastating effect on these countries' economies. In both cases, a halving of agricultural prices would reduce their respective GDPs by at least 5 percent. Thus, while there is clear patterns of methods for increasing agricultural productivity emerge from the Brazilian and Argentine cases, their overall viability in terms of environmental and social sustainability are far from clear.

Given the current heavy reliance on initial phases of production, relatively limited diversification of market outlets, and lack of local capital in all stages of production, policies must evolve quickly to capture new markets (Bisang et. al. 2009). Both Argentina and Brazil will need to focus on continuing to move up global value chains to maintain and build upon competitive advantages already in place, increase and ensure a sustainable process of diffusion of technologies used to add value, and increase the efficiency of distribution and logistics to shorten the distance between all parties involved. Public policy will need to focus on strengthening innovation at the base of the production pyramid of agribusiness, developing and strengthening local capital and its position in global value chains to increase the presence of local firms in international and domestic agro-industry, and developing strategies for negotiation with multinationals to make production processes more local (Bisang et. al. 2009).

CONCLUSION

This chapter examined the sources of the recent food shortage and why supply and demand pressures, particularly related to petroleum prices, are likely to continue to be concerns over the long-term. We then looked at why increasing agricultural productivity is the key to increasing global supply of food. However, there are a number of obstacles to improving agricultural productivity, all of which pose quite significant challenges. Huge Northern subsidies are the greatest impediment to reducing global poverty, which is concentrated in the rural areas of the South. Inequality in land ownership in the South and lack of clear property rights mean that productivity increases tend to be towards exports, ie Northern consumers, with inadequate attention to domestic food markets. The end result is a bifurcation of the agricultural markets, with small and family farmers increasingly marginalized and unable to compete and large agribusinesses operating in global markets dominated by large multinationals. Lastly, we see that the incapacity of Southern states to engage in agricultural R&D, to regulate markets, and to provide extension and other services, is a major policy failure not just for them but for the world.

Through the cases of Brazil and Argentina, we see rare examples of the success of Southern institutions first in promoting Green Revolution techniques and next in scaling up to global export markets. We note the crucial role of government institutions in these successes, leading to a thriving source of revenue and growth with important spillover effects throughout the economy. However, we also see that such improvements have not addressed the problems of sustainability and equity. We suggest that further institutional reforms are needed, as well as a recognition of market opportunities for improving food quality through organic and sustainable farming. The latter provide an opportunity for a new supply market that, in requiring in its labour intensiveness and limited use of costly inputs, offers another way to thrive through helping small and family farmers. Such transformation requires further investment and transformation into government support institutions. Thus, we offer a dual market solution- a cheap food one represented by Argentina and Brazil's success, and a high quality organic and sustainable one that has yet to be harnessed.

REFERENCES

Alston, Julian M., Jason M. Beddow, and Philip G. Pardey. 2009. Mendel versus Malthus: Research, Producivity and Food Prices in the Long-Run. Revised- Sept. Staff Paper P09-1. University of Minnesota.

BBC News. 2008. Q&A: World Food Prices. Oct. 15. Found at news.bbc.co.uk

Bisang, Roberto, Guillermo Anlló, Mercedes Campi and Ignacio Albornoz. (2009). "Cadenas de Valor en la Agroindustria." In Bernardo Kosacoff and Ruben Mercado (eds.). *La Argentina ante la nueva internacionalización de la producción: Crisis y oportunidades.* CEPAL.

CEPAL. (2005). *El Nuevo Patrón de Desarrollo de la Agricultural en América Latina y el Caribe.* Santiago, Chile.

Chaddad, Fabio R. and Marcos S. Jank. 2006. "The Evolution of Agricultural Policies and Agribusiness Development in Brazil." *Choices,* 21(2): 85-90.

Cline, William R. 2007. *Global Warming and Agriculture: Impact Estimates by Country.* Washington: Center for Global Development and Peterson Institute for International Economics.

Cooke, Bryce and Miguel Robles. 2009. *Recent Food Prices Movements: A Time Series Analysis.* Dec. IFPRI Discussion Paper No. 00942. Washington: International Food Policy Research Institute.

Economist Intelligence Unit. 2008. *Country Profile: Argentina*. London: EIU.

Economist Intelligence Unit. 2008. *Country Profile: Brazil*. London: EIU.

Ekboir, Javier M. 2003. "Research and Technology Policies in Innovation Systems: Zero Tillage in Brazil." *Research Policy*, 32(4): 573-586.

FAO (Food and Agriculture Organization of the United Nations). 2002. *World agriculture: towards 2015/2030: summary report*. Rome: FAO

FAO. 2009. *How to Feed the World in 2050:executive summary*. High-level expert forum, Rome, Oct. 12-13, 2009.

Helfand, Steven M. and Gervasio Castro de Rezende. 2001. *Brazilian Agriculture in the 1990s: Impact of the Policy Reforms*. Discussion Paper No. 785, Instituto de Pesquisa Econômica Aplicada, Ministério do Planejamento, Orçamento e Gestão, Brazil.

Hira, Anil. Reassessing ISI and Neoliberalism in Historical Context, accepted at *Revista de Economia Politica*. (Brazil), vol. 27, no.3, (107: July-Sept) 2007: 345-56.

Hira, Anil, and Jared Ferrie. Fair Trade: Three Key Challenges to Reaching the Mainstream, *Journal of Business Ethics*, 63, 2 (2006): 107-118.

James, Jennifer S., Philip G. Pardey and Julian M. Alston. 2008. *Agricultural R&D Policy: A Tragedy of the International Commons*. Staff Paper P08-08. Sept. Dept. of Applied Economics and INSTEPP, University of Minnesota.

Knox, Paul, John Agnew, and Linda McCarthy. 2008. *The Geography of the World Economy*. 5th ed. London: Hodder Education.

Lattuada, Mario and Eduardo Moyano Estrada. 2001. "Crecimiento económico y exclusión social en la agricultural familiar argentina." *Economía Agraria y Recursos Naturales*, 1(2): 171-193.

OECD-FAO. 2009. *OECD-FAO Agricultural Outlook 2009-2018*. Paris: OECD.

Pardey, Philip G., Nienke Beintema, Steven Dehmer, and Stanley Wood. 2006. *Agricultural Research: A Growing Global Divide?* Aug. Washington: International Food Policy Research Institute.

Reca, Lucio G. and Gabriel H. Parellada. 2001. "La agricultural argentina a comienzos del milenio: Logros y desfíos." *Desarrollo Económico*, 40(16): 707-737.

Reboratti, Carlos. 2004. *La Argentina rural entre la modernización y la exclusión*. Latin American Studies Association Meeting, Las Vegas, Nevada, Oct. 7-9, 2004.

Schnepf, Randall D., Erik Dohlman and Christine Bolling. 2001. *Agriculture in Brazil and Argentina: Developments and Prospects for Major Field Crops*. Dec. Agriculture and Trade Report No. WRS013, Economic Research Service, USDA.

Stads, Gert-Jan and Nienke M. Beintema. 2009. *Public Agricultural Research in Latin America and the Caribbean: Investment and Capacity Trends*. Synthesis Report, Agricultural Science and Technology Indicators, International Food Policy Research Institute and the Inter-American Development Bank.

Trigo, Eduardo, Eugenio Cap, Valeria Malach and Federico Villarreal. 2009. *The Case of Zero-Tillage Technology in Argentina*. Nov. IFPRI Discussion Paper No. 00915. Washington: International Food Policy Research Institute.

Tweeten, Luther and Stanley R. Thompson. 2008. Long-term Global Agricultural Output Supply-Demand Balance and Real Farm and Food Prices. Dec. Working Paper AEDE-WP 0044-0-8 Dept. of Agricultural, Environmental, and Development Economics. The Ohio State University.

UNCTAD (United Nations Commission on Trade and Agriculture). 2008. *Addressing the Global Food Crisis: Key trade, investment and commodity policies in ensuring sustainable food security and alleviating poverty.* UNCTAD/OSG/2008/1 Advance Unedited Version. The High-Level Conference on World Food Security: The Challenges of Climate Change and Bioenergy. June 3-5, Rome, Italy.

INDEX

A

access, 167, 176, 211, 244, 260, 261, 264, 266
accountability, 39
accounting, ix, 133, 136, 259, 266
acetic acid, 36, 44, 199
acetone, 11, 13, 16, 198
acidic, 16, 37, 53, 55, 60, 188, 216
acidity, 37, 49, 126, 185, 249
acidosis, x, xii, 165, 166, 177, 178, 210, 216, 243, 244, 249, 252
activated carbon, 16, 19, 28
active compound, 20, 21, 26, 27, 130
active oxygen, 187, 206
active site, 204
acute lymphoblastic leukemia, 197
acylation, 81
adaptation, 167, 210, 244, 246, 264
adaptations, 206
additives, 227
adenocarcinoma, 83, 88, 185
adhesion, 83
adjustment, 15, 210
ADP, 241
advancements, 263, 266
Africa, ix, 102, 113, 147, 152, 257, 259, 261
age, 22, 43, 70
agencies, 93, 98, 260
agricultural exports, 263
agricultural market, 261, 268
agricultural producers, 265, 267
agricultural sector, 90, 262, 263, 265, 266
agriculture, viii, x, xiii, 59, 60, 61, 62, 63, 65, 89, 90, 93, 107, 134, 145, 151, 152, 155, 161, 162, 164, 253, 254, 256, 259, 260, 261, 262, 263, 264, 265, 266, 269
Agro-processing development, viii, 89
alanine, 203

albumin, xi, xii, 106, 209, 211, 243, 246, 249, 250
alcohols, 2, 3, 6, 8, 9, 15, 24, 188
Alcohols, 4, 9, 10, 29, 122
aldehydes, 2, 3, 6, 8, 9, 12, 21, 122
Aldehydes, 4, 7, 10, 12, 21
alfalfa, 54, 66, 152, 153, 158, 159, 160, 161, 211, 217, 223, 229, 249, 250
algae, xi, 195, 196, 197, 202, 203, 206, 207
algorithm, 141
alkalinity, 216
alkaloids, 157
allergic reaction, 77, 120
allergy, 77, 78, 79
American Heart Association, 77
amines, 9
amino, x, 9, 16, 28, 66, 75, 77, 116, 122, 165, 203, 206, 222, 223, 224, 227
amino acid, x, 9, 16, 28, 66, 75, 77, 116, 122, 165, 203, 206, 222, 223, 224, 227
amino acids, x, 16, 66, 116, 122, 165, 203, 206, 222, 223, 224, 227
ammonia, 197, 198
ammonium, xi, 58, 60, 195, 198, 201, 204
amylase, 115, 169, 239, 245
androgen, 71
angiogenesis, 69, 72, 82
Angola, 258
anhydrase, 44
ANOVA, 199, 200, 201, 202
antagonism, 43
anthocyanin, 117, 118, 123, 124, 127, 128, 129
anticancer activity, 69
antigen, 72, 83
antioxidant, 26, 33, 68, 75, 119, 128, 129, 185, 196, 202, 204, 205, 206, 207
APC, 91, 92, 93, 96, 97
apex, 58
apoptosis, 72, 74, 82, 83, 185, 186
apples, 102

272 Index

appropriate technology, 260
aqueous solutions, 8
Arabidopsis thaliana, 207
Argentina, v, xiii, 133, 144, 146, 147, 148, 149, 253, 258, 262, 265, 266, 267, 268, 269, 270
arginine, 69, 116
Aristotle, 153
aromatic compounds, 123
aromatic hydrocarbons, 12, 28
arrhythmia, 85
arteriosclerosis, 126
Artificial Neural Networks, v, 133, 140
ascorbic acid, 58, 83, 185, 225, 227, 229
aseptic, 90, 107, 109
Asia, viii, ix, 65, 70, 113, 129, 257, 258, 259, 261
Asian countries, 68, 70, 78
aspartate, 203
aspartic acid, 116, 197, 199, 203, 204
aspirate, 170
assessment, 59, 145, 235
assimilation, x, 46, 49, 56, 58, 165, 210, 216, 249
astringent, 3, 8, 14, 16, 22, 180
atherosclerosis, 69
atherosclerotic plaque, 75, 117, 129
athletes, 66
ATP, 166, 237

B

Bacillus subtilis, 203, 207
backlash, 261
bacteria, 16, 65, 68, 86, 203, 206
Bangladesh, 25
barriers, 23, 239, 260, 263, 266
barriers to entry, 266
base, x, 57, 85, 110, 151, 168, 187, 198, 204, 219, 245, 256, 260, 267
beans, vii, viii, xi, 1, 2, 7, 8, 10, 12, 13, 14, 15, 21, 22, 24, 25, 26, 27, 29, 30, 31, 32, 41, 65, 66, 67, 76, 78, 81, 83, 84, 88, 152, 160, 221, 224, 225, 226, 229
bedding, 167
beef, 29, 173, 217, 263, 265
Belgium, 207
beneficial effect, 72, 74, 77, 78, 127, 227
benefits, ix, 23, 31, 56, 73, 76, 77, 79, 82, 126, 151, 153, 154, 158, 161, 162, 175, 196, 204, 260, 262
benign, 71, 72
benign prostatic hyperplasia, 72
benign prostatic hypertrophy, 71
benzene, 4, 19
beverages, 20, 32, 90, 98, 102
bias, 257
bicarbonate, 168, 212, 227, 245

bile, 68, 75, 185
bile acids, 68, 75
bioassay, 26
bioavailability, 67, 68, 78, 85, 130, 222, 227, 229
biodegradation, 29
biodiversity, xi, 195, 196, 203, 204
biogas, 46, 108
biological activity, xi, 179, 185
biological systems, 27
biomarkers, 71
biomass, 136, 152, 154, 155, 156, 157, 162, 198, 205, 206, 241
biomedical agents, 73
biomolecules, 199, 202
biosynthesis, xii, 202, 204, 206, 216, 225, 231, 235, 236, 238
biotechnology, 267
blood, xi, xii, 30, 66, 69, 75, 77, 79, 83, 86, 87, 88, 118, 166, 207, 209, 210, 213, 214, 216, 217, 218, 243, 244, 247, 248, 249, 250, 251
blood pressure, 77, 79
body composition, 83
body weight, xii, 71, 210, 213, 243, 244
bonds, 7, 15, 78, 166
bone, 66, 68, 73, 74, 77, 78, 80, 81, 82, 86, 87, 88, 210, 216
bone form, 74
bone mass, 73, 74, 87
bone mineral content, 74, 81
bone resorption, 74, 87, 88, 210, 216
Brazil, xiii, 100, 102, 253, 256, 257, 258, 259, 262, 263, 264, 265, 266, 267, 268, 269
breakdown, 3
breast cancer, 68, 70, 77, 78, 80, 81, 82, 83, 85, 88
bun, 65
Burma, ix, 113
business environment, 92
by-products, 3, 108, 257

C

CAD, 210, 212, 216
caffeine, 25
calcium, x, xi, 65, 66, 67, 73, 74, 79, 80, 82, 88, 120, 165, 168, 209, 210, 211, 213, 216, 217, 218, 224, 244, 245
calcium carbonate, 211
calibration, 30
cancer, xii, 2, 66, 68, 69, 70, 71, 72, 77, 78, 79, 80, 81, 82, 83, 84, 85, 86, 87, 88, 118, 129, 185, 186, 197, 221
capillary, 18, 19, 28, 72, 183, 184
carbohydrate, 83, 114, 152, 180, 198, 199, 203, 210, 212, 218, 223, 230, 235, 236, 239, 240

carbohydrate metabolism, 240
carbohydrates, xi, 2, 9, 32, 66, 69, 79, 156, 168, 196, 218, 221, 222, 223, 224, 228
Carbohydrates, 8, 115, 223
carbon, 16, 19, 28, 37, 48, 49, 68, 87, 136, 141, 145, 146, 149, 154, 166, 202, 235, 240, 258
carbon dioxide, 48, 49, 68, 87
carboxyl, 187
carcinogen, 72
carcinogenesis, 72, 87, 196
carcinoma, 71, 87, 88, 185, 197, 208
cardiac arrhythmia, 85
cardiac muscle, 87
cardiovascular disease, xii, 66, 75, 76, 77, 79, 83, 221
cardiovascular risk, 77
carefulness, 249
Caribbean, 259, 269
carotene, vii, xi, 24, 121, 195, 196, 197, 198, 199, 200, 201, 202, 203, 204, 205
carotenoids, 196, 199, 202, 206, 225, 228, 236
casein, 73, 77, 106
cash, 54, 153
cash crops, 54
catalysis, 198
catchments, 107, 109
cation, xi, 168, 209, 210, 212, 217, 218, 219, 224, 249
cattle, 100, 104, 106, 172, 173, 175, 176, 177, 178, 211, 217, 218, 251, 252
Caucasians, 86
cell culture, xi, 195, 200, 201, 202, 203, 204
cell cycle, 185, 186
cell death, 185, 187
cell differentiation, 81
cell division, 235
cellular homeostasis, 204
cellulose, 222
central nervous system, 76
chakiya, viii, 89
challenges, vii, 1, 2, 61, 176, 249, 250, 263, 268
chelates, 47, 48, 50
chemical, ix, xi, 2, 3, 8, 9, 17, 24, 27, 29, 39, 59, 73, 80, 109, 113, 127, 128, 130, 136, 158, 159, 167, 168, 169, 173, 179, 180, 185, 207, 212, 228, 229, 238, 245, 262, 265, 266, 267
chemical properties, 17, 228
chemical reactions, 3
chemical structures, 29
chemicals, 59, 157, 222, 227, 235
chemometric techniques, 17
chemopreventive agents, 185
China, ix, 100, 113, 254, 256, 257, 258, 259, 261

chitosan, 207
chloroform, 12
chlorophyll, 46, 49, 56, 201, 203
cholesterol, 66, 67, 68, 72, 75, 76, 77, 78, 82, 118, 119, 120, 129, 185, 211, 213, 216
chromatography, 2, 17, 20, 23, 24, 25, 26, 27, 30, 32, 33, 80, 81, 129, 170, 182, 183, 198
chronic diseases, 66, 76, 77
chymotrypsin, 78, 221
classes, 49, 66, 223
cleaning, 91, 107
cleavage, 119
climate, ix, xii, 37, 107, 133, 137, 144, 146, 147, 253, 254, 256, 257, 262
climate change, xii, 146, 253, 254, 256, 257
clinical interventions, 77
cloning, 239
closure, 152
clover, viii, x, 51, 52, 65, 66, 77, 151, 153, 154, 155, 156, 157, 158, 159, 160, 161, 162, 163, 164
clusters, 186
coastal region, 156
cobalt, 168, 245
coffee, 25, 101, 107, 256, 261
cognitive function, 76, 77, 81
colon, xii, 2, 68, 69, 77, 83, 88, 185, 188, 221
colon cancer, 2, 68, 77, 83, 185
color, iv, ix, 2, 113, 122, 124, 125, 126
combined effect, 70
commercial, xi, 16, 21, 87, 101, 107, 123, 182, 183, 195, 202, 203, 204, 246, 250
commercial crop, 101
commodity, 256, 257, 265, 266, 270
communication, 107
communities, 153
community, 85, 91, 161, 197
companion cell, 233
competition, x, xii, 73, 93, 151, 153, 154, 155, 156, 157, 164, 235, 243, 264
competitive advantage, 267
complementarity, 154
complex carbohydrates, 66
complex interactions, 140
complexity, 79, 141, 154
composition, vii, ix, x, 1, 2, 7, 18, 25, 27, 32, 75, 78, 80, 83, 84, 113, 114, 115, 127, 128, 142, 165, 167, 168, 169, 172, 177, 180, 186, 199, 206, 212, 215, 228, 229, 230, 245, 246, 247, 249, 250
compost, 46, 47, 48, 108
compounds, vii, ix, xi, 1, 2, 3, 4, 6, 7, 8, 9, 10, 11, 12, 13, 15, 16, 17, 18, 19, 20, 21, 22, 23, 24, 25, 26, 27, 28, 29, 30, 31, 32, 33, 46, 68, 76, 77, 113,

117, 119, 120, 122, 123, 127, 129, 130, 166, 179, 180, 183, 188, 196, 197, 202, 203, 204, 205, 206

conditioning, 145

consensus, 45, 77, 257

conservation, ix, 137, 145, 146, 147, 151, 153, 162, 262

constituents, 8, 25, 28, 31, 65, 117, 119, 224, 228

construction, 95, 139

consumers, viii, 20, 22, 89, 90, 98, 257, 259, 262, 268

consumption, vii, xi, 1, 2, 6, 14, 23, 66, 67, 70, 71, 74, 75, 76, 77, 79, 81, 83, 84, 87, 90, 97, 99, 100, 221, 223, 254, 256, 257

consumption habits, 257

containers, 90, 102, 107, 109

contaminated food, 227

contamination, 169, 227

control group, 118, 125

cooking, 22, 27, 29, 32, 72, 90, 91, 122, 123, 128, 222, 228, 229

cooperation, 93

copper, vii, 35, 36, 56, 57, 58, 59, 60, 120, 168, 207, 245

coronary heart disease, xii, 75, 77, 87, 129, 221, 228

correlation, 74, 85, 118, 122, 147, 171, 203, 238

correlation coefficient, 118

cortical bone, 73

cosmetic, 124

cosmetics, 107

cost, 20, 42, 94, 95, 96, 97, 100, 102, 108, 154, 159, 175, 197, 204, 235, 258, 260, 266

costochondral junction, 170

cotton, 58, 101, 222, 260

crises, xii, 253

crop production, 145, 146, 147, 152, 156, 159, 161, 163, 164, 227

crop residue, vii, 35, 136, 137, 156, 157

crop rotations, 134, 161, 163, 262

cross-sectional study, 86

crystalline, 45, 116

crystals, 17

cultivars, vii, 2, 35, 36, 48, 49, 52, 53, 59, 84, 114, 119, 121, 128, 129, 134, 138, 155, 156, 157, 226, 233, 236, 239, 240

cultivation, viii, 36, 49, 54, 61, 101, 110, 113, 127, 145, 148, 180, 239

cultural practices, 162

culture, xi, 16, 51, 56, 87, 88, 195, 198, 201, 202, 204, 208

culture conditions, 201

cuticle, 234

CVD, xii, 75, 84, 221

cycles, 154

cycling, 147

cyclohexanone, 13

cysteine, 33, 196, 204

D

data analysis, ix, 22, 133

data collection, 244, 246

decomposition, ix, 12, 17, 23, 48, 123, 137, 151, 156

defects, 22, 67, 81

defence, 100, 197

deficiencies, vii, 35, 36, 38, 39, 40, 42, 43, 46, 53, 56

deficiency, vii, xii, 35, 36, 37, 39, 40, 41, 42, 43, 44, 46, 47, 48, 49, 50, 51, 52, 53, 54, 55, 56, 57, 58, 59, 60, 61, 62, 63, 76, 121, 231, 236

deficit, 135, 143

deforestation, 258, 260, 261

degradation, xii, 3, 6, 7, 13, 15, 123, 130, 166, 171, 173, 174, 178, 210, 216, 224, 226, 236, 243, 244, 249, 261, 267

degradation rate, xii, 166, 171, 174, 210, 243, 249

dehydration, 91, 102, 108

Denmark, 79, 169, 246

Department of Agriculture, 67, 87

depression, xii, 76, 243

deprivation, 261

depth, ix, 133, 136, 138, 139, 140, 142, 148

deregulation, 241, 264

derivatives, 19

desorption, 18, 19, 24

destruction, 24, 258

detectable, 21, 23, 118

detection, 17, 20, 180, 182, 183, 188

detection system, 17

developed countries, xi, 221, 263

developing countries, xi, 32, 221, 259, 260, 262

deviation, 40, 146

diabetes, xii, 2, 67, 68, 77, 78, 79, 82, 118, 221

diet, x, xi, xii, 2, 66, 67, 69, 71, 72, 73, 75, 76, 77, 78, 80, 81, 82, 84, 86, 87, 88, 118, 152, 159, 165, 167, 171, 174, 175, 176, 209, 211, 212, 213, 214, 215, 217, 225, 227, 243, 244, 247, 248, 249, 250

dietary fiber, 66, 67, 75, 76, 78, 79, 84, 113, 116, 124, 178, 219

dietary intake, 70, 71, 74, 81

digestibility, x, xii, 77, 87, 88, 165, 169, 170, 175, 176, 177, 178, 212, 215, 216, 219, 222, 223, 225, 227, 228, 229, 243, 244, 245, 247, 248, 250

digestion, 67, 78, 173, 176, 177, 178, 216, 252

dihydroxyphenylalanine, 157

dioxygenase, 3

direct investment, 265

diseases, xi, xii, 2, 57, 58, 66, 76, 77, 78, 79, 154, 155, 195, 196, 221

Index 275

disorder, 47

distillation, 15, 24

distilled water, 170, 198, 247

distribution, 24, 26, 36, 59, 60, 62, 75, 93, 98, 100, 171, 235, 245, 247, 265, 267

diversification, 267

diversity, 69, 101

DNA, 69, 72, 81, 119, 191

docosahexaenoic acid, 67

domestic policy, 262

dominance, 232, 263, 264

dough, 125, 130

drainage, 163

drought, 53, 152, 253, 256

drug resistance, 85

dry matter, x, 39, 43, 49, 51, 155, 159, 165, 169, 170, 212, 214, 215, 240, 246, 248

drying, viii, 89, 91, 108, 153

dysplasia, 71

E

ECM, 174, 175, 215

ecology, 162, 164

economic growth, 254, 262

economic liberalization, 264

economics, 94, 96, 128, 155

economies of scale, 260

ecosystem, 54, 59, 158

edema, 213, 214, 215

education, 91, 93, 259, 261

EEA, 143, 145, 147

eicosapentaenoic acid, 67

election, vii, 1, 2

electricity, 92, 107

Electronic Nose, 17

employment, viii, 89, 90, 91, 93, 95, 100, 101, 102, 107, 108, 257

employment opportunities, 101, 102

encoding, 236, 239

endosperm, xii, 119, 121, 166, 231, 233, 235, 236, 237, 238, 239, 240, 241

energy, vii, x, xi, xii, 1, 23, 44, 66, 99, 100, 125, 165, 174, 176, 184, 196, 204, 209, 210, 211, 215, 216, 217, 218, 223, 224, 226, 227, 231, 237, 244, 247, 248, 250, 252

energy consumption, 99

energy density, 218

energy supply, 217

England, 212

entrepreneurs, 92

environment, 39, 69, 92, 101, 107, 154, 160, 161, 173, 196, 204, 249, 259, 260

environmental conditions, 134

environmental control, 136

environmental degradation, 261

environmental stress, xi, 195, 196, 197

environmental stresses, xi, 195, 196, 197

enzyme, 3, 6, 21, 26, 27, 28, 32, 44, 71, 75, 176, 185, 197, 198, 201, 203, 204, 224, 236, 240, 241

enzymes, xii, 6, 7, 13, 15, 27, 44, 56, 58, 69, 72, 79, 173, 188, 202, 221, 223, 225, 231, 235, 236, 238, 239, 240

epidemiologic studies, 70, 83

epinephrine, 170

epithelial cells, 70, 85

epithelium, 73, 185

equilibrium, 18, 19

equipment, viii, 18, 89, 90, 91, 92, 93, 94, 98, 107, 108, 109, 256, 259, 264

equity, 257, 259, 260, 268

erosion, ix, 136, 146, 151, 152, 153, 154, 158, 160, 162, 262

ESI, 183, 184, 190

estrogen, 69, 70, 71, 73, 76, 80, 84, 85

estrogen receptor modulator, 80

ethanol, 14, 15, 24, 27, 182, 223

ethyl acetate, 120

ethylene, 235

etiology, 217

EU, 77, 113, 260

Europe, viii, 26, 65, 102, 106, 156

European Union, 263

evacuation, 178

evaporation, 26, 183

evidence, 28, 70, 77, 83, 154, 174, 185, 256

evolution, 263

exclusion, 77, 269

excretion, 66, 74, 75, 79, 82, 86, 185

exercise, 40, 167, 245

experimental design, 138, 158, 197, 204

expert systems, 145

exploitation, 72

export market, 101, 102, 104, 107, 263, 268

exports, 102, 104, 105, 110, 255, 256, 260, 261, 263, 265, 266, 267, 268

exposure, 12, 70, 85, 173, 203

expulsion, xi, 209, 213, 214, 215

extraction, 15, 16, 18, 19, 20, 23, 24, 27, 28, 32, 33, 36, 38, 99, 100, 101, 109, 180, 181, 182, 188, 198

extracts, 15, 19, 24, 28, 77, 118, 119, 122, 128, 129, 157, 181, 182, 185, 188, 198

extrusion, 222, 228

F

FAI, 63

families, 2, 9, 65, 152

famine, 253
farmers, viii, ix, 36, 44, 49, 89, 97, 109, 133, 134,
    138, 153, 154, 256, 259, 260, 261, 262, 266, 267,
    268
farmland, 263, 264, 265
farms, x, 161, 165, 211, 256, 260
fat, x, xi, xii, 3, 7, 14, 23, 27, 66, 67, 75, 76, 78, 87,
    166, 169, 175, 176, 209, 211, 212, 215, 216, 224,
    243, 244, 246, 247, 248, 249
fatty acids, x, 3, 6, 7, 9, 12, 21, 24, 29, 32, 67, 75, 82,
    85, 87, 116, 117, 165, 169, 174, 210, 218, 250
FDA, 72, 75, 77
feedstock, 257
fermentation, x, xi, 16, 75, 99, 165, 167, 170, 173,
    174, 176, 188, 205, 209, 216, 222, 230, 244, 249,
    250, 251
ferrous ion, 117
fertility, ix, 45, 62, 133, 135, 137, 144, 158
fertilization, viii, 35, 36, 39, 45, 137, 138, 140, 141,
    146, 147, 149
fertilizers, vii, 35, 46, 65, 134, 137, 142, 145, 256,
    259, 263, 264, 265, 266, 267
fever, 210, 213, 217
fiber, 66, 67, 69, 75, 76, 78, 79, 84, 86, 113, 116,
    124, 173, 177, 178, 211, 212, 219, 251
fiber content, 113, 116
fibers, 24, 75
fibre, vii, xi, 1, 2, 18, 19, 23, 25, 69, 78, 80, 221,
    222, 229
Fibre Solid Phase Microextraction, 18
fibroblast growth factor, 72
fibroblasts, 71, 81
field crops, 49, 54
financial, 92, 101, 148, 254
financial performance, 148
fish, 12, 81, 87, 104, 107
fish oil, 81, 87
fixation, 40, 58, 65, 66, 152, 153, 164
flank, 71
flatulence, 68, 221, 230
flavonoids, 16, 69, 83, 87, 88, 119, 120
flavor, 24, 25, 26, 27, 28, 29, 30, 31, 32, 33, 122,
    124, 126, 130
flavour, vii, 1, 2, 3, 4, 6, 7, 8, 9, 10, 11, 12, 13, 14,
    15, 16, 18, 20, 22, 23, 24, 25, 27, 28, 29, 31, 107
Flavour, v, vii, 1, 2, 3, 4, 8, 9, 14, 15, 16, 17, 23, 32
flora, 72, 84
flour, vii, viii, 1, 2, 6, 14, 24, 28, 71, 74, 89, 93, 94,
    96, 97, 100, 108, 115, 116, 122, 124, 125, 126,
    130, 131, 182
flowers, 13, 66, 107
fluctuations, 246, 249, 262
fluid, 72, 88, 129, 169, 170, 174, 177, 218, 237

food industry, 28, 98
food processing industry, 110
food production, xii, 196, 253, 258, 262
food products, 103, 257
food safety, 25
food security, vii, 146, 253, 270
forage crops, 155
foreign direct investment, 265
foreign investment, 99, 103, 104
forest ecosystem, 54
formamide, 198
formation, vii, 1, 2, 3, 6, 7, 8, 9, 12, 14, 15, 16, 23,
    24, 29, 32, 40, 48, 53, 74, 80, 88, 119, 173, 185,
    188, 196, 224, 239
formula, 260
fragments, 183
France, 176
free radicals, 202
freedom, 171, 213, 233, 247
freezing, 9
freshwater, 196, 197, 198, 207
frost, 12, 159
fructose, 118, 129
fruits, viii, 3, 20, 30, 33, 90, 99, 101, 102, 107, 108,
    109, 110, 198
functional food, ix, 32, 79, 113, 117, 127
funding, 264
fungi, x, 179, 203, 205
furan, 4, 29, 123

G

gametophyte, 240
gastrointestinal tract, 67, 86
GATT, 107
GDP, 262, 265, 266
gel, 182, 198
gene expression, 232
gene regulation, 238
gene transfer, 196, 207
genes, 236, 238, 239
geography, 261
geometry, 85
Germany, 169, 213
germination, 6, 28, 53, 124, 130, 156, 157, 161, 222,
    223, 224, 225, 226, 227, 228, 229, 230, 240
gestation, 217, 251
ginger, 110
ginseng, 183
gland, 80, 85, 217
glioblastoma, 186
global climate change, 146
global economy, 261
global markets, 256, 268

gluconeogenesis, 218
glucose, xi, 78, 79, 83, 87, 88, 177, 203, 209, 211, 213, 216, 217, 228, 241, 244, 246, 249
glucose tolerance, 78
glucoside, 15, 117, 118, 119, 123, 124, 128, 130
glutamate, 196, 203, 204, 206
glutamic acid, 116, 203, 223
glutamine, xi, 195, 198, 199, 200, 201, 202, 203, 205, 206, 207
glutathione, xi, 33, 118, 120, 195, 196, 197, 198, 201, 203, 204, 205, 206, 207
glycerol, 216
glycine, 69, 196
glycoside, 81, 180
glycosylation, 180
government policy, 263, 264, 265, 266
grades, 99
grading, 90, 91, 102, 107, 109
grass, 41, 160
grasses, 152
grasslands, 134, 149
grazing, 146
Greece, ix, 113, 151, 207
green alga, xi, 195, 196, 197, 198, 203, 204
Green Revolution, 257, 258, 259, 260, 268
greenhouse, 38, 146
gross domestic product, viii, 89
groundwater, 148, 258
growth factor, 69, 70, 72, 80, 85
growth rate, 102, 106, 107
guessing, 256
guidelines, 167, 211, 245

H

habitat, 198
hardness, 115, 122, 124, 125, 126
harvesting, 3, 6, 9, 23, 101, 153, 256, 264
Hawaii, 81, 87
health, ix, xi, 23, 29, 31, 67, 68, 73, 74, 75, 76, 77, 78, 79, 80, 81, 82, 84, 87, 113, 119, 127, 158, 174, 195, 196, 202, 204, 205, 209, 217, 218, 222, 229
health care, 196
health effects, 29, 78, 80, 84
health risks, xi, 195, 196
heart disease, xii, 2, 75, 77, 87, 129, 221, 228
heavy metals, 207
hemagglutinins, 65, 72
hemicellulose, 116
hepatocytes, 218
herbicide, 154, 155, 158, 159, 164, 264
hexane, 15, 27
high fat, 67

higher education, 259
high-yielding crop, vii, 35
histamine, 120
histogram, 187
history, 42, 77, 136, 180, 263
HIV, 82, 187
HIV-1, 82, 187
homeostasis, 79, 204, 206, 216
homogeneity, 171, 247
Hong Kong, 65
hormone, 72, 75, 76, 85, 86
human, vii, ix, xi, xii, 1, 2, 14, 17, 20, 32, 63, 66, 68, 70, 71, 72, 74, 77, 78, 79, 80, 81, 82, 83, 84, 85, 86, 87, 88, 113, 119, 134, 152, 183, 186, 187, 188, 195, 196, 198, 202, 203, 204, 205, 218, 221, 222, 225, 227, 230, 243
human body, 222, 225
human health, 87, 196, 202, 204, 205
human immunodeficiency virus, 187
human subjects, 66, 79
humidity, 6
Hunter, 122
hydrocarbons, 12
Hydrocarbons, 4, 10, 12
hydrogen, 13, 66, 68, 88, 117
hydrogen peroxide, 88, 117
hydrolysis, 7, 21, 76, 115, 181, 187, 197, 223, 225, 226, 237
hydroperoxides, 6, 12, 13, 27, 28, 32
hydrophobicity, 180, 185
hydroquinone, 38
hydroxide, 252
hydroxyl, 117, 119, 123, 183, 187
hydroxyl groups, 123
hygiene, 107
hypercholesterolemia, 69
hyperglycemia, 128
hyperlipidemia, 118, 126, 129
hyperplasia, 72, 86
hypertension, 69, 79
hypertrophy, 71, 87
hypocotyl, 222
hypoglycemia, 210
hypothesis, 70, 72, 75, 78, 178, 210, 256
hypothesis test, 178

I

identification, 26, 127, 183
imagination, 257
immersion, 18, 19, 124
immobilization, 198, 207
immune function, 166
immunity, 79, 166, 216

immunodeficiency, 187, 218
imports, 260, 266
in vitro, 69, 70, 71, 72, 73, 74, 82, 83, 85, 86, 88, 118, 123, 186, 225
in vivo, xii, 70, 72, 83, 85, 87, 88, 166, 243
incidence, 36, 37, 40, 54, 56, 70, 71, 74, 75, 76, 77, 86, 88, 154
income, viii, 66, 89, 91, 93, 100, 107, 259, 261
increased access, 264
independence, 85
Independence, 98, 110
independent variable, ix, 133, 139, 143
India, v, vii, xi, 35, 36, 37, 38, 44, 49, 53, 54, 56, 57, 58, 59, 60, 62, 63, 66, 89, 90, 91, 93, 98, 99, 100, 101, 102, 103, 104, 105, 106, 110, 155, 163, 195, 196, 198, 204, 206, 221, 229, 231, 232, 233, 240, 254, 256, 257, 259
Indian soils, vii, 35, 36, 38, 60
individuals, 68, 77, 87, 91
Indonesia, 106
inducer, 81
induction, 86, 129, 185
industrial sectors, 264
industrialized countries, 155
industries, viii, 90, 91, 99, 102, 103, 106, 107, 108, 196, 262, 263
industry, vii, 1, 2, 15, 17, 28, 63, 93, 98, 99, 107, 109, 110, 259, 262, 267
infants, 67
infection, 170
inflammation, 72, 88, 169, 246
inflation, 262
influenza, 187
influenza a, 187
information technology, 62
infrastructure, 107, 258, 260, 262
ingestion, 216
ingredients, vii, 1, 2, 21, 25, 57, 117, 167, 168, 245
inhibition, 71, 79, 85, 87, 117, 118, 119
inhibitor, 67, 69, 78, 79, 80, 82, 83, 84, 87, 88, 130, 186, 187, 223, 226
inoculum, 198, 199
inositol, 67, 87
institutional reforms, 268
institutions, xiii, 253, 260, 265, 268
insulin, 67, 69, 75, 86, 118, 128, 129, 166, 228
insulin dependent diabetes, 86
insulin resistance, 118, 129
insulin sensitivity, 128
integrity, 77, 126
integument, 235
intellectual property, 260
intellectual property rights, 260

interaction effect, 201, 204, 206
Inter-American Development Bank, 266, 269
interest rates, 262, 264
interface, 17, 234
interference, 92, 156, 158, 159, 183, 198
international standards, 107
intestinal flora, 84
intestine, 68, 72, 86
introspection, 38
In-Tube Solid Phase Microextraction, 19
investment, 91, 95, 99, 103, 107, 109, 259, 261, 262, 263, 265, 266, 267, 268, 270
investments, 104, 263, 266
iodine, 115, 168, 245
ion-exchange, 16
ionization, 18, 183
Iran, 165, 167, 171, 177, 205, 209, 211, 217, 243, 245, 246, 251
Ireland, 183, 189
iron, vii, 3, 35, 38, 44, 46, 47, 48, 49, 56, 58, 60, 61, 63, 66, 67, 78, 120, 168, 224, 230, 245
irrigation, vii, 35, 51, 134, 258, 262
Islam, 146
isoflavone, 15, 31, 32, 68, 69, 70, 73, 76, 77, 78, 80, 85
isoflavonoids, 81, 85
isolation, 18, 180, 182, 205
isoleucine, 224
isomers, 117, 119, 121
isoprene, 202
isozyme, 29
isozymes, 6
issues, 63, 90, 107, 147, 254, 260
Italy, ix, 60, 113, 169, 247, 270
iteration, 141

J

janta, viii, 89
Japan, ix, 70, 71, 76, 80, 82, 83, 106, 113, 189, 190, 259
Japanese women, 73, 85
Jordan, 258

K

K^+, 168, 212
keratinocytes, 130
ketones, 2, 3, 8, 9
Ketones, 4, 11, 13, 21, 122
kidney, 16, 67, 78, 80, 86
Korea, ix, 113, 117, 127, 128, 130

Index

279

L

lactation, x, xi, 166, 167, 173, 209, 210, 211, 217, 218, 244, 251

lactic acid, 16, 203

lactose, 106, 169, 175, 212, 215, 246, 248

landscape, 135

Laos, ix, 113

large intestine, 67

latency, 70, 71

Latin America, 258, 259, 261, 269

LDL, 72, 75, 77, 119

leaching, 36, 37, 49, 54, 55, 226

lead, viii, 9, 12, 32, 43, 90, 109, 135, 185, 204, 226, 227, 257

learning, 140, 141, 145

learning process, 141

legume, vii, viii, ix, xii, 1, 2, 3, 6, 13, 14, 15, 16, 21, 23, 31, 53, 65, 66, 69, 75, 78, 79, 81, 82, 151, 152, 153, 154, 155, 156, 157, 159, 160, 161, 163, 180, 221, 222, 228, 229

Legumes, v, vi, vii, viii, ix, 1, 2, 3, 8, 9, 14, 15, 17, 23, 29, 65, 66, 68, 75, 78, 84, 151, 152, 153, 154, 161, 188, 191, 221, 229

lentils, vii, viii, x, xi, 1, 2, 7, 8, 22, 29, 65, 66, 76, 152, 179, 221, 223, 229, 230

lesions, 56, 72, 83

leucine, 223, 224

leukemia, 81, 83, 197, 203, 206

leukoplakia, 80, 87

liberalization, 99, 260, 264, 265

life cycle, 154, 158

light, 37, 53, 54, 55, 58, 153, 154, 157, 163, 182, 183, 196, 198, 227, 250, 265

light scattering, 182, 183

linoleic acid, 3, 6, 7, 24, 27, 32, 67, 81, 116, 117, 119

lipases, 3, 7, 15, 26

lipid metabolism, 81

lipid oxidation, 7, 13, 24, 26

lipid peroxidation, 120, 236

lipids, 2, 7, 9, 12, 26, 30, 33, 79, 83, 86, 87, 115, 118, 196, 198, 203

lipoproteins, 75, 79

Lipoxygenase (LOX), 3

liquid chromatography, 25, 26, 33, 182

liver, 120, 210, 217, 218, 250

liver damage, 120

livestock, 106, 155, 264

loans, 94, 261

local conditions, 259, 264

localization, 239

locust trees, viii, 65

low-density lipoprotein, 75, 83

LSD, 46, 48, 50, 51, 55, 57, 199, 200

lumbar spine, 74

lung metastases, 71

lupins, viii, 65, 66

lymphocytes, 218

lysine, 77, 224

M

machinery, 93, 95, 99, 100, 263

magnesium, 168, 224, 245

magnitude, 49, 55, 57, 69, 74, 175

Maillard reaction, 8, 9, 14, 16, 33

majority, vii, 1, 2, 186, 265

Malaysia, 102, 106, 229

malignant cells, 197

malnutrition, 49

man, 95, 152

management, ix, 36, 49, 58, 61, 77, 91, 92, 93, 133, 134, 135, 137, 138, 144, 145, 146, 147, 149, 151, 153, 155, 156, 159, 162, 163, 164, 177, 210, 218, 250, 262

manganese, vii, 35, 49, 50, 51, 53, 59, 60, 61, 62, 63, 120, 168, 245

manipulation, x, 165, 238

mannitol, 39

manufacturing, 101, 106, 261, 262

manure, 46, 47, 48, 66, 153, 159, 164

mapping, 62

marginalisation, 257

marine environment, 204

marketing, vii, viii, 1, 22, 89, 93, 97, 99, 107, 127, 260, 261, 265

markets need, vii, 1, 2

Marx, 144

masking, 14, 16

mass, 18, 29, 30, 32, 33, 73, 74, 75, 80, 86, 87, 129, 182, 183, 184, 198, 238, 239

mass spectrometry, 18, 29, 30, 80, 129

mastitis, 167, 169, 246

materials, viii, 9, 19, 36, 90, 92, 93, 104, 108, 126, 128, 131, 134, 246

mathani, viii, 89

matrix, 18, 19

measurement, 24, 93, 169, 187, 197

measurements, 247, 251

meat, viii, 2, 16, 29, 65, 104, 105, 106, 254, 256, 257

Mediterranean, 156, 160, 162, 229

melanoma, 86, 186

menopause, 76, 84, 87

mesquite, viii, 65

meta-analysis, 217

metabolic acidosis, 210

metabolic changes, 75, 222
metabolic disorder, 217
metabolic disorders, 217
metabolism, 2, 9, 46, 49, 58, 66, 73, 75, 80, 81, 82, 85, 119, 120, 174, 176, 196, 203, 209, 210, 218, 224, 240, 250, 252
metabolites, x, xii, 73, 75, 179, 188, 210, 213, 214, 218, 243, 244, 248, 249, 251
metabolized, 68, 188, 223
metabolizing, 236
metalloenzymes, 46
metals, 207, 224, 227
metastasis, 87, 186
meter, 136, 169, 170, 212, 247
methanol, 9, 182
methodology, 28, 143
Mexico, 100, 146, 155, 159, 160, 239, 260
mice, 71, 72, 83, 84, 88, 128, 129, 130
micronutrients, vii, 35, 36, 37, 38, 39, 41, 49, 58, 59, 60, 62, 63, 66, 78, 227
microorganisms, 188, 197, 203, 205
Middle East, 102, 105, 258, 259
migration, 257, 261
military, 263
military dictatorship, 263
mineralization, 136, 148
mini oil ghanis/kolhus, viii, 89
model system, 28, 32, 128, 188
modelling, 141, 146
models, 80, 137, 139, 140, 141, 142, 143, 146, 147, 171, 213, 219
modernisation, 63
modernization, 92, 99, 101, 109
modifications, 39
moisture, ix, 36, 108, 122, 126, 133, 135, 136, 139, 140, 142, 143, 154, 156, 157, 163, 166, 171, 251
moisture content, 122, 135, 139, 140, 143, 171, 251
mold, 230
molecular biology, 207
molecular oxygen, 7
molecular weight, 7, 9
molecules, x, 179, 188, 225
molybdenum, 60
morphology, xii, 207, 231
mortality, 68, 70, 71, 72, 78
mortality rate, 68, 70, 71, 78
motivation, 259
mucosa, 68, 217
multinational companies, 265
multinational corporations, 261
multiple regression, 139, 143
multiplication, 197
mung bean, 228, 229

N

Na^+, 168, 212
NaCl, xi, 168, 195, 198, 199, 200, 201, 202, 203, 204, 205, 245
National Research Council, 66, 85, 177, 218, 251
natural compound, 12
negative relation, 134
neoliberalism, 260
nervous system, 76
Netherlands, 164, 170, 207
neural network, ix, 133, 134, 140, 141, 142, 144, 145, 146, 147, 148, 149
neural networks, ix, 133, 134, 140, 146, 147, 148
neurons, 141, 142
neutral, 22, 27, 38, 178, 185, 211, 212, 219
nitrates, 139, 140, 142
nitric oxide, 120, 128
nitric oxide synthase, 120
nitrogen, ix, xi, 13, 40, 46, 49, 58, 60, 65, 66, 77, 122, 133, 136, 137, 138, 140, 142, 145, 148, 149, 151, 152, 153, 154, 159, 160, 164, 166, 176, 197, 198, 199, 200, 201, 203, 205, 206, 209, 250, 259
nitrogen compounds, 166
nitrogen fixation, 40, 58, 65, 66, 152, 153, 164
nitrogenase, 58
NMR, 191
nodules, 65, 104
non-insulin dependent diabetes, 86
non-polar, 18, 19
North Africa, 258, 259
North America, 76, 87, 207
nucellus, 233, 237
Nuclear Magnetic Resonance, 189
nucleic acid, 46
nucleotide sequence, 207
null, 14, 23
nutraceutical, 81, 117, 205
nutrient, x, xii, 37, 40, 42, 43, 46, 47, 53, 63, 78, 127, 158, 165, 172, 173, 197, 212, 215, 222, 227, 228, 229, 243, 244, 250
nutrients, vii, 35, 36, 37, 40, 43, 50, 60, 62, 73, 153, 154, 159, 221, 223, 227, 229
nutrition, viii, xi, 2, 31, 32, 35, 43, 58, 60, 63, 67, 78, 83, 120, 134, 160, 178, 219, 221, 230
nutritional imbalance, 203
nutritional status, 38

O

obesity, 67, 69, 78, 79
obstacles, viii, 89, 268

Index

oil, viii, ix, 7, 10, 11, 12, 13, 15, 28, 59, 60, 81, 87, 89, 90, 91, 98, 99, 100, 101, 107, 108, 109, 119, 120, 129, 133, 137, 138, 140, 142, 144, 146, 152, 153, 254, 256, 257, 260
oligosaccharide, 68
omega-3, 87
open policy, 100
operations, 91, 110
operon, 207
opportunities, 90, 101, 102, 268
optical fiber, 24
optimization, 44, 197, 204, 206, 263
organ, 235
organic compounds, 12, 19, 30
organic food, 113
organic matter, 36, 38, 40, 53, 56, 134, 136, 138, 144, 146, 147, 149, 152, 153, 159, 170
organic polymers, 17
organic solvents, 15
organism, 198, 203
organize, 91, 256
organoleptic quality, vii, 1, 2, 222
organs, 40, 235
ornithine, 185, 203
ovariectomy, 73
oxidation, 3, 7, 9, 10, 12, 13, 24, 26, 46, 49, 75, 119, 185, 187, 223
oxidative reaction, 7
oxidative stress, 118, 120, 197, 204
oxygen, 7, 15, 117, 128, 187, 196, 197, 202, 204, 205, 206

P

p53, 86
Pacific, 83, 129, 259
Pakistan, 66
palpitations, 76
parallel, 148, 170
paralysis, 260
parenchyma, 233
pasture, 54, 66, 134, 152
pastures, 134, 138, 140, 142
pattern recognition, 17
peanuts, viii, 31, 65, 66, 152
peas, vii, viii, x, xi, 1, 2, 6, 7, 8, 9, 12, 13, 14, 24, 26, 27, 28, 29, 30, 31, 33, 65, 66, 102, 177, 179, 221, 222
peptides, 15, 25, 29, 79, 224
peripheral blood, 250
peroxidation, 117, 119, 205, 236
peroxide, 29, 88, 117, 205
Peru, 260
pesticide, 227, 229, 260

pests, 154, 159
petroleum, 198, 254, 255, 256, 257, 259, 268
Petroleum, 254, 255
pH, x, xi, xii, 15, 19, 36, 37, 38, 39, 40, 48, 56, 72, 148, 165, 166, 169, 170, 174, 177, 187, 198, 199, 209, 212, 214, 215, 216, 224, 243, 247, 248, 249, 250
pharmaceutical, 66, 196
pharmacology, 27
phenolic compounds, ix, 8, 16, 26, 28, 113, 119, 123, 129
Phenolic Compounds, 8
phenylalanine, 223
Philadelphia, 169, 245
Philippines, 114, 241
phloem, xii, 231, 233, 235, 236, 237
phosphatidylcholine, 6, 7
phospholipids, 3, 7, 31
phosphorous, 211, 213, 228
phosphorus, ix, 44, 56, 58, 133, 138, 140, 142, 145, 148, 168, 245
phosphorylation, 71, 74, 81
photosynthesis, 40, 44, 46, 49, 205, 240
phycocyanin, 207
physical properties, 2, 114, 144, 173, 180
physicochemical properties, 128
Physiological, xii, 128, 217, 231, 235, 241
physiological factors, 234
physiology, 41, 78, 148, 233
phytohaemagglutinin, 86
phytosterols, 69, 84
piezoelectric crystal, 17
placebo, 77, 85
placenta, xi, 209, 210, 213, 215, 216
plant diseases, 155
plant growth, 53
plant type, 206
plants, xi, 6, 12, 23, 26, 37, 38, 39, 40, 43, 44, 46, 49, 50, 53, 54, 56, 58, 59, 65, 66, 101, 106, 107, 154, 156, 157, 158, 160, 195, 196, 197, 202, 203, 204, 206, 207
plaque, 117, 129
polar, x, 18, 19, 179
polarity, 185
policy, 63, 100, 104, 254, 262, 263, 264, 265, 266, 267, 268
policy initiative, 263, 265
policy issues, 63
policy makers, 267
politics, 256, 260
pollen, 53
pollination, 53, 135
pollutants, 30, 32

polyamine, 185
polycyclic aromatic hydrocarbon, 28
polydimethylsiloxane, 19
polyphenols, 25, 119, 221
polypropylene, 126
polythene, 95, 102, 108, 109
polyunsaturated fat, 3, 6, 82, 85
polyunsaturated fatty acids, 3, 82, 85
pools, 154, 197, 207
population, viii, xi, xii, 82, 91, 99, 100, 110, 113, 153, 158, 195, 198, 253, 254, 258, 259, 261, 262
population density, 153
population growth, xii, 253, 254, 258, 261
Portugal, 170
positive correlation, 74, 203
potassium, 58, 169, 210, 212, 246
potato, 11, 12, 54, 56, 59, 102
potential benefits, 73, 153
poultry, 46, 47, 98, 101, 105, 106
poverty, 257, 260, 261, 268, 270
precipitation, 25, 53, 55, 156, 158
predators, x, 179
pregnancy, 212, 215
preparation, iv, 18, 19, 20, 30, 32, 103, 183
pressure gradient, xii, 231, 234
prevention, viii, xii, 56, 76, 77, 78, 79, 80, 87, 89, 217, 221, 228
private investment, 267
producers, 143, 166, 256, 259, 260, 264, 265, 267
production technology, 134
profit, viii, 89, 92, 93, 96
profitability, x, 151
project, 39, 62, 77, 92, 238
proliferation, 67, 69, 70, 71, 80, 81, 84, 85, 86, 87, 185, 210
proline, xi, 195, 197, 198, 201, 203, 204, 205, 207
promoter, 88
propagation, 141
property rights, 260, 268
proposition, 236
prostate cancer, 68, 71, 72, 78, 82, 83, 85, 86, 87
prostate carcinoma, 71
prostate specific antigen, 72, 83
protease inhibitors, 67, 69, 77
protection, 78, 203, 204, 264, 266
protectionism, 259
protein components, 75
protein hydrolysates, 25
protein kinase C, 185
protein kinases, 69, 79
protein synthesis, 44, 74, 175, 216, 224

proteins, xi, xii, 2, 8, 9, 14, 15, 16, 24, 31, 65, 66, 71, 72, 76, 77, 79, 81, 82, 83, 152, 196, 209, 222, 223, 224, 230, 238, 239, 243, 250
Protein-Saponin Interaction, 8
proteinuria, 69
proteolytic enzyme, 15, 27, 225
Pseudomonas aeruginosa, 205
public sector, 259
puffing pan, viii, 89
pulp, 102, 167, 168, 252
purification, 28, 81, 182, 188, 205, 207
purity, 182
P-value, 172, 174, 175, 214, 215, 247, 248
PVP, 198
Pyrazines, 4, 11, 13, 29
pyridoxine, 225
pyrite, 47, 48
pyrolysis, 32
pyrolysis reaction, 32
pyrophosphate, 38, 202

Q

quality control, 17, 22
quality standards, 107
quantification, 127, 182, 183
quantitative estimation, 25
quercetin, 120
questioning, 136
quotas, 260, 265

R

radiation, 136, 154
radicals, 72, 118, 119, 129, 187, 196, 202
rainfall, ix, 53, 59, 133, 134, 135, 136, 139, 140, 142, 143
rancid, 3, 7, 11, 14
raw materials, viii, 90, 93, 108
reaction time, 256
reactions, 3, 7, 27, 29, 32, 33, 46, 77, 120, 207, 254
reactive oxygen, 117, 128, 196, 197, 202, 204
reality, 263
reasoning, 40, 73
receptors, 75, 76, 84, 85
recession, 255, 256
recovery, 19, 91, 99, 100, 101, 108, 182, 203, 256
rectal temperature, 213
rectum, 169, 212, 247
recycling, vii, 35
reducing sugars, 223, 227
regression, ix, 133, 139, 140, 141, 142, 143, 146, 149
regression analysis, 146
regression method, ix, 133, 139

regression model, ix, 133, 139, 140, 141, 142, 143
regrowth, 156
regulations, 235
relevance, 36, 43, 72, 87
requirements, 45, 143, 145, 177, 197, 251
researchers, 38, 266
reserves, 156, 222
residuals, 171, 213, 247
residues, vii, 24, 35, 68, 136, 157, 227, 229
resins, 16, 103
resistance, 6, 80, 85, 118, 125, 129, 236
resource utilization, ix, 151
resources, xi, 14, 91, 154, 157, 158, 159, 195, 196, 203, 204, 264
respiration, 224, 225
response, viii, ix, 17, 21, 28, 35, 42, 44, 47, 51, 53, 54, 57, 61, 69, 80, 83, 87, 88, 123, 133, 137, 139, 145, 147, 148, 158, 162, 164, 177, 185, 187, 196, 203, 206, 207, 251, 256, 266
retinol, 128, 185
retinopathy, 69
revenue, 252, 257, 264, 265, 266, 267, 268
reverse transcriptase, 82
riboflavin, 225, 227, 229
ribosome, 65, 83
risk factors, 67, 82, 84
risks, xi, xii, 176, 195, 196, 243
rodents, 70, 72, 98
room temperature, 182, 186
root, 49, 53, 65, 161
roots, 43, 66, 152, 222
rotations, 134, 157, 161, 163, 262
runoff, ix, 151, 154, 162
rural areas, viii, 90, 91, 92, 93, 99, 100, 101, 108, 109, 256, 257, 268
rural development, viii, 89
rural people, 91, 92, 100, 107

S

safety, 25, 85, 227
salinity, 196, 197, 201, 203, 204, 205, 206, 207
salts, xi, 47, 209, 210, 211, 216, 217, 218
SAP, 62
saponin, 16, 25, 30, 72, 180, 185
SARA, 166, 174, 176, 244, 249
saturated fat, 75, 76, 78
savannah, 152
scaling, xi, 195, 268
scope, viii, 89, 90, 100, 107, 109
secretion, 72, 75, 174
security, vii, 146, 253, 270
sediment, 134
sedimentation, 126

seed, vii, 1, 2, 6, 8, 26, 27, 32, 48, 50, 59, 97, 134, 152, 155, 157, 159, 164, 180, 222, 223, 224, 233, 236, 237, 240, 241, 257, 261
seeding, 51, 155, 156, 159, 161
seedlings, 157
selectivity, 20
selenium, 120, 168, 245
semiconductor, 17
semiconductors, 17
serum, 69, 70, 71, 72, 74, 75, 79, 83, 183, 211
services, iv, 90, 107, 108, 109, 256, 257, 259, 264, 266, 267, 268
sewage, 46, 47
shape, 186
shelf life, 102, 108, 109
shoot, 49
shoots, 222
shortage, 77, 268
showing, 35, 40, 52, 69, 71, 141, 216, 234, 235, 238
shrimp, 104
shrubs, 66, 154
sialic acid, 186
sieve element, 233
sieve tube, xii, 231
signal transduction, 68, 69, 78
signaling pathway, 81, 87, 186
signals, 86, 218, 238
signs, 40, 211, 213
silbatta, viii, 89
silica, 18, 24, 183, 198
Singapore, 70, 83, 102, 106, 179
small intestine, 67, 72
smooth muscle, 216
social benefits, 262
sodium, 38, 54, 55, 58, 88, 168, 169, 227, 245, 252
sodium hydroxide, 252
software, 146
soil erosion, ix, 136, 151, 153, 154, 158, 160, 262
soil type, vii, 35, 37, 40, 55, 56, 137, 138, 158
solid phase, 18, 19, 24, 26, 182
Solid Phase Dynamic Extraction, 19
solubility, 25, 26, 54, 115
solution, 47, 48, 51, 53, 55, 57, 58, 170, 199, 257, 262, 268
solvents, 15, 27, 33, 182
South Africa, 24, 102, 258
South America, 257
South Asia, 258, 261
sowing, 42, 47, 48, 51, 54, 56, 134, 138, 140, 142, 143, 156, 164
soybeans, vii, 1, 2, 6, 7, 8, 14, 15, 23, 26, 28, 29, 31, 32, 67, 69, 70, 77, 78, 82, 83, 84, 152, 223, 225, 226, 229, 263, 265

284 Index

soymilk, 6, 14, 15, 16, 23, 27, 28, 32, 71
Spain, 155
specialists, 256
species, vii, viii, ix, 35, 42, 49, 52, 66, 71, 113, 114, 117, 128, 151, 152, 153, 154, 155, 156, 157, 158, 160, 163, 196, 197, 202, 204, 205, 206
speculation, 73, 254
spending, 173, 259, 266
spillover effects, 268
stability, 28, 72, 123, 125, 126, 134, 145, 148, 154, 188, 196, 262, 265
stabilization, 117, 129
standard deviation, 40
standard error, 175
standard of living, 196
standardization, 265
starch, x, xi, xii, 25, 87, 88, 115, 127, 128, 165, 166, 173, 174, 176, 177, 178, 209, 210, 215, 216, 217, 223, 227, 228, 229, 231, 235, 236, 237, 238, 239, 241, 243, 244, 249, 251
starch biosynthesis, xii, 231, 235, 236, 238
starch polysaccharides, 128
starvation, 198, 204, 261
state, 49, 110, 205, 260, 264, 265, 266, 267
state intervention, 261
states, 35, 37, 46, 49, 54, 56, 106, 259, 268
statistics, 62
steel, 19, 98, 138, 170, 171
sterile, 56, 240
steroids, 84
stir bar sorptive extraction (SBSE), 19
storage, vii, x, 1, 2, 3, 6, 7, 8, 9, 12, 23, 27, 28, 32, 33, 90, 98, 101, 107, 108, 109, 115, 126, 136, 140, 142, 179, 222, 227, 228, 233, 235, 237, 239, 246
stress, 42, 118, 120, 156, 161, 196, 197, 201, 202, 203, 204, 205, 206, 207
stroke, 228
structure, x, 3, 8, 33, 68, 73, 76, 91, 140, 141, 153, 161, 171, 179, 180, 185, 187, 232, 233, 235, 240
subacute, x, xii, 165, 166, 177, 178, 243, 244, 249, 252
subcutaneous injection, 74
sub-Saharan Africa, 258, 259
subsistence, 260, 261
subsistence farming, 261
substitutes, 2
substitution, 15, 50, 76
substrate, 3, 26
substrates, 7
sucrose, xii, 25, 83, 223, 226, 231, 233, 235, 236, 237, 238, 239, 240
sugarcane, 101, 256

sulfate, 35, 45, 46, 47, 57, 88
sulfur, 29, 31, 168, 245
sulfuric acid, 170
sulphur, 9, 12, 13, 46, 66
Sulphur Compounds, 12
Sun, viii, 89, 108, 203, 207
supa, viii, 89
supplementation, 36, 77, 81, 82, 84, 120, 210, 222
suppliers, x, 165
suppression, x, 25, 151, 153, 155, 156, 157, 158, 159, 160, 161, 162, 163, 237, 240
surplus, 91, 105, 106, 110, 263
survival, 240
susceptibility, 42, 61, 225
sustainability, ix, 151, 153, 164, 257, 260, 264, 267, 268
sustainable development, 262
symptoms, 40, 42, 43, 44, 52, 53, 56, 58, 66, 74, 76, 77, 87
synergistic effect, 70
synthesis, 44, 46, 56, 58, 72, 74, 75, 175, 185, 197, 202, 203, 204, 216, 217, 224, 225, 235, 236, 238, 241, 250

T

T cell, 218
Taiwan, 59
tamoxifen, 74
tannins, 222, 223, 225, 226, 227
tariff, 260, 264
tax breaks, 260
taxes, 263, 265, 267
taxpayers, 259
technical support, 266
techniques, x, 14, 15, 17, 18, 20, 23, 62, 102, 108, 109, 140, 142, 143, 147, 159, 165, 178, 222, 229, 259, 260, 261, 262, 264, 266, 268
technological advancement, 263
technologies, 264, 265, 266, 267
technology, vii, 28, 49, 62, 93, 101, 102, 107, 108, 109, 127, 128, 130, 131, 134, 147, 182, 256, 260, 262, 263, 264, 265, 266
temperature, 6, 18, 19, 28, 90, 102, 107, 109, 115, 123, 126, 134, 136, 154, 157, 163, 182, 183, 186, 187, 205, 213
testing, 22, 31, 40, 93, 147, 178, 249
testosterone, 71, 72
textiles, 107
textural character, 125
texture, ix, 22, 36, 37, 54, 107, 124, 125, 126, 133, 134, 138, 140, 142
TGF, 69, 74, 82
Thailand, ix, 113

therapeutic benefits, 126
therapy, 68, 76, 78, 202, 205
thermal degradation, 123, 130
thermal stability, 123
thermostability, 72
timber production, 66
tissue, xii, 9, 39, 71, 76, 81, 87, 88, 197, 231, 233, 235, 237, 239
tobacco, 101, 207
tocopherols, 121
tofu, 70, 72
toluene, 12, 19, 123
tones, 98, 99
total cholesterol, 118, 120
total factor productivity, 263
total parenteral nutrition, 67, 83
total product, 101, 134
toxicity, 120, 207
TPA, 36
trade, 42, 127, 137, 256, 259, 260, 261, 264, 270
trade liberalization, 260, 264
trafficking, 261
training, 21, 40, 42, 139, 140, 141, 142, 147
traits, vii, ix, 1, 151, 158
transduction, 68, 69, 78
transformation, 58, 72, 82, 185, 265, 268
transforming growth factor, 69, 85
transition period, xi, 209, 217
translocation, 43
transplantation, 71
transport, 53, 102, 148, 232, 234, 239, 241
transportation, 91, 254
treatment, xi, xii, 14, 15, 26, 27, 48, 50, 58, 71, 72, 73, 76, 79, 80, 87, 129, 155, 167, 171, 174, 176, 177, 186, 187, 195, 196, 197, 199, 200, 203, 206, 211, 212, 213, 216, 217, 221, 249
triggers, 75
triglycerides, 7, 75, 77
Trinidad, 163
trypsin, 78, 81, 82, 88, 221, 223, 226, 227
tumor, 70, 71, 72, 73, 81, 88, 185, 186, 196, 197, 202, 207
tumor cells, 72, 186
tumor development, 71
tumor growth, 71, 72, 73
tumor metastasis, 186
tumorigenesis, 185
tumors, 67, 70, 71, 80, 81
tumour growth, 86
turgor, 237
Turkey, 100
turnover, 223
type 2 diabetes, 69

Typic Haplustepts of Punjab, 45
tyrosine, 69, 71, 74, 79, 80, 81, 84, 87, 223
Tyrosine, 116

U

UK, 62, 86, 145, 146, 159, 176, 205, 212
ulcerative colitis, 72, 88
ultrasound, 182
underwriting, 262
United, ix, 66, 70, 82, 113, 263, 269, 270
United Nations, 269, 270
United States, ix, 66, 70, 82, 113, 263
urban, viii, 80, 89, 93, 100, 104, 198, 257, 260, 263
urban areas, viii, 80, 89, 93, 100, 104
urban settlement, 198
urbanization, 257, 258
urea, xi, 46, 204, 209, 211, 213, 216, 217
urine, x, xi, xii, 165, 170, 209, 216, 218, 243, 247, 248, 249
Uruguay, 147
USA, 75, 82, 102, 113, 145, 146, 147, 148, 155, 171, 176, 207, 229, 241, 246, 251
USDA, 129, 269
usual dose, 73
UV, 182, 183, 190, 211

V

vacuum, 27
validation, 139, 140, 141, 142
valine, 224
variables, ix, 133, 137, 139, 140, 141, 142, 143, 144, 148, 202
variations, 49, 71, 85, 147
varieties, vii, ix, 1, 23, 40, 41, 49, 53, 61, 104, 106, 110, 113, 114, 115, 116, 117, 118, 119, 120, 121, 124, 126, 127, 131, 163, 228, 240, 262
vascular bundle, 233, 235, 240
vascular diseases, 79
vascular wall, 76
vasomotor, 76, 87
vegetable oil, 152
vegetables, 3, 12, 13, 14, 20, 22, 26, 30, 57, 58, 67, 90, 98, 99, 101, 102, 107, 108, 109, 110, 162, 163
vegetation, 134, 162
viscosity, 115, 126
vitamin A, 121, 168, 196, 202, 225, 245
vitamin B1, 121
vitamin C, 225, 228
vitamin D, 168, 210, 245
vitamin E, 119, 168, 245
vitamins, xi, 2, 168, 196, 221, 222, 223, 245
volatility, 259

volatilization, 17
voluntary organizations, 91
vulva, 169, 211, 247

W

Washington, 85, 146, 177, 189, 218, 251, 269, 270
waste, 47
wastewater, 26
water, ix, x, 3, 6, 7, 11, 14, 15, 19, 22, 38, 39, 40, 44, 47, 49, 55, 68, 80, 92, 115, 120, 122, 125, 126, 128, 129, 133, 135, 136, 139, 140, 142, 144, 147, 148, 151, 152, 153, 154, 156, 158, 159, 167, 170, 179, 180, 182, 198, 203, 205, 211, 222, 226, 240, 247
water absorption, 125
workers, viii, 15, 23, 42, 90, 224, 262

worldwide, 134, 136, 143, 154, 253, 256
WTO, 107, 127

X

xerophthalmia, 196

Y

yeast, 207

Z

zinc, vii, 35, 44, 45, 46, 53, 58, 59, 60, 61, 62, 67, 78, 85, 86, 120, 168, 224, 230, 245
zinc oxide, 45
zinc sulfate, 35, 45, 46
ZnO, 44, 45